T0248700

Circuits and Systems for the Internet of Things

CAS4IoT

EDITOR

João Goes
Universidade NOVA de Lisboa,
NOVA School of Science and Technology,
Portugal

Tutorials in Circuits and Systems

For a list of other books in this series, visit www.riverpublishers.com

Series Editor

Franco Maloberti

President IEEE CAS Society
University of Pavia, Italy

LONDON AND NEW YORK

Published 2017 by River Publishers
River Publishers
Alsbjergvej 10, 9260 Gistrup, Denmark
www.riverpublishers.com

Distributed exclusively by Routledge
4 Park Square, Milton Park, Abingdon, Oxon OX14 4RN
605 Third Avenue, New York, NY 10017, USA

Circuits and Systems for the Internet of Things CAS4IoT / by João Goes, Franco Maloberti.

Routledge is an imprint of the Taylor & Francis Group, an informa business

ISBN 978-87-93519-90-9 (print)

Table of contents

Introduction

Internet-of-Things (IoT) can be envisaged as a dynamic network of interconnected physical and virtual entities ("things"), with their own identities and attributes, seamlessly integrated in order to e.g. actively participate in economic or societal processes, interact with services, and react autonomously to events while sensing the environment. By enabling things to connect and becoming recognizable, while providing them with intelligence, informed and context based decisions are expected in a broad range of domains spanning from health and elderly care to energy efficiency, either providing business competitive advantages to companies, either addressing key social concerns. The level of connectivity and analytical intelligence provided by the IoT paradigm is expected to allow creating new services that would not be feasible by other means.

CAS4IoT book targets post-graduated students and design engineers, with the skills to understand and design a broader range of analog, digital and mixed-signal circuits and systems, in the field of IoT, spanning from data converters for sensor interfaces to radios, ensuring a good balance between academia and industry, combined with a judicious selection of worldwide distinguished authors.

CHAPTER 1

Ultra-Low-Voltage and Micro-Power Analog Design for Internet of Things

Franco Maloberti
University of Pavia, Pavia, ITALY

Designing low power and low voltage analog circuits for IoT systems is challenging. Many applications require analog circuits working with sub 1-V supply voltages and with power consumption in the hundreds of nW range while preserving good performances. The present target is to have analog circuits operating at 0.5-V or less. With that supply voltage it is required to design basic building blocks like operational amplifiers, comparators and voltage reference generators. For each of them, the presentation briefly reviews the state of the art and presents some recently published examples of implemented solutions. The analog-to-digital converter is essential for every modern system. Ensuring very low power while providing a medium-high resolution (9-14 bit) over a signal bandwidth up to about 1 MHz is a typical goal. For these specifications, SAR converters and sigma-delta modulators are typically used. Techniques and recent circuit implementations capable of satisfying diverse requests are presented. A 10-bit, 200-kS/s, 250-nA ADC operating down to 0.7-V and voltage references operating with supply voltages in the 0.4-0.6-V range will be presented.

Introduction

➢ **What is INTERNET OF THINGS?**
- **"Internet of Objects"**
- **"Machine-to-Machine Era"**
- **"Machine-to-Object Era"**

Intelligent System of Systems

The Internet of Things can be considered as a very large network connecting things. These things can be computing devices, mechanical machines, objects, or even people, having the ability to collect and exchange data over a network without needing a human-to-human or human-to-computer interaction. With the ever growing number of devices connected together collecting data, an intelligent system of systems is created with the ability to transform businesses and lives in many ways.

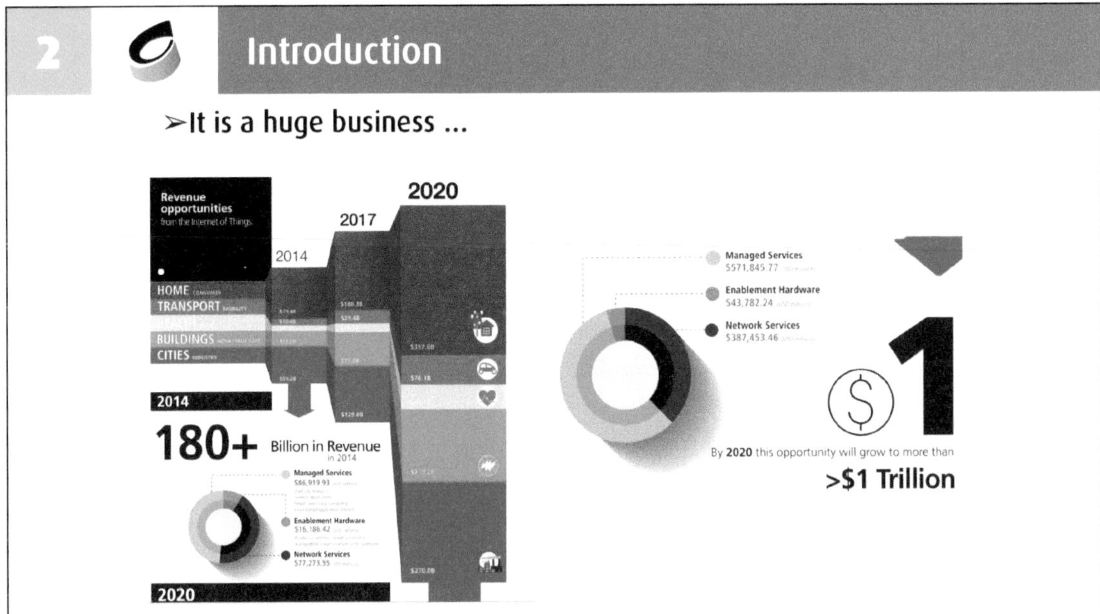

Internet of Things is important for many reasons, one of them is that it is a huge business. In 2014, the total revenue of the Internet of Things was above 180 billion dollars and it is expected to grow to more than one trillion dollars by 2020. A big fraction of the revenue will come from managed and network services and a good fraction from hardware, i.e., devices and systems integrated together.

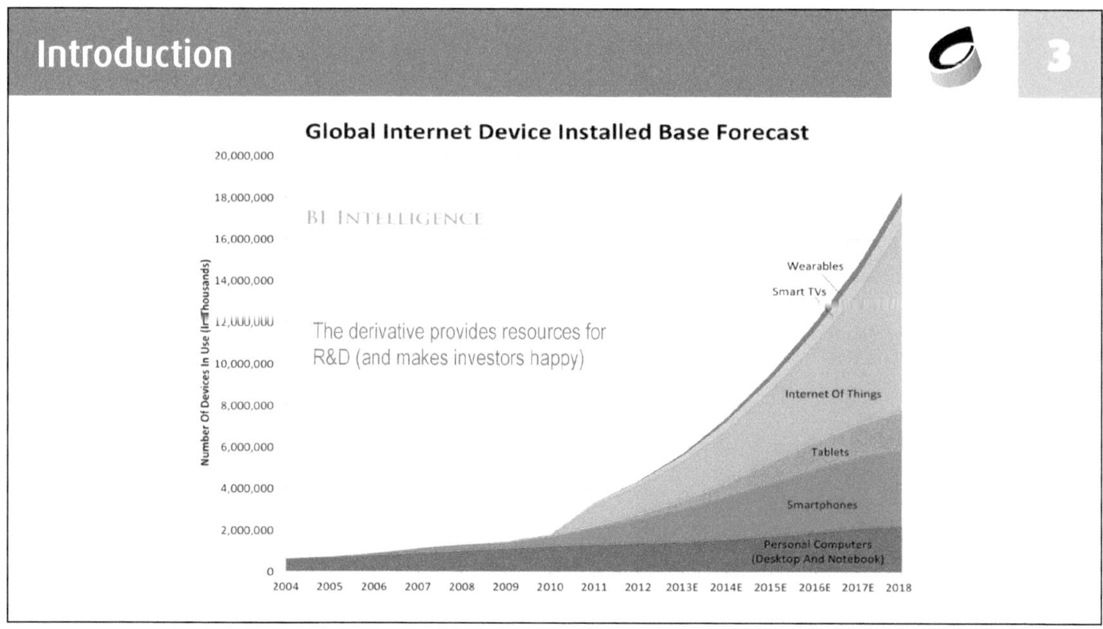

The global internet device installed base forecast predicts that, by 2018, between 40 and 45% of objects should be connected together using an IoT type connection. By then, Internet of Things sensors and devices are expected to exceed mobile devices as the largest category of connected devices as connections move beyond computing devices and start to power everyday devices from thermostats to smart outlets.

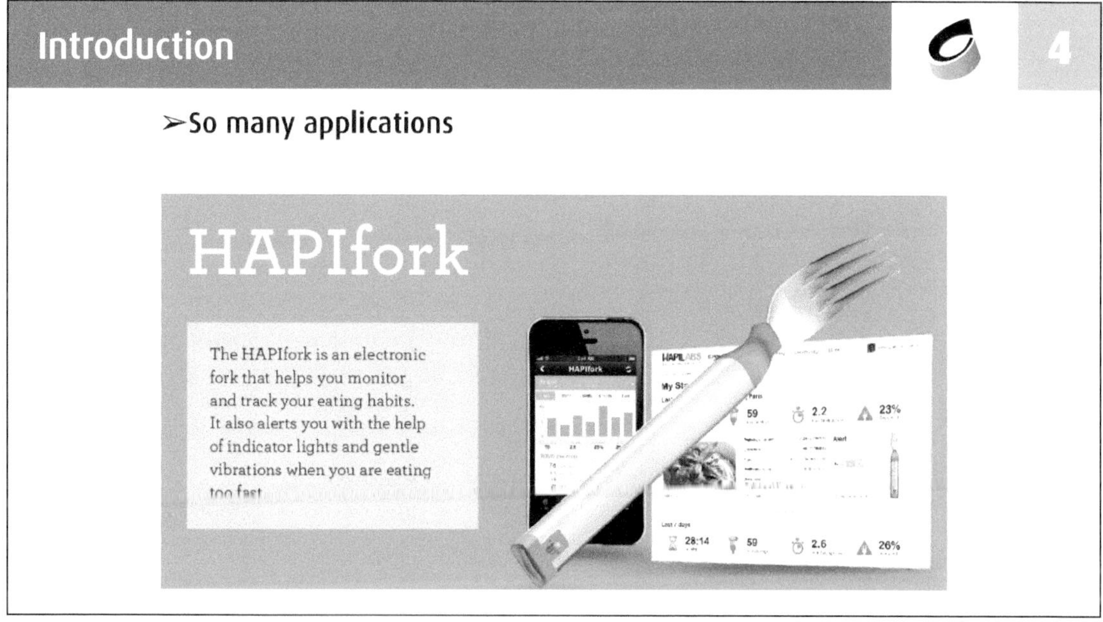

Internet of Things can be used in many applications. Several examples can be found on the Internet, one of these examples is the HAPIfork which helps people collect data on their eating habits, and thus helping them to improve them. The benefits include helping people to reduce the speed at which they eat meals, improving their digestion and helping people to eat at the right time so that they eat at optimal times for their bodies.

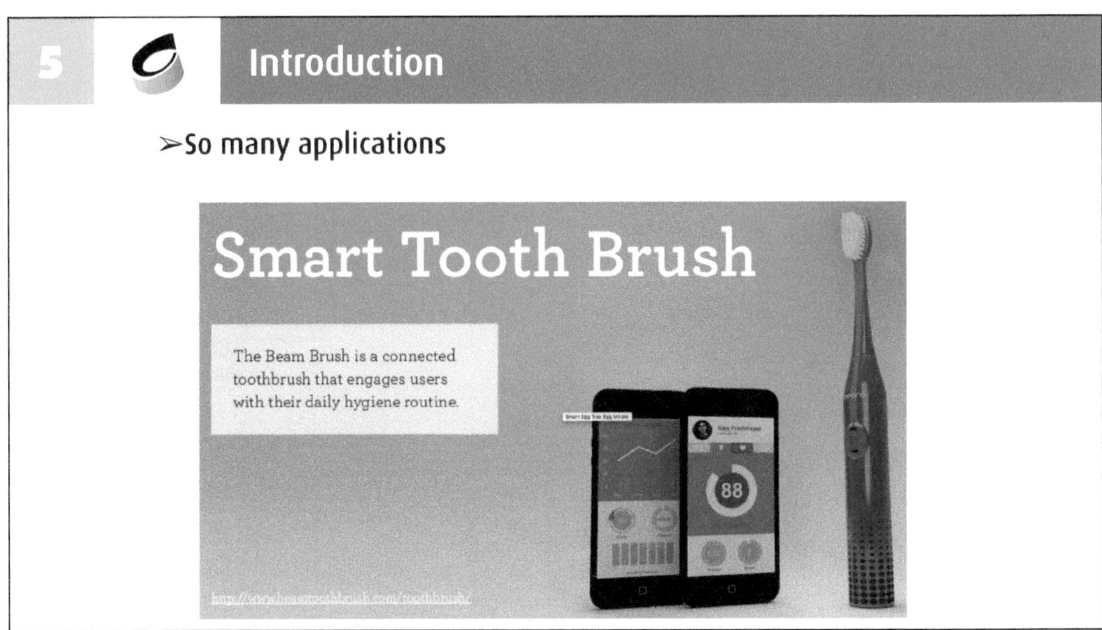

Another example is the smart tooth brush. This device uses an embedded accelerometer to track the user's brushing schedule. The sensor measures how long a person brushes their teeth and how often, sending this information via Bluetooth to a Smartphone so that the users can see their brushing habits over time. To encourage users to brush their teeth for a longer period of time (two minutes), the brush can play music during this time. This device/application is particularly useful for tracking children's brushing habits.

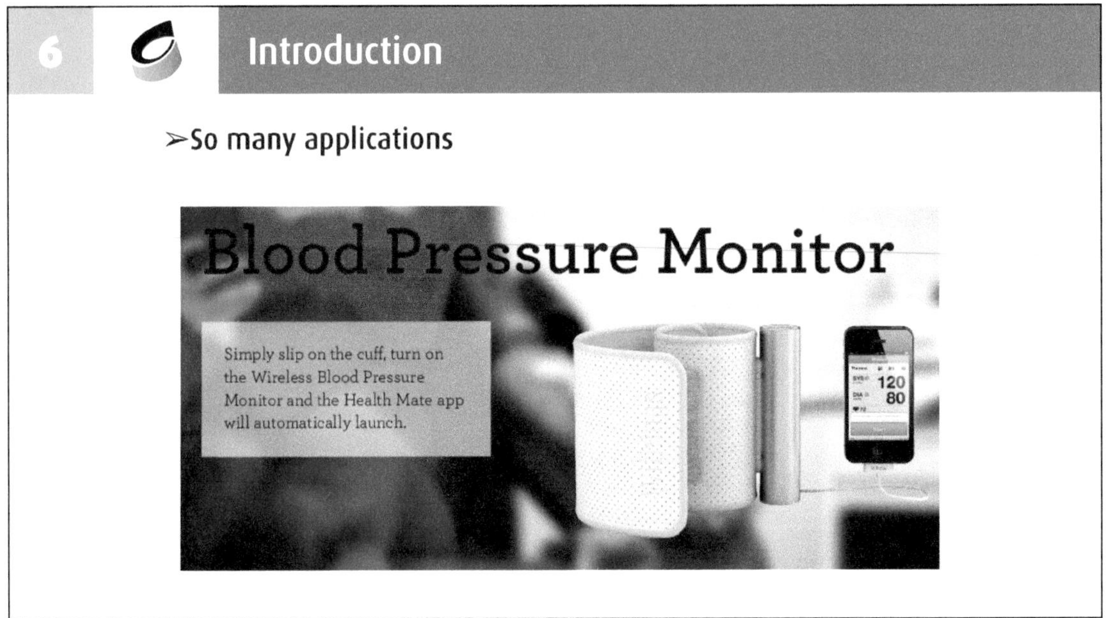

The blood pressure monitor is yet another example of an Internet of Things application. This application gives instant color-coded feedback based on ESH (European Society of Hypertension) and the AHA (American Heart Association) recommendations for hypertension. The device stores the data in a cloud and creates easy to understand charts to better understand the user's heart health. The data can also be forwarded to the user's physician.

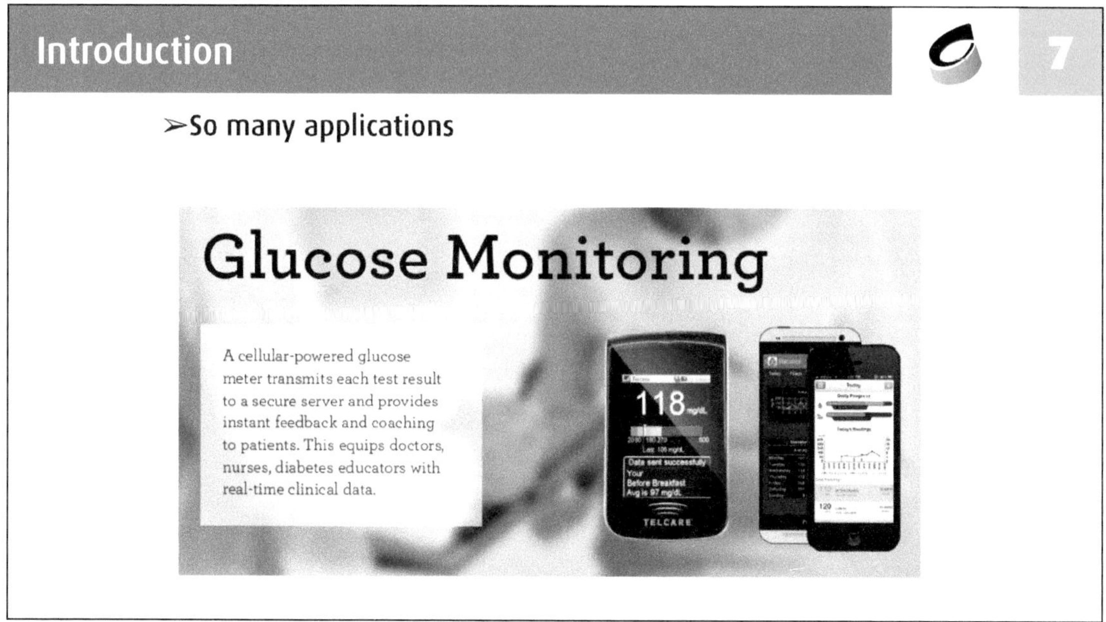

One final example is the glucose monitoring tool. Similarly to the blood pressure monitor, this device collects data and stores it in a secure server allowing patients to monitor blood glucose levels in a non-invasive way. This data can be used to deliver personalized alerts and to forecast immediate trends in the user's blood glucose levels, allowing them to adjust their diet and medication according to the current levels.

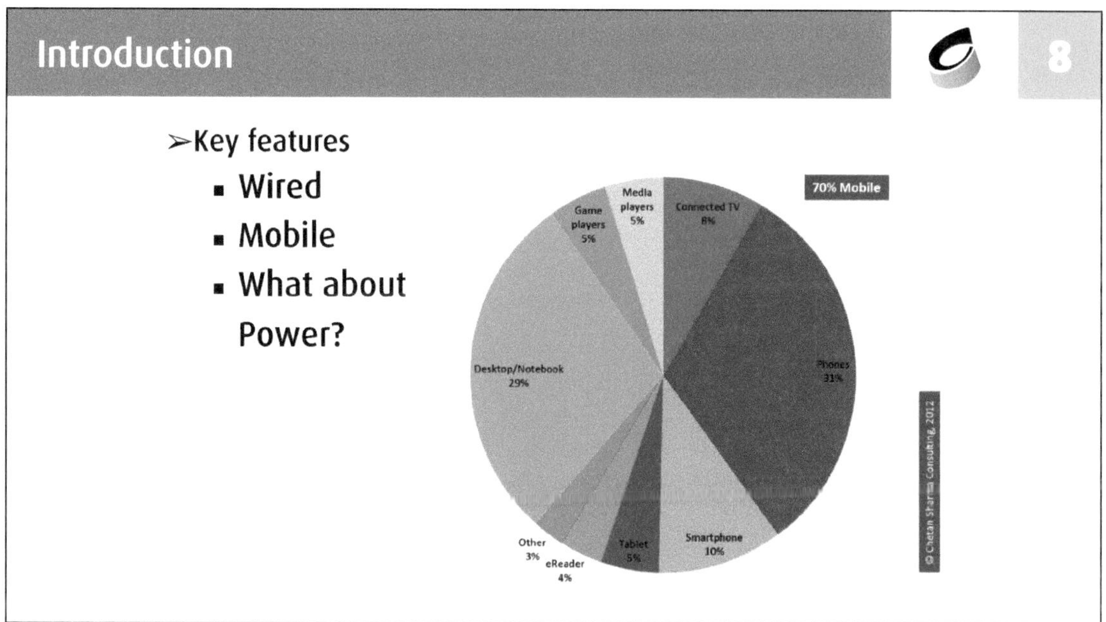

Devices can be divided into two categories: wired devices or mobile devices. In mobile devices the power consumption is critical since it determines the amount of time it can be used before needing to be recharged or replaced. In 2012 mobile devices already represented 70% of all connected devices.

9  **Introduction**

➢**Key features**

- Refueled
- Autonomous
- How harvesting power?
- How optimizing power?
- Where storing power?

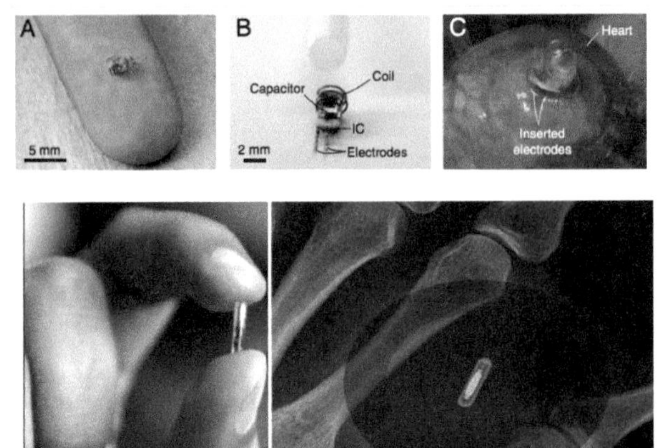

The mobile devices can be recharged or autonomous, where the latter gathers power from the environment to recharge the devices. The questions are, how to perform the energy harvesting, how to minimize the power consumption of the devices, and how to store the power harvested from the environment.

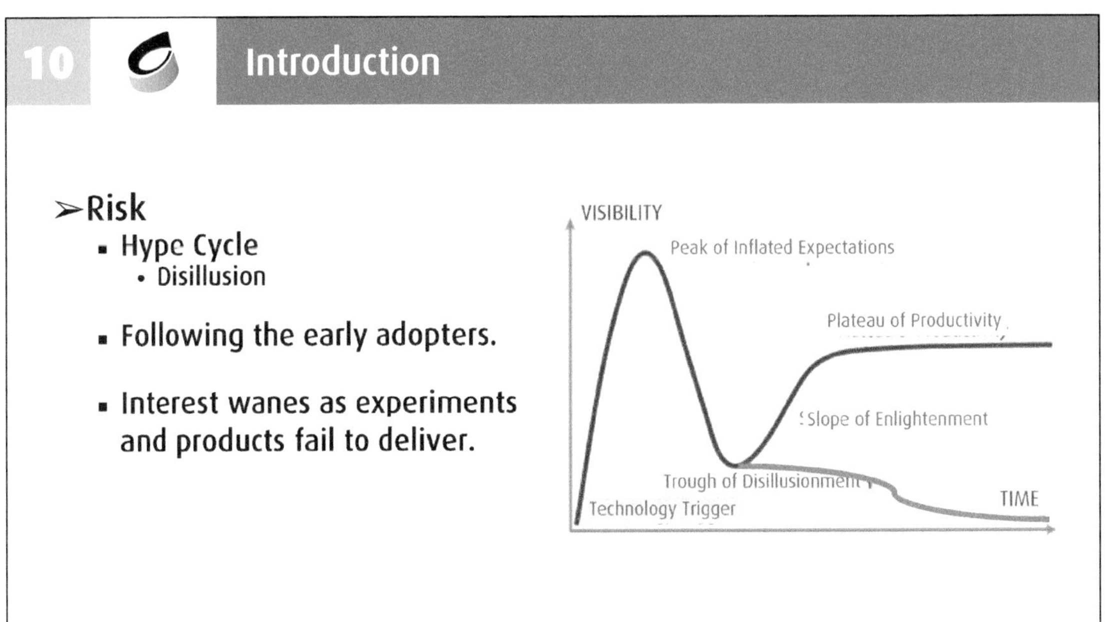

The Hype cycle provides a graphical representation of the five phases of maturity emerging technologies go through. Among them, "Trough of Disillusionment" is the most critical phase since it is the phase where the media's interest wanes as experiments fail to deliver. Here, the producer needs to improve their product/technology to the satisfaction of the early adopters, otherwise they risk losing the consumers trust.

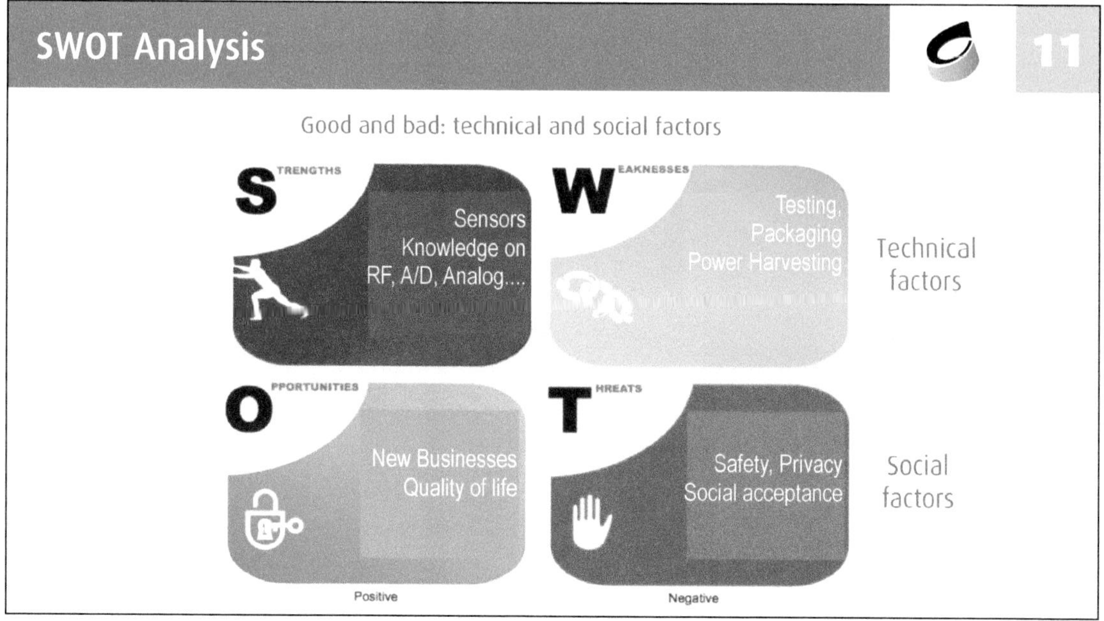

What is important to is understand is what are the possible limits of a given application/device. This can be done using a SWOT analysis, determining what are the strengths, weaknesses, opportunities, and threats for the application/device. The first two can be classified as technical factors and the last two as social factors. The next slides will give some details on each of these four groups in the context of the Internet of Things.

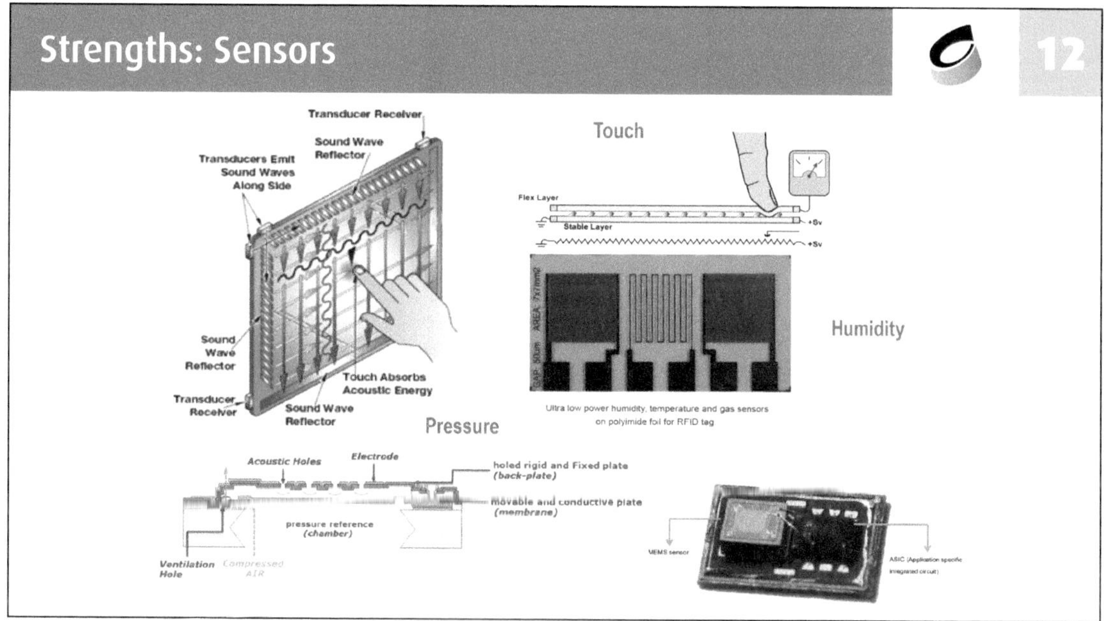

The main driving force of the Internet of Things are the sensors. The collection of data from the environment can be performed using combination of different types sensors or a network of sensors, which can then be used to create historical records and quality management of products or be used as triggers for alarms or change a system's behavior depending on the data collected, allowing the creation of smart systems.

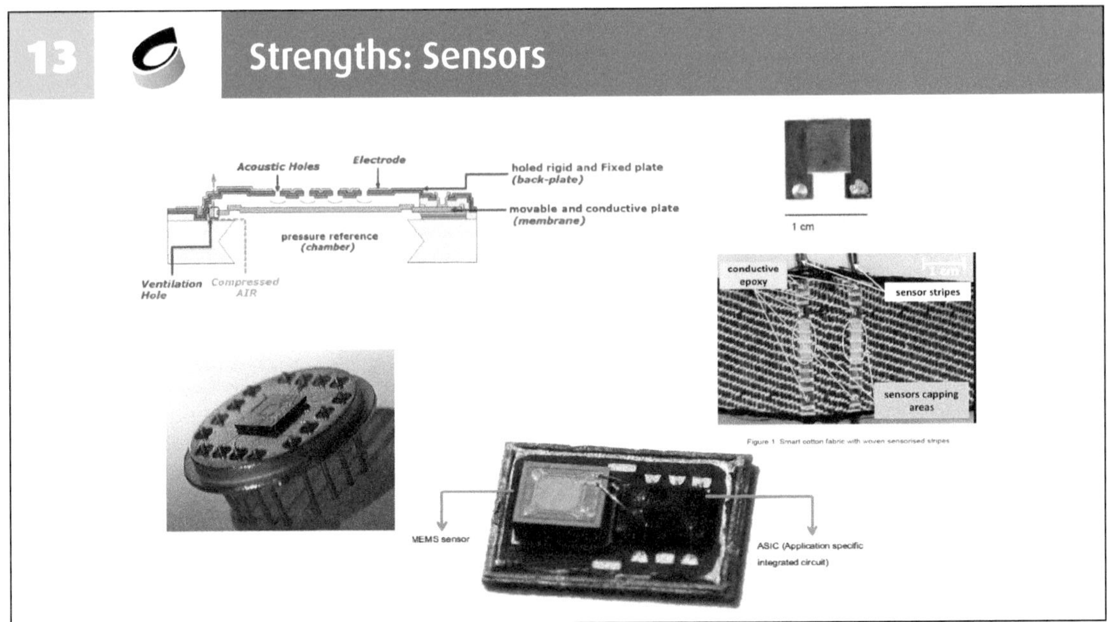

Sensors can also be incorporated into fabric, for example an intelligent T-shirt that can be used to monitor a person's status. Recently, the microelectromechanical systems (MEMs) have gained interest to perform sensing, mainly due to their low power performances, good reliability, and lower price when compared with standard sensors.

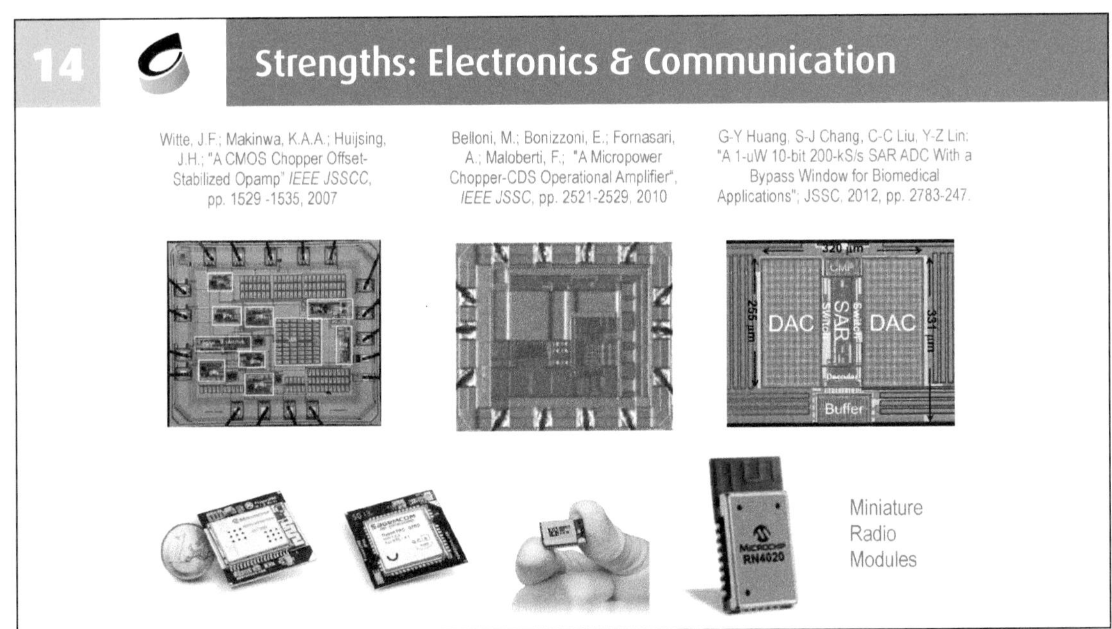

Especially in sensors that are integrated in silicon, the level of the signal can be very small. Many solutions can be found in the literature to solve this problem, for example amplifiers that are very sensitive and able to reject the offset, or converters with reasonable resolution and consuming very little power. Some examples of low power consumption circuits will be discussed later. For communication there are already miniature radio modules that are very small and maybe in the future they can be further miniaturized.

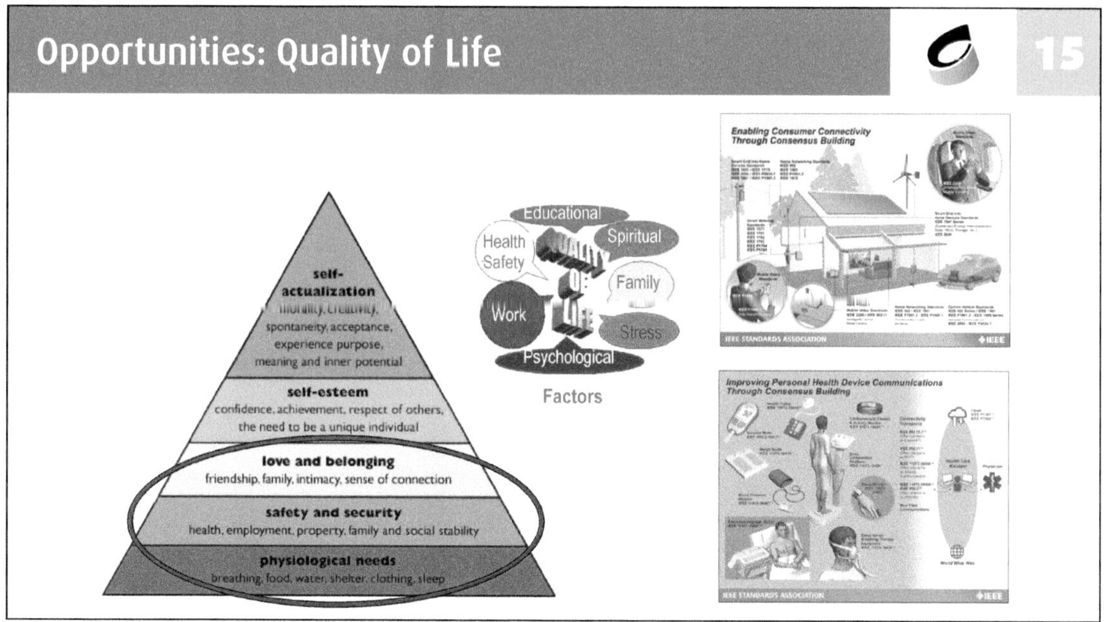

The opportunities are mainly on the side of the social effect. With the Internet of Things we have the possibility to improve the quality of life. The pyramid shows the different levels of needs of a person and the Internet of Things can help improve the bottom three levels. In the previous slides some examples were already given on how Internet of Things applications can help monitor and improve a person's health, although there are many more examples that improve communication between people or applications to improve water quality or food production.

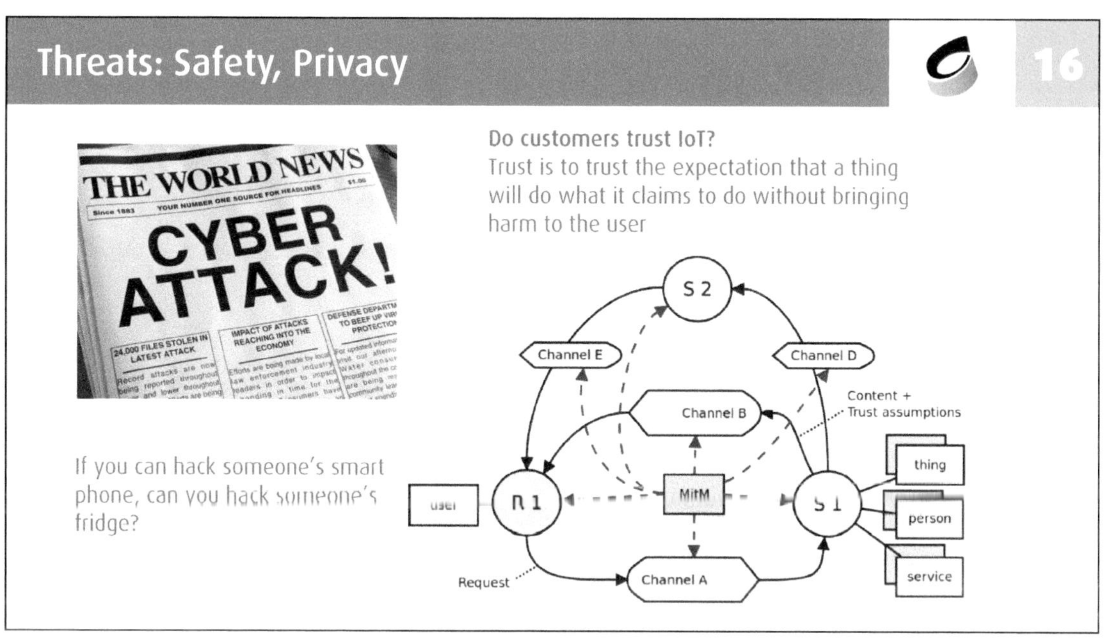

With the incorporation of new devices into IoT systems, concerns about safety and privacy of information can arise, leading to the question of whether or not consumers truly trust these systems to keep their information safe. Security concerns have also increased, since badly configured gadgets can provide a backdoor for hackers to break into private/corporate networks. Since many IoT devices are very small in size, are low cost and disposable, and have limited processing power when connected together, the use of encryption and other security measures is very limited.

17 

Threats: Machine dependency

Billion of connected things are very very powerful

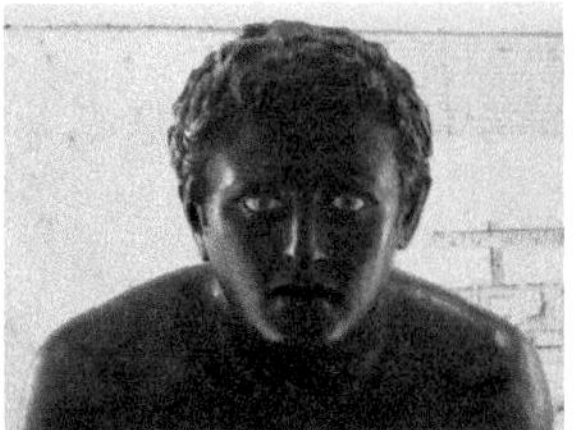

Good or bad?
Man against robots?
Or man against man?
Peace or conflict?

Easier life ...
Safe control of persons and things ...
but
Less intellectual trouble (brain use) ...
Evolution faster than custom and tradition changes ...
Supposition to beat the nature (fatal conceit) ...

Another question that can be made is: is it good or bad to have many systems connected together, creating a very powerful system? Although such systems can make a person's life easier, it reduces brain use, which can lead to reduced intelligence. Although technology is evolving very fast and with the intention of surpassing nature, at present nature still surpasses technology.

18 

Weaknesses: Packaging Testing

Testing systems for IoT is quite different from testing VLSI circuits.

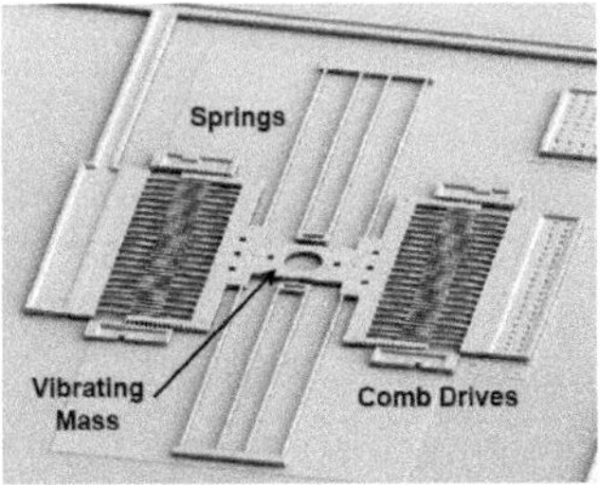

It is necessary to test all things physical, chemical, biological, electrical. It is necessary to test all sensory reactions. Interaction with humans is an incredibly crucial part of the test.

One of the most important weaknesses is the packaging and testing. With a system made of sensors, actuators, and electronics it becomes very difficult and complex to perform the testing of the system. In electrical systems it is only necessary to consider the electrical behaviors between circuits but in these more complex systems it is also necessary to consider the physical, chemical, biological and electrical behaviors between these elements.

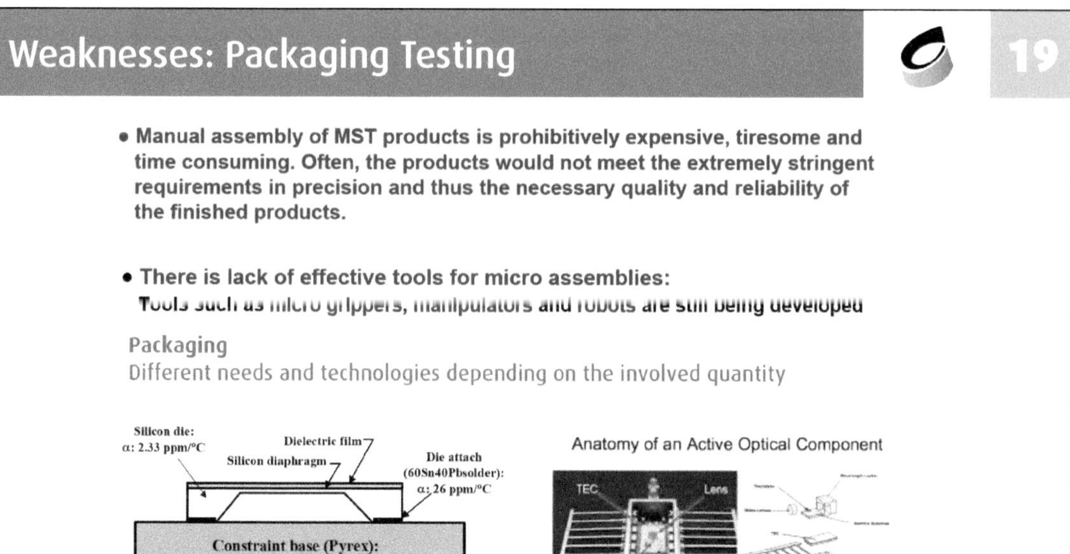

Weaknesses: Packaging Testing
19
• Manual assembly of MST products is prohibitively expensive, tiresome and time consuming. Often, the products would not meet the extremely stringent requirements in precision and thus the necessary quality and reliability of the finished products.
• There is lack of effective tools for micro assemblies:
Tools such as micro grippers, manipulators and robots are still being developed
Packaging
Different needs and technologies depending on the involved quantity
Silicon die:
α: 2.33 ppm/°C
Dielectric film
Silicon diaphragm
Die attach
(60Sn40Pbsolder):
α: 26 ppm/°C
Constraint base (Pyrex):
α: 7 ppm/°C
Anatomy of an Active Optical Component
TEC
Lens
Laser
Thermistor
Wavelength Locker

One of the most important weaknesses is the packaging and testing. With a system made of sensors, actuators, and electronics it becomes very difficult and complex to perform the testing of the system. In electrical systems it is only necessary to consider the electrical behaviors between circuits but in these more complex systems it is also necessary to consider the physical, chemical, biological and electrical behaviors between these elements.

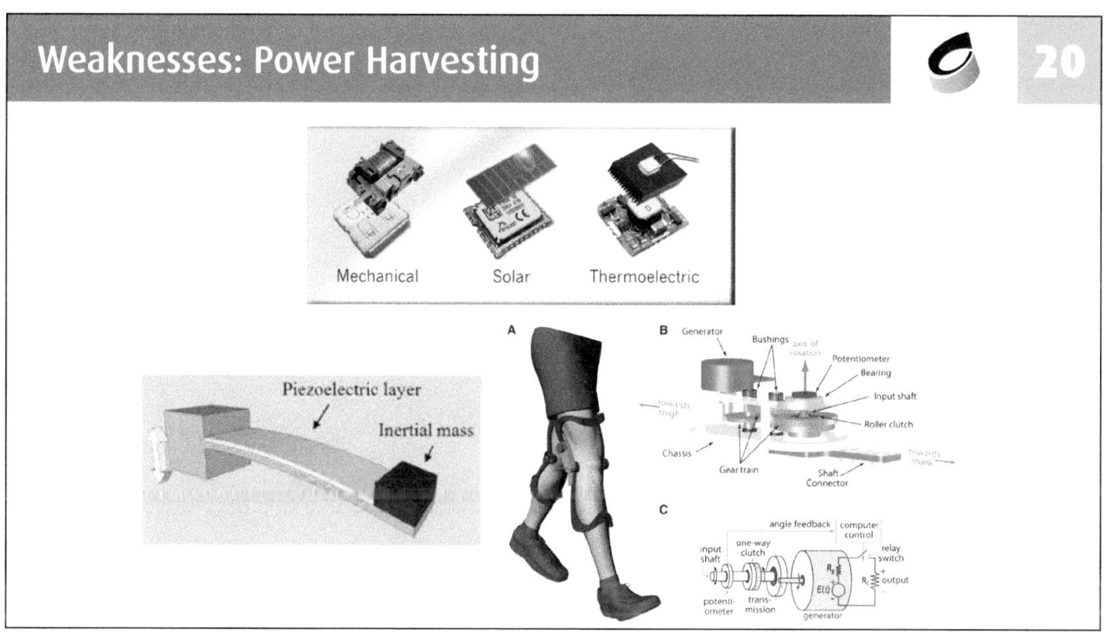

Power harvesting can be achieved with a different number of external sources (solar power, thermal energy, wind energy, etc). The choice of external source to be used to harvest power is dependent on the location on which the device/system is going to be used.

21 Weaknesses: Power Harvesting

POWER DENSITY OF ENERGY HARVESTING TECHNIQUES.

Energy harvesting technique	Power density
Photovoltaic	Outdoors (direct sun): 15 mW/cm^2 Outdoors (cloudy day): 0.15 mW/cm^2 Indoors: <10 μW/cm^2
Thermoelectric	Human: 30 μW/cm^2 Industrial: 1 to 10 mW/cm^2
Piezoelectric	250 μW/cm^3 330 μW/cm^3 (shoe inserts)
Electrostatic	50 to 100 μW/cm^3
Electromagnetic	Human motion: 1 to 4 μW/cm^3 Industrial: 306 μW/cm^3
RF	GSM 900/1800 MHz: 0.1 μW/cm^2 Wifi 2.4 GHz: 0.01 μW/cm^2
Wind	380 μW/cm^3 at the speed of 5 m/s

The table shows the power density achieved by different types of energy harvesting techniques. The best technique to be used is dependent on the location in which it is going to used. For example, the power produced by photovoltaic panels is directly proportional to the amount of sunlight it receives so this type of device should be used in sunny locations. Now if we change to an indoors location, the amount of power produced by the photovoltaic panel greatly decreases, so it is no longer efficient to be used in this location and an alternative technique should be considered. The piezoelectric technique is not commonly used since is relies on vibrations to harvest energy.

22 Weaknesses -> Strength: Energy Storage

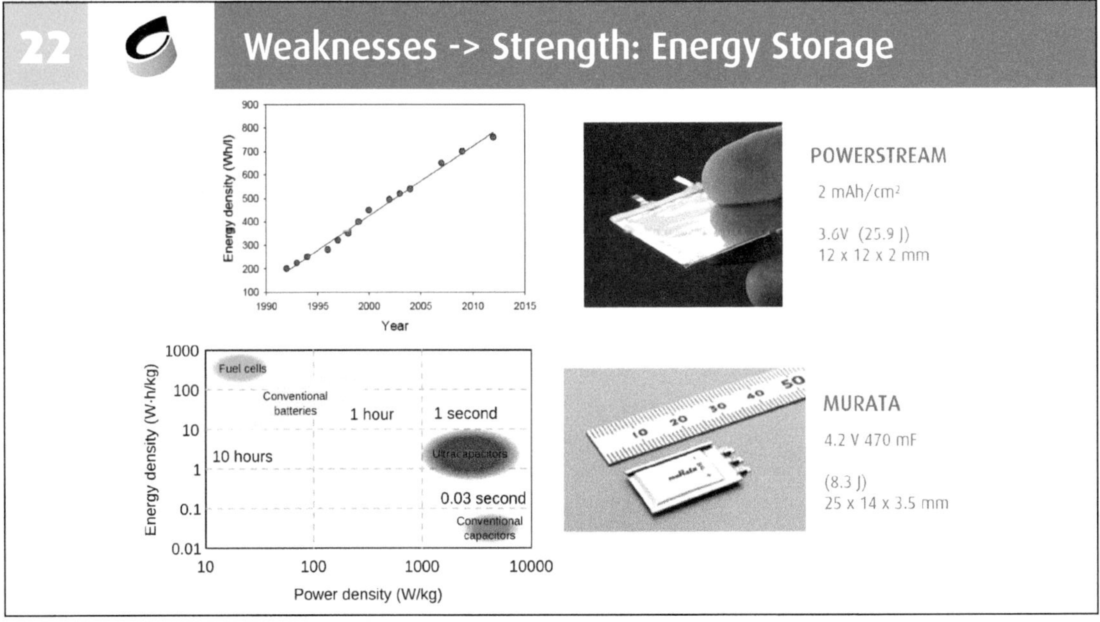

Energy density in storage devices has been improving at a steady pace over the last two decades, however, the increase of energy stored in small devices can be dangerous due to the risk of explosion. The powerstream is an example of a rechargeable battery capable of achieving 2mA per cm2 and with a very small area (12x12 mm). Another possibility as energy storage devices are supercapacitors, which are also becoming smaller and capable of quick recharging with time.

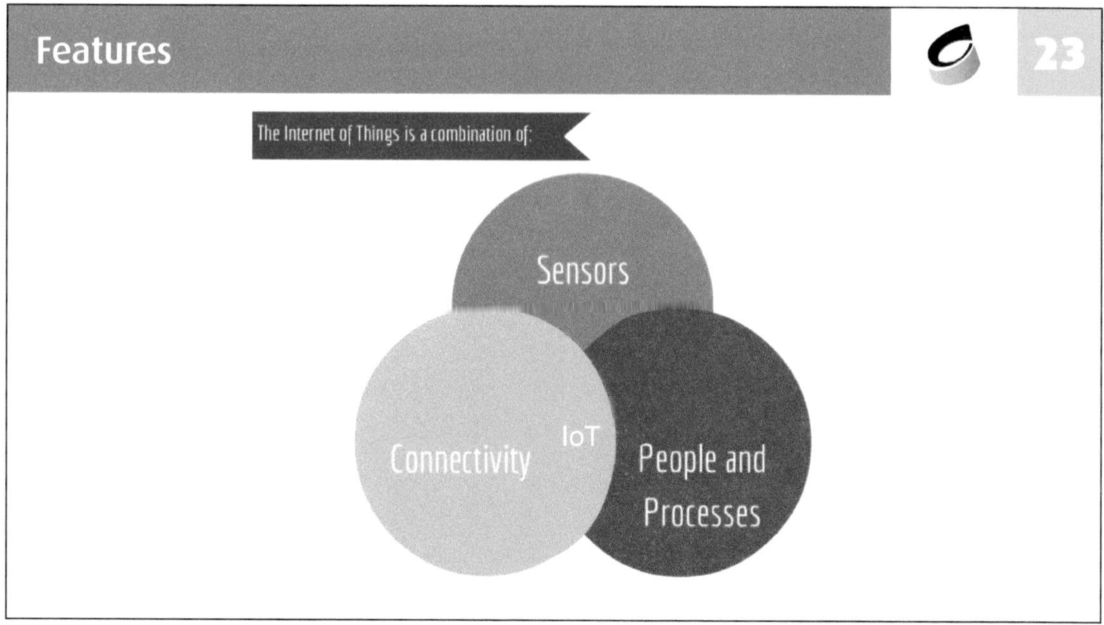

The Internet of Things is a combination of sensors, connectivity, and people and processes. Sensors represent the digital nervous system of the IoT, where location data can be acquired using GPS, and the eyes and ears come from cameras and microphones spread throughout the IoT network. Once the data is acquired and digitized, it can be placed into networks allowing the communication between devices. The data obtained from the sensors can be used in bidirectional systems that integrate people and processes for better management of the IoT network.

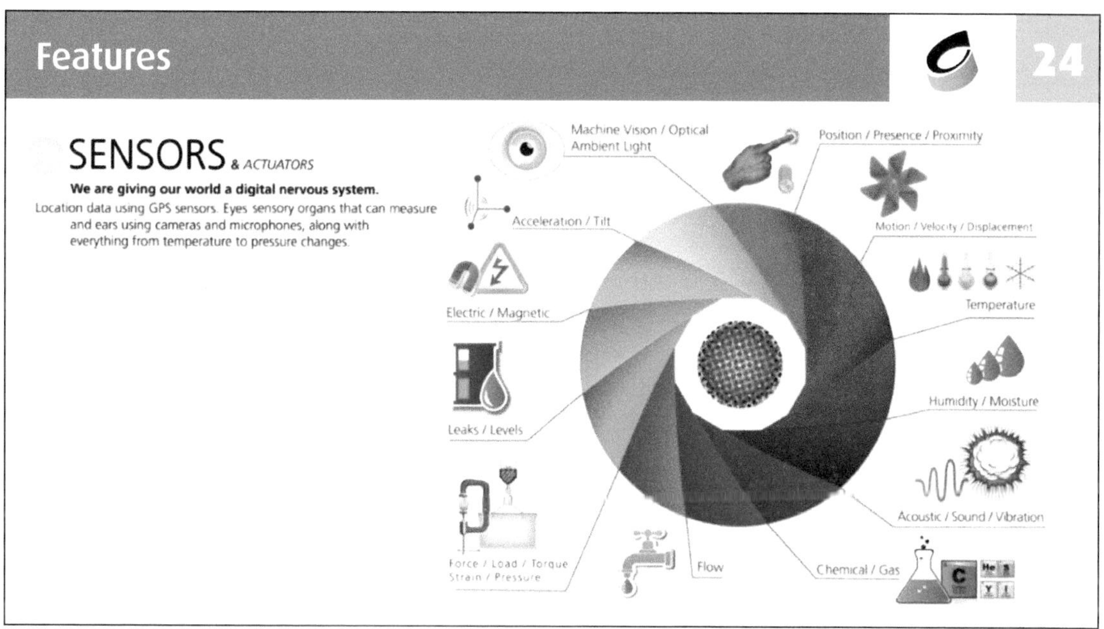

The digital nervous system of the IoT network is achieved by using many different types of sensors which range from: location data using GPS; eyes and ears using cameras and microphones; and sensory organs than can measure anything (temperature, humidity, velocity, pressure, electric and magnetic fields, etc).

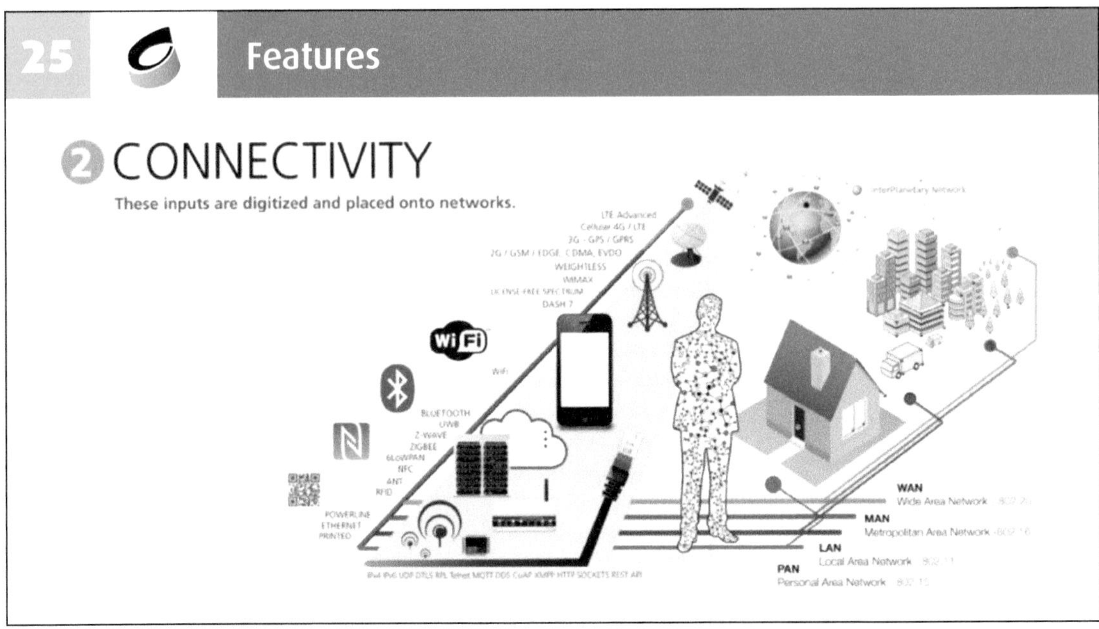

Once digitized, the information can be transmitted between devices using different methods/protocols of transmission (Wi-Fi, ZigBee, Bluetooth), and is able to reach any type of network (PAN, LAN, MAN, WAN).

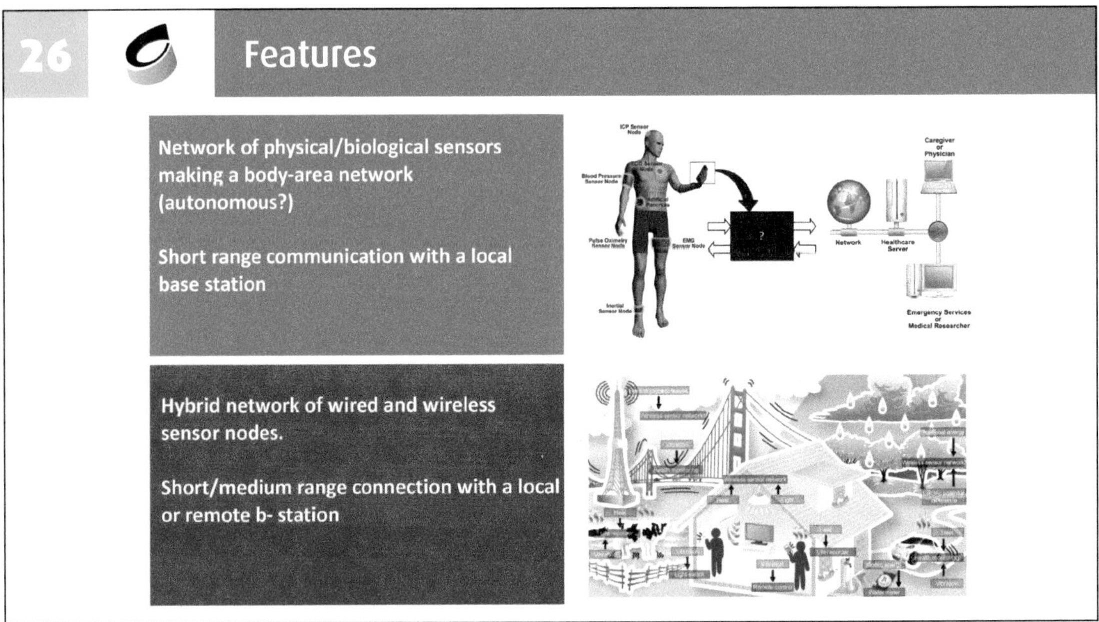

For the Internet of Things there are two possibilities. One is to have a network of sensors, ideally autonomous, in a small area communicating with a local base station (for example a Smartphone) which can then communicate with any part of the world. Another possibility is to have a hybrid network of wired/wireless sensors where there is a mix of wired and wireless communication between devices.

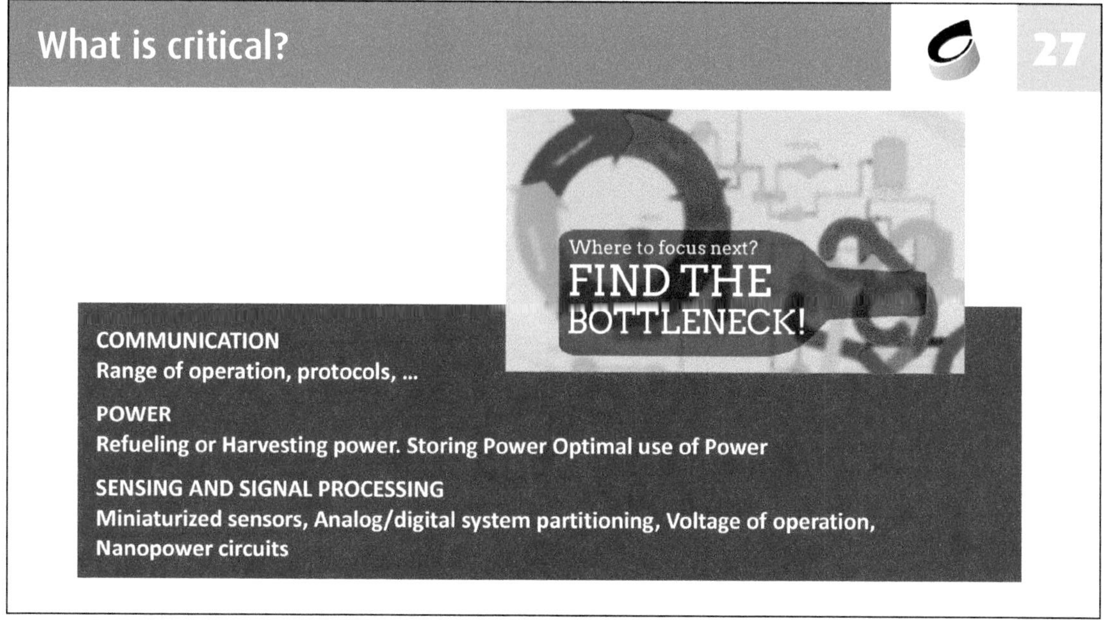

To avoid the green line shown in the hype cycle graph from slide 11, i.e., the loss of interest, certain things needs to be considered, such as the communication (range of operation, protocols, ...), reducing the power consumption while maintaining effective operation, using energy harvesting solutions and storing elements to increase the life-time of the devices, and sensing and signal processing.

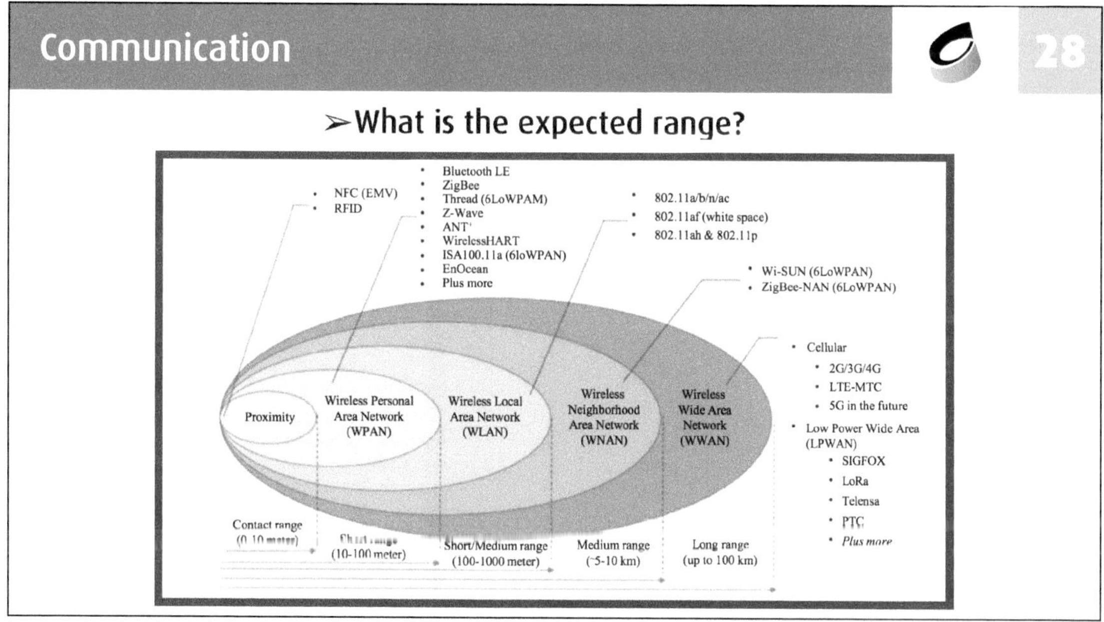

One way to characterize communication networks is by their communication range. Depending on this range and the intended application, different types of communication protocols can be used. For example, Bluetooth is commonly used to transfer information between two or more devices that are close together and when the communication speed is not an issue (low bandwidth applications). Wi-Fi on the other hand, is better suited to operate full-scale networks due to the faster connection, better range from the base station and due to having better security features when compared with Bluetooth.

29 

Communication

- For short to medium range communication Wireless Personal and LocalArea Network technologies (WPAN\LAN) such as: Bluetooth, ZigBee, 6LowPAN, and Wi-Fi are used.

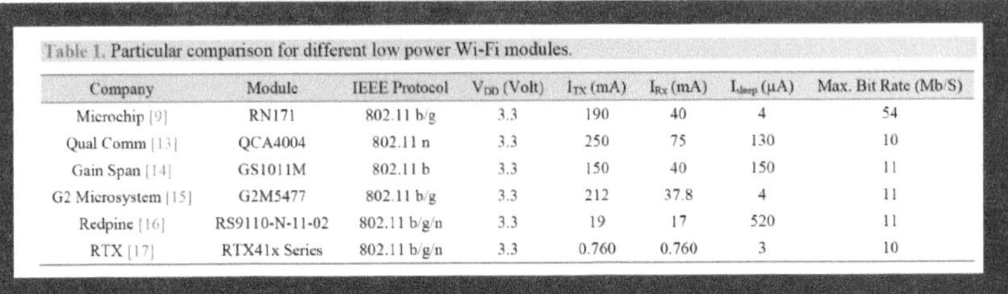

Table 1. Particular comparison for different low power Wi-Fi modules.

Company	Module	IEEE Protocol	V_{DD} (Volt)	I_{TX} (mA)	I_{RX} (mA)	I_{sleep} (μA)	Max. Bit Rate (Mb/S)
Microchip [9]	RN171	802.11 b/g	3.3	190	40	4	54
Qual Comm [13]	QCA4004	802.11 n	3.3	250	75	130	10
Gain Span [14]	GS1011M	802.11 b	3.3	150	40	150	11
G2 Microsystem [15]	G2M5477	802.11 b/g	3.3	212	37.8	4	11
Redpine [16]	RS9110-N-11-02	802.11 b/g/n	3.3	19	17	520	11
RTX [17]	RTX41x Series	802.11 b/g/n	3.3	0.760	0.760	3	10

- Power lower by a factor 10 for dedicated modules.
- Very low duty cycle and very low current in the sleep mode.

In the examples presented in the table, it can be seen that the current consumption for transmitting and for receiving is decreasing. However the power level is high and much more than is normally harvested. In order to achieve low power consumption, we need a storing element and low duty cycle, so it is not possible to be transmitting all the time, however, the sleep mode, i.e., the time in which the circuit is neither transmitting nor receiving data can be still consuming too much current. Power is also consumed to switch in between operation modes.

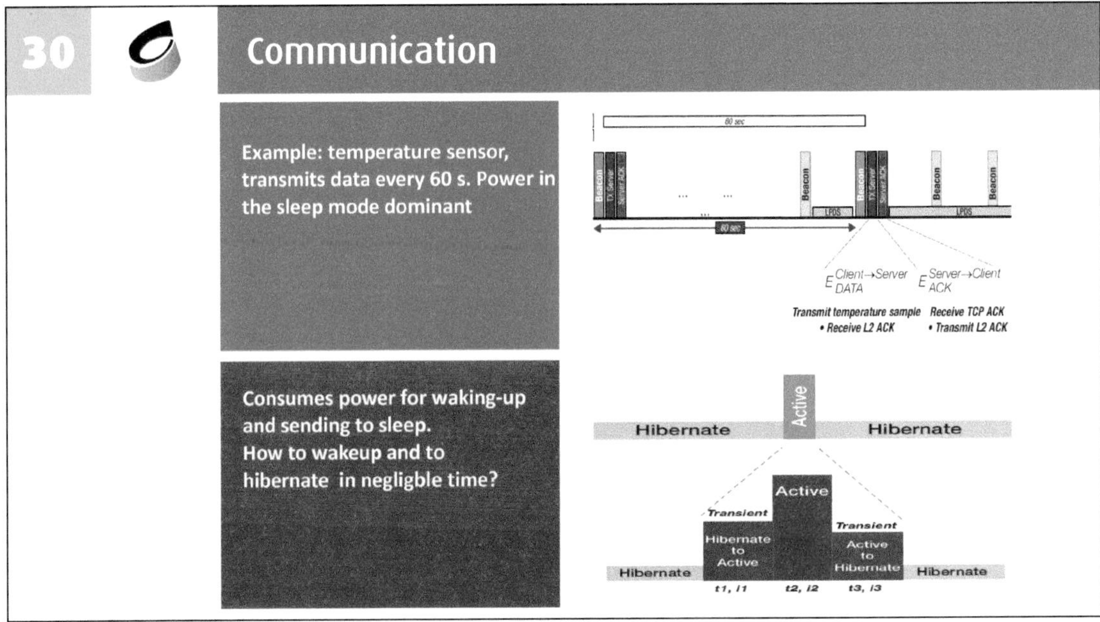

In low power applications, it becomes increasingly important to reduce the power consumption in sleep mode, and to reduce the time it takes for the circuit to switch between states (hibernate to active and active to hibernate). By doing this it is possible to increase the energy efficiency of the devices. An example on how this can be done is given in the next slide.

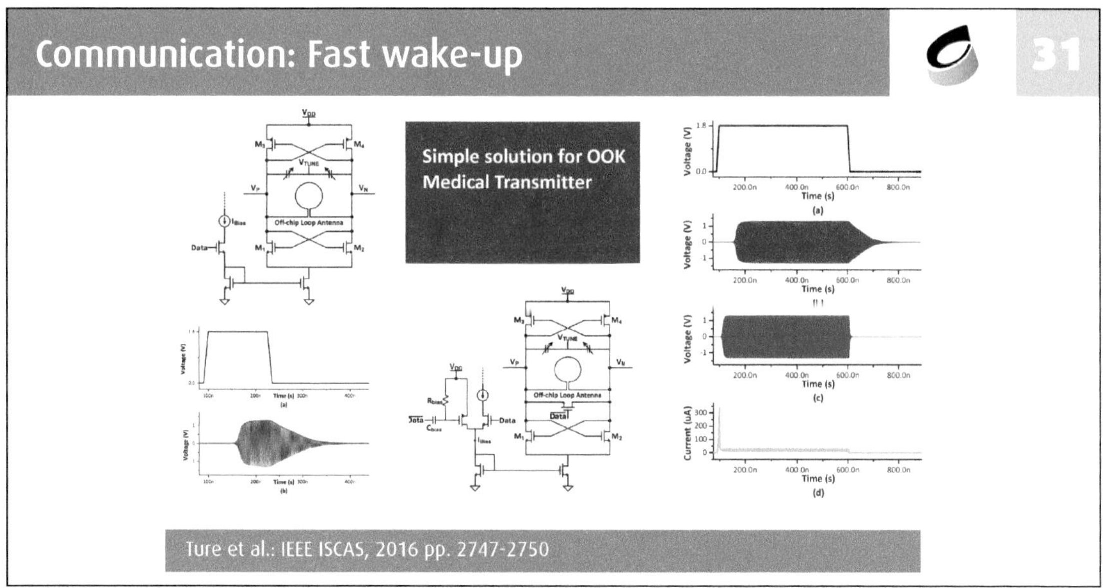

Simple solution for OOK Medical Transmitter

Ture et al.: IEEE ISCAS, 2016 pp. 2747-2750

To minimize power consumption, the cross-coupled voltage controlled oscillator completely turns off the transmitter during the transmission of '0' bits and only oscillates when transmitting '1' bits, i.e., the data bit controls the bias current and the operation of the oscillator. To decrease the time it takes for the circuit to turn-on and to turn-off, which limits the data communication rate, and additional current source is used during the build-up of the oscillator and a switch with small ON resistance is connected to the differential output pins of the oscillator. Using this approach, the turn-on time can be reduced by more than 3 times and the turn off time by more than 16 times, increasing the maximum achievable communication data rate by more than 6 times.

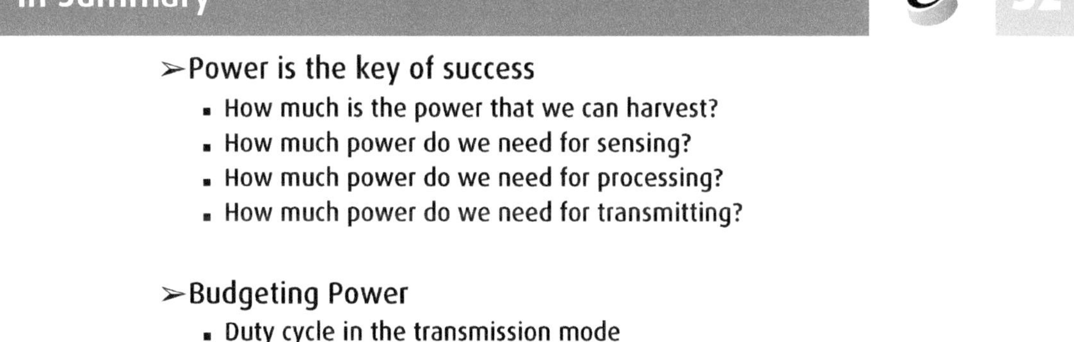

In Summary
32
➢Power is the key of success
How much is the power that we can harvest?
How much power do we need for sensing?
How much power do we need for processing?
How much power do we need for transmitting?
➢Budgeting Power
Duty cycle in the transmission mode
Power in the sleep mode
Power for analog and digital processing comparable with transmission sleep mode

In summary, one of the key elements to have success in the Internet of Things, or in the Internet of Everything, is power. What is important is to know is: How much power can be harvested? How much power is needed to sense? How much power is needed for the processing of the signal? How much power is needed for the transmission and for the sleep mode?

The budget will determine the duty cycle that is possible to have in the transmission mode and the system needs to be able to provide the necessary power for the sleep mode, which should be a fraction or at least the same power that is needed for the sensing and the analog to digital processing.

33 

Therefore

Many IoT applications need ultra low power (and voltage)

It is the key of success

IoT applications involve the use of sensors or adding connectivity to devices in hard to reach places where power is not available. In these situations, where the devices cannot be directly connected to a power outlet, a battery-operated solution is typically the preferred solution. This leads to the need to implement ultra low power devices in order to increase their life time. When a power outlet is not available the need of autonomous operation exacerbate the problems.

Low Voltage Low Power Analog

➢Early times

The Electronic Watch and Low-Power Circuits.
Eric Vittoz, CSEM founded 1962

The total current drain was less than 100nA at 32kHz, almost constant for a supply voltage ranging from 0.8 to 3 volts.
(1977)

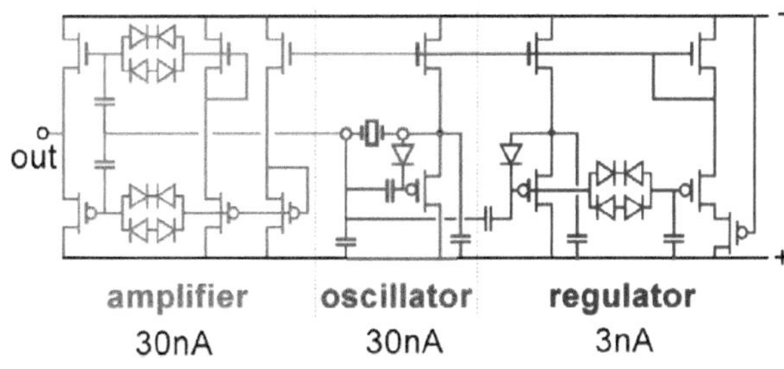

In the 1970's it was already possible to design circuits, in CMOS technology, with low power consumption. Eric Vittoz, working at CSEM in Switzerland, was a pioneer on this field. For example in 1977, a quartz oscillator was designed, capable of operating with a current smaller than 100 nA and a supply voltage ranging from 0.8 to 3 V, which was the voltage range of the watch's battery depending on its status.

Low Voltage Low Power Analog

35

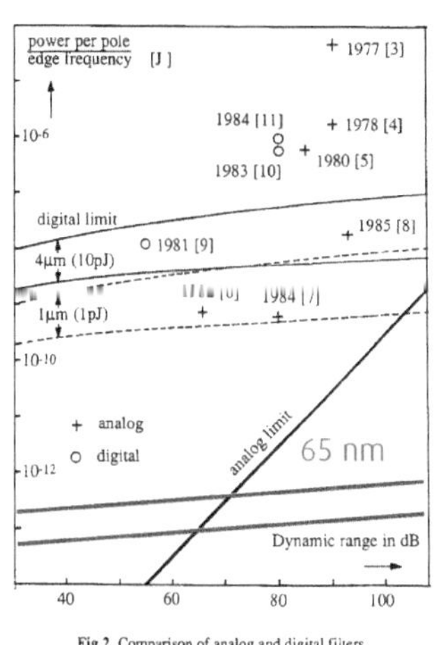

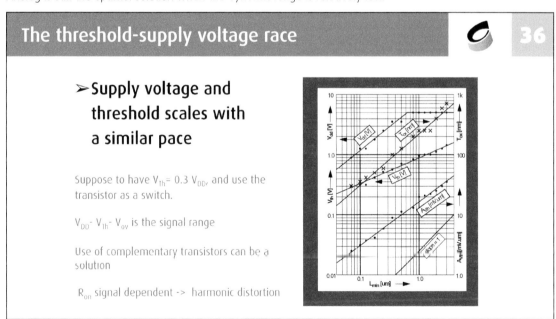

Fig.2. Comparison of analog and digital filters

Vittoz, ISCAS 1990

Minimum analog power consumed per pole $P = 8kTV_S^2$

Minimum noise power $V_N^2 = kT/C$

Minimum energy per cycle and pole $P/f = 8kTV_S^2/V_N^2$

It is estimated that a digital filter requires 50 n² operations per cycle with a clock frequency 10 times f_p

Energy per operation 2-4 fJ for 1V 65 nm CMOS
Plus the energy for the S&H and the A/D.

For low dynamic range analog is still competitive

In 1990 a paper was presented at the ISCAS conference comparing analog and digital solutions, using a filter as an example. Equations for the minimum analog power consumed per pole, the minimum noise power and the minimum energy per cycle and pole were estimated. At the time analog solutions were better than digital in terms of power consumption. Currently, with the reduction of the channel's length, digital solutions are better than the analog counterparts when the dynamic range of the system is relatively large, above 65 or 70 dB. Analog is still the optimal solution when the dynamic range is relatively low.

The threshold-supply voltage race

36

➢ **Supply voltage and threshold scales with a similar pace**

Suppose to have $V_{Th} = 0.3\ V_{DD}$, and use the transistor as a switch.

$V_{DD} - V_{Th} - V_{ov}$ is the signal range

Use of complementary transistors can be a solution

R_{on} signal dependent -> harmonic distortion

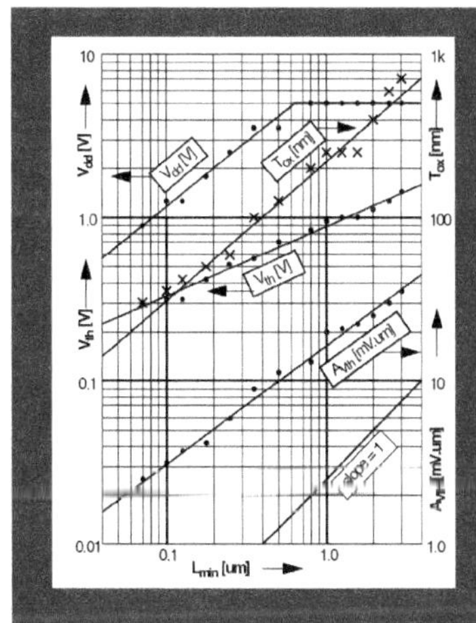

With the scaling down of nanometer CMOS technology, the oxide's thickness has been decreasing, requiring the use of smaller supply voltages to avoid destroying the gate of the transistors. With the reduction of the supply voltage, the performance of transistors operating as switches is degraded, since the ON resistance becomes increasingly dependent on the input signal, increasing the harmonic distortion introduced by these switches. A possible solution to overcome this drawback is the use of complementary transistors (transmission gates).

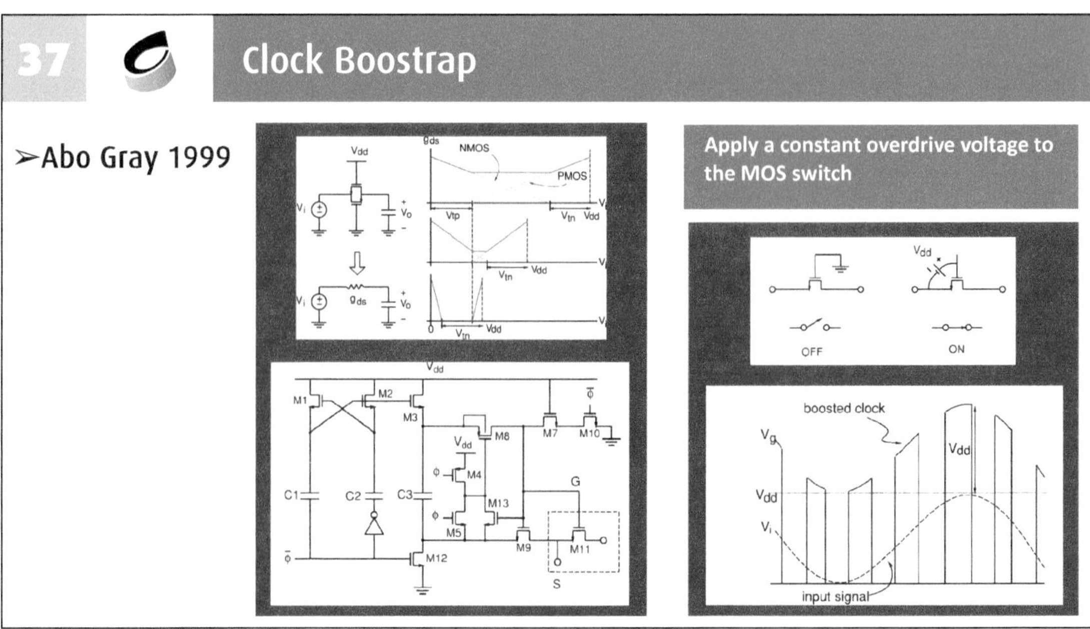

With the reduction of the supply voltage, the resistance of CMOS switches becomes increasingly non-linear, increasing the harmonic distortion that is introduced by these switches. When the supply voltage is low, clock boost circuits are typically used, increasing the voltage that is applied to the gate of the transistors, reducing the switches resistance and its nonlinearity.

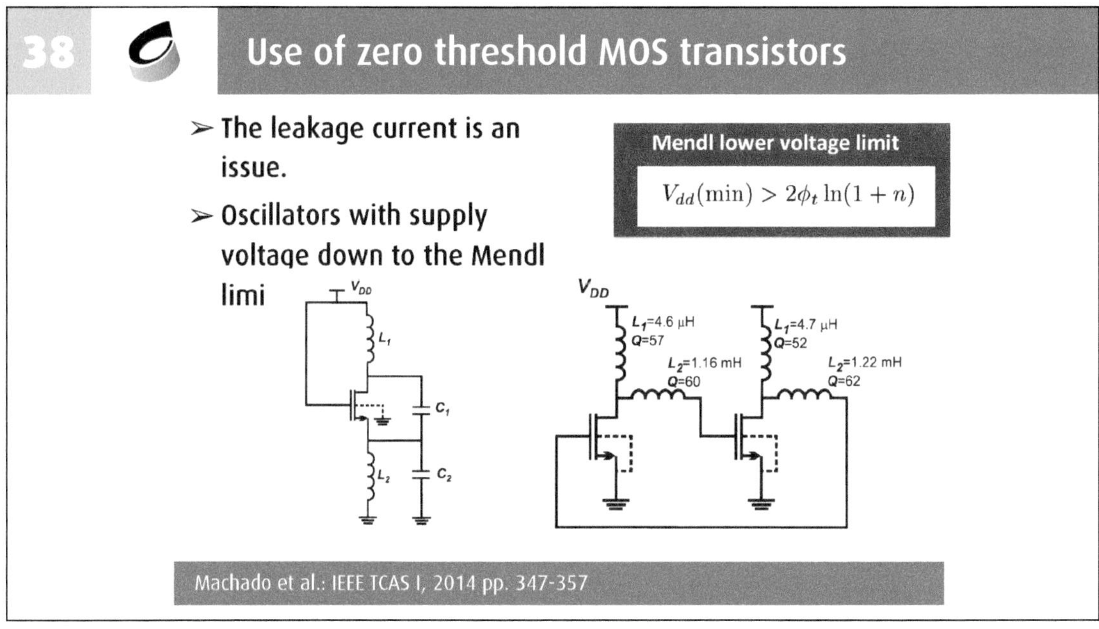

The enhanced swing Colpitts oscillator (common-gate) and the enhanced swing inductive-load ring oscillator (common-source) are two examples of circuits capable of operating with a supply voltage well below (ln 2) *k* T/q, by using a common-gate configuration, which always has an intrinsic gain higher than unity, or a transistor in common-source configuration, boosting the voltage gain with an appropriate inductive load. To achieve ultra-low-voltage voltage, zero-VT MOSFETs and high-quality-factor inductors should be used to operate with supply voltages below the thermal voltage.

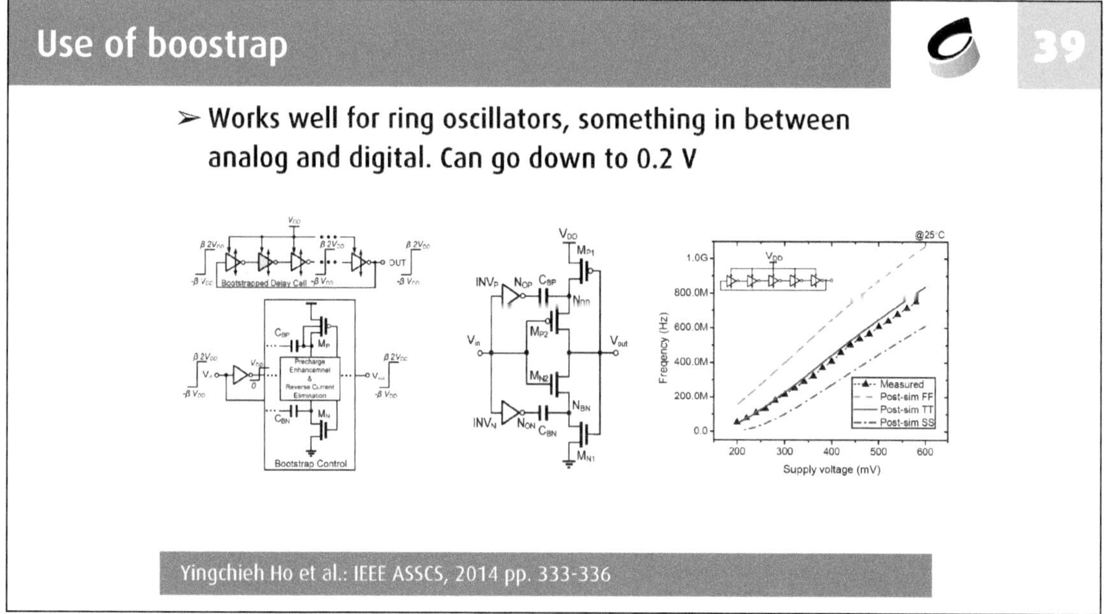

Use of boostrap

Yingchieh Ho et al.: IEEE ASSCS, 2014 pp. 333-336

The bootstrapped inverter-based ring oscillator is capable of operating with supply voltages between 200 and 600 mV. The operation in this voltage range is achieved by using a delay cell which provides a boosted voltage swing that enhances the driving capability and lowers the large process variations that occur in the low voltage region. The figure on the right shows the measured transfer curve of the operating frequency as a function of the supply voltage, which ranges from 200 to 600 mV.

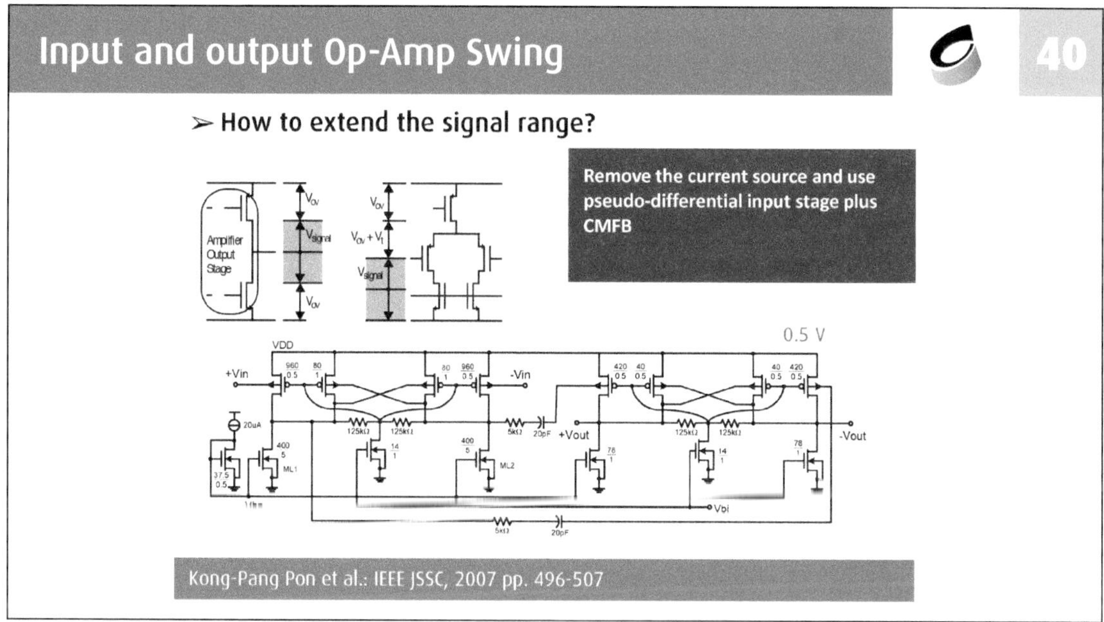

Input and output Op-Amp Swing
40
➢ How to extend the signal range?
Remove the current source and use pseudo-differential input stage plus CMFB
Amplifier Output Stage
0.5 V
VDD
+Vin
-Vin
+Vout
-Vout
Vbi
Kong-Pang Pon et al.: IEEE JSSC, 2007 pp. 496-507

Low voltage operation can be achieved using internal voltage boosting techniques, however, this approach increases the risk of destroying the transistor's gate, affecting the reliability of nanoscale CMOS implementations. Another option is to use low-threshold voltage options, although it increases fabrication costs since it involves using additional masks. The presented OTA achieves ultra-low-voltage operation without these approaches by using the transistors in an unconventional way, i.e., input signals are injected in the bulk.

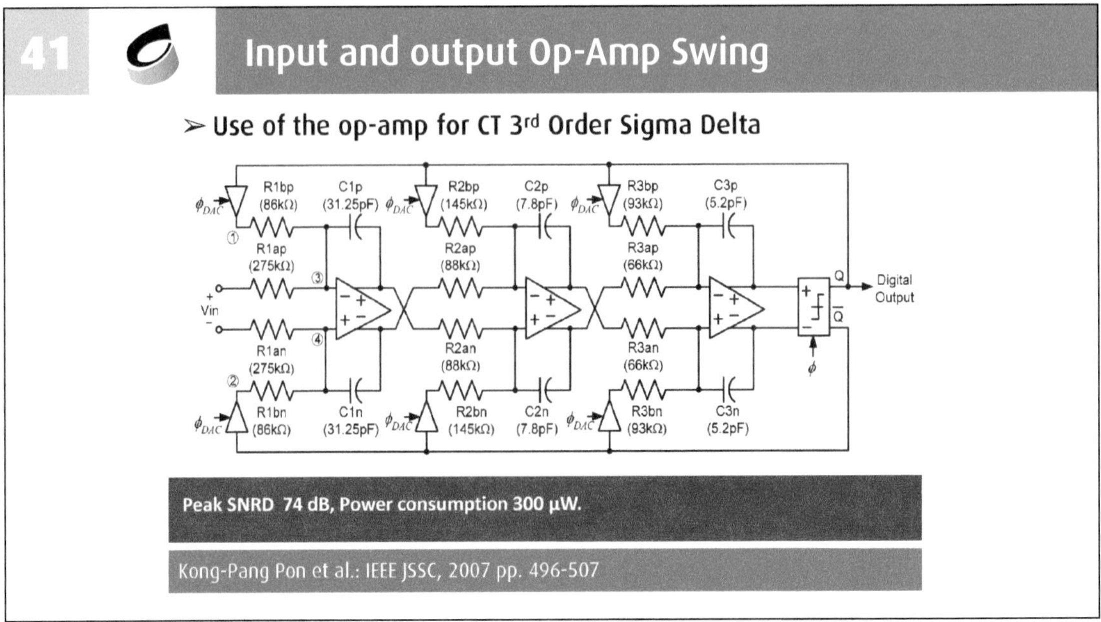

The same body-input technique is used in the comparator to achieve low-voltage operation. To avoid coupling between the multiple feedback paths when the DAC output floats, the DAC is split into several DACs, one for each feedback path. These DACs are appropriately sized so that there is no size or power penalty with this change. The sigma-delta modulator achieves a SNDR of 74 dB for a supply voltage of 500 mV and has a total power consumption (including the output buffers) of 370 μW.

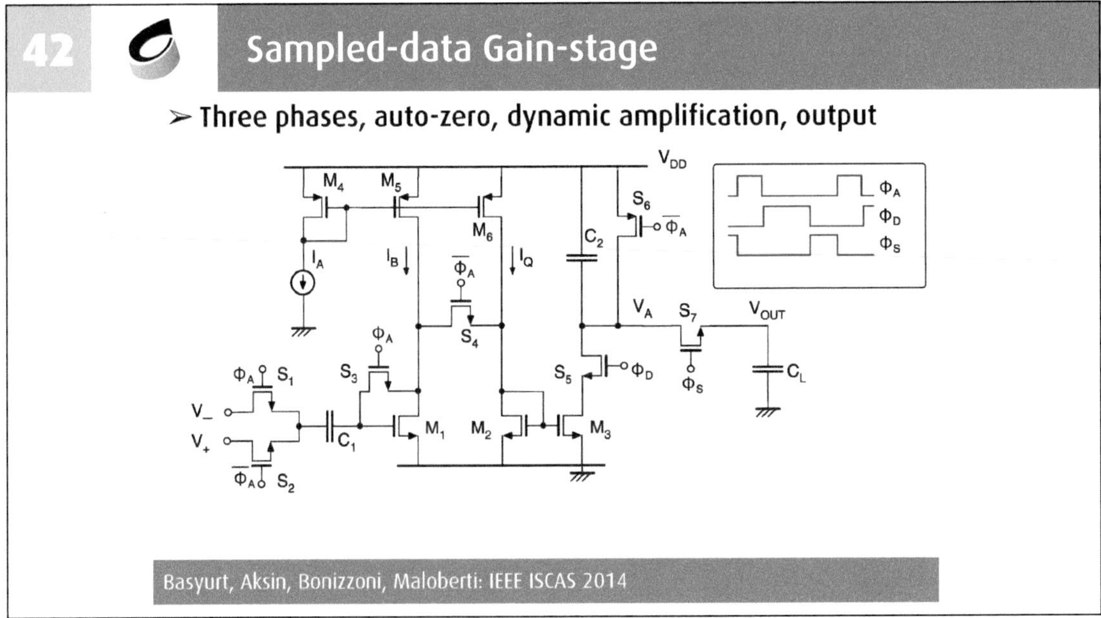

42
Sampled-data Gain-stage
➢ Three phases, auto-zero, dynamic amplification, output
Basyurt, Aksin, Bonizzoni, Maloberti: IEEE ISCAS 2014

This is an attempt to secure low voltage for a medium gain amplifier. During phase Φ_A, transistor M_1 is in diode configuration to charge capacitor C_1 with the difference between V_ and the V_{GS} of transistor M_1. During the amplification phase (Φ_D), the signal current flows through switch S4 to transistor M2 and is mirrored by a given factor to M_3, charging capacitor C_2, where current I_Q sets the quiescent discharge voltage. In the last phase (Φ_S), a new sample is available at the output.

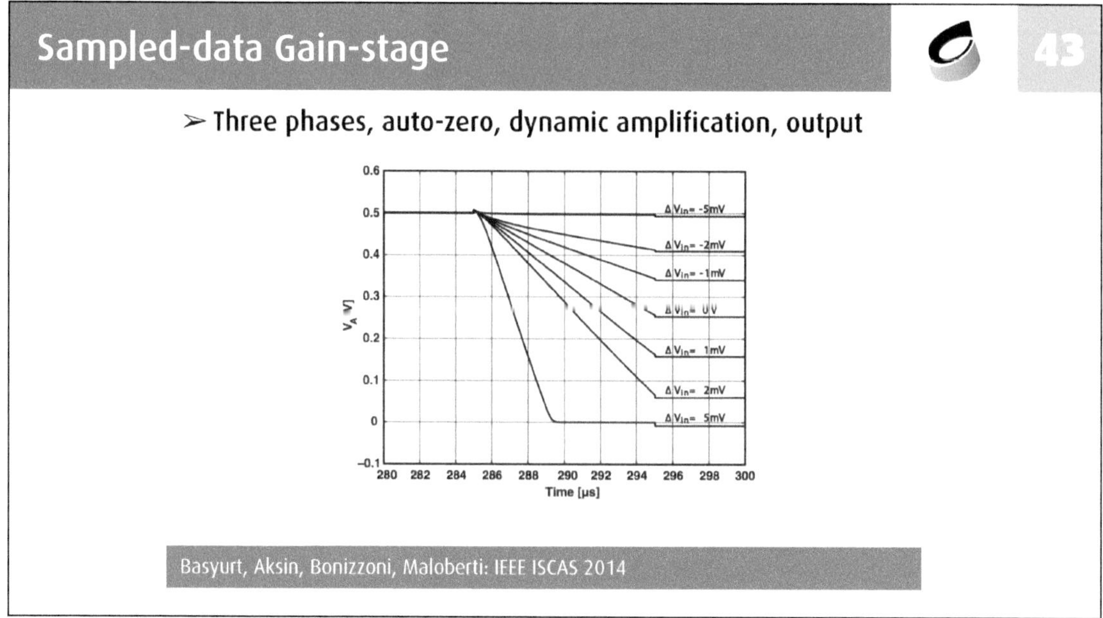

The plot shows how the voltage in node V_A changes for different differential input voltages. Due to the clock feedthrough introduced when switch S_6 opens and switch S_7 closes there is a small input referred offset. For differential input voltages higher than 3.5 mV the output begins to saturate to the supply levels. The circuits' behavior is also slightly asymmetrical.

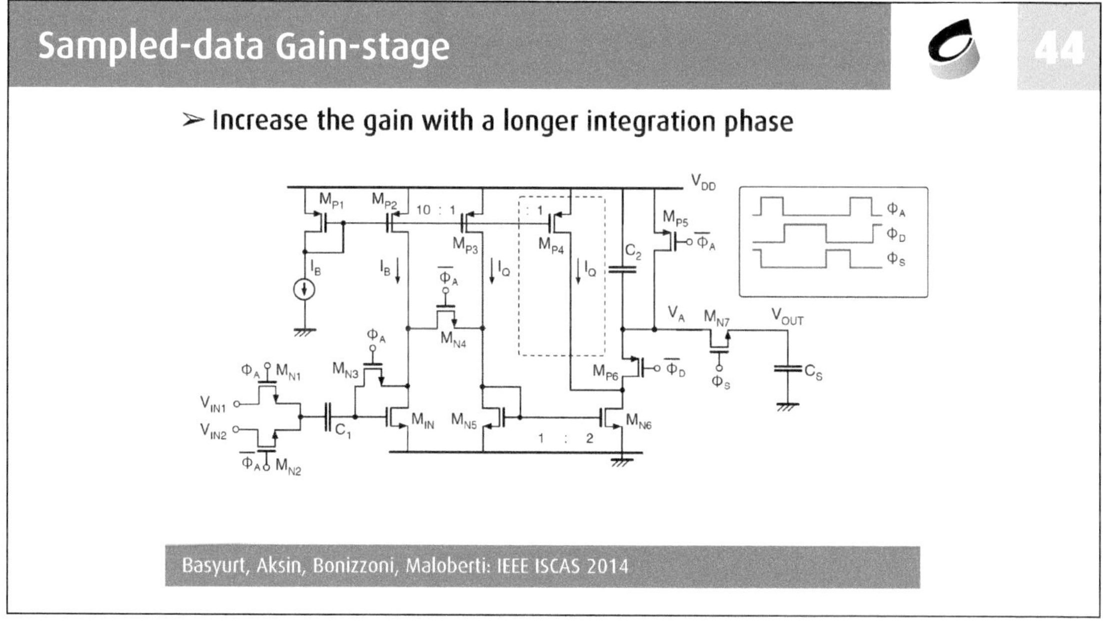

In order for the circuit to operate properly using a low supply voltage and a common-mode voltage of $V_{DD}/2$, the input transistors/switches (S_1/M_{N1} and S_2/M_{N2}) need to be driven by clock boosted phases.

The accuracy of the DC gain is dependent on accuracy of phase Φ_D, since it controls the charging time of capacitor C_2. The DC gain of the sampled-data gain stage can be improved by increasing the mirror factor between transistors M_2/M_{N5} and M_3/M_{N6} and by using transistor M_{P4} to supply the necessary current to set the quiescent current used to charge C_2 to the expected value I_Q.

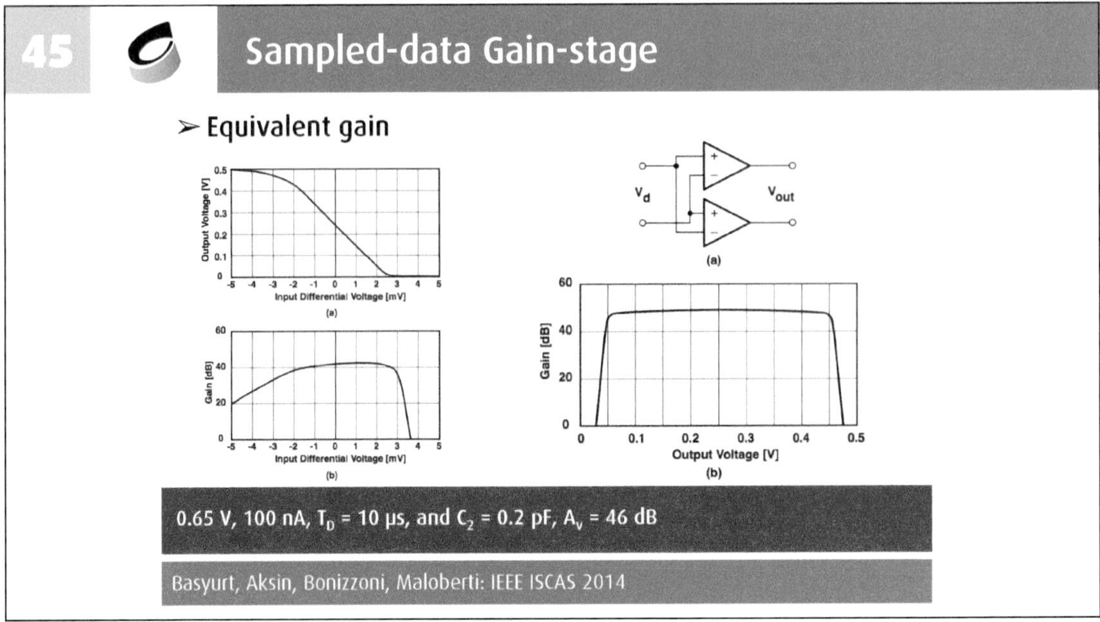

The figures on the right show the output voltage and the gain curves as a function of the input differential voltage. Due to the asymmetry of the gain curve, a pseudo differential configuration is used, increasing the sampled-data gain stage linearity in the output voltage range, at the cost of doubling the power consumption.

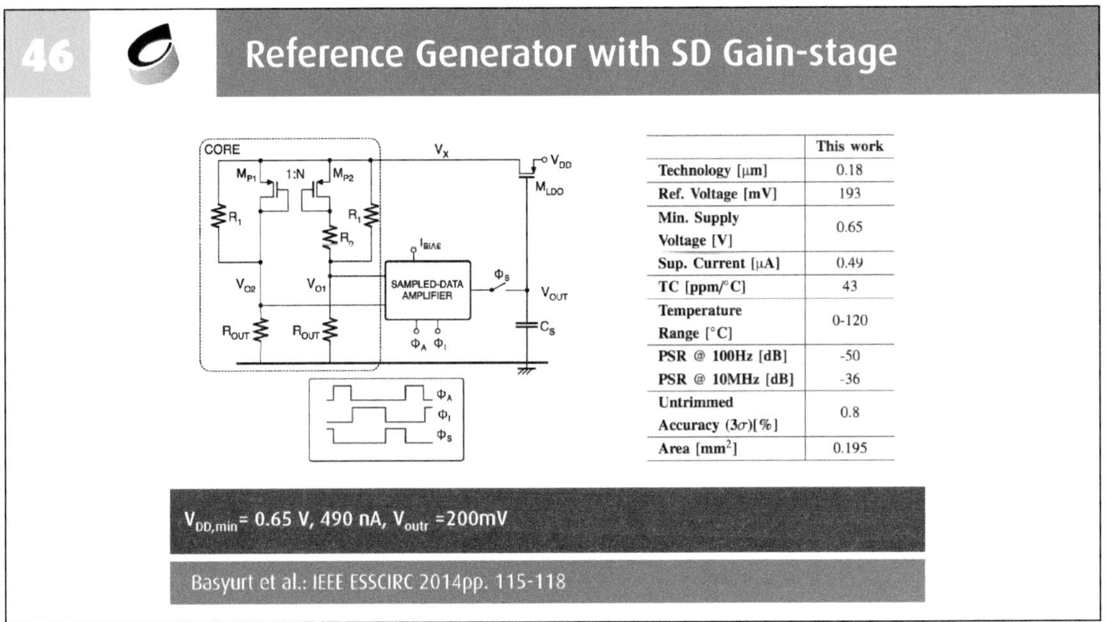

	This work
Technology [μm]	0.18
Ref. Voltage [mV]	193
Min. Supply Voltage [V]	0.65
Sup. Current [μA]	0.49
TC [ppm/°C]	43
Temperature Range [°C]	0-120
PSR @ 100Hz [dB]	-50
PSR @ 10MHz [dB]	-36
Untrimmed Accuracy (3σ)[%]	0.8
Area [mm²]	0.195

The non-conventional amplifier has been used in a low-supply voltage reference generator. The reversed current-mode bandgap circuit generates the reference voltage across the output resistors ROUT, where transistors MP1 and MP2 operate in the sub-threshold region to provide a diode-like voltage-current response to generate PTAT and CTAT currents, avoiding the use of current mirrors which are often source of inaccuracy. An LDO is used to generate the voltage Vx. The circuit works to a minimum supply voltage of 650 mV.

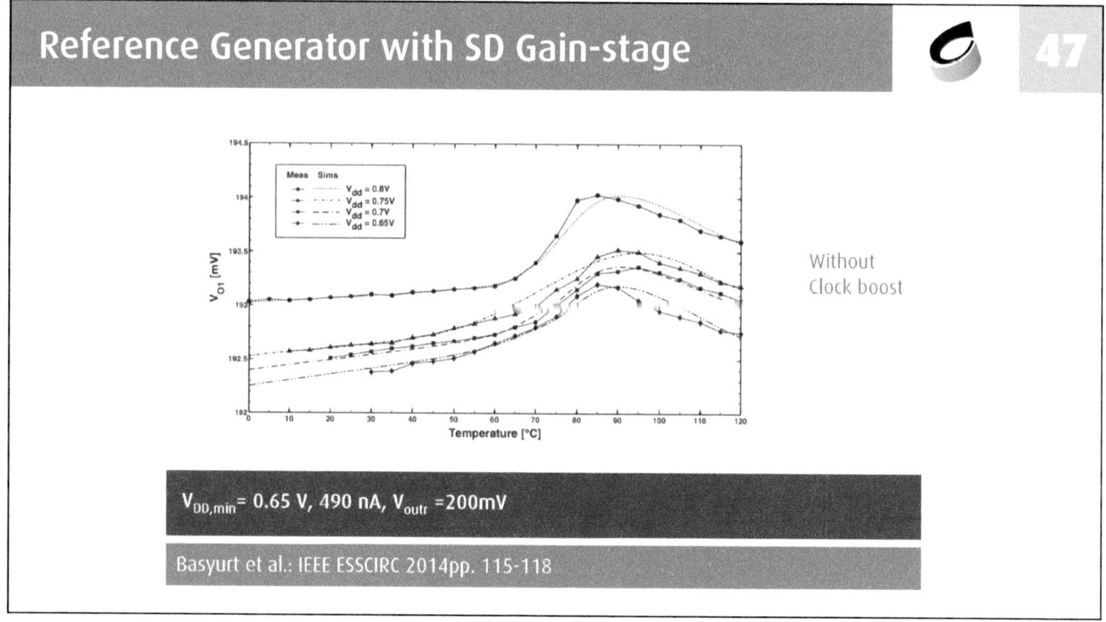

For lower supply voltages and low temperatures, the loop control may not work properly due to the threshold voltage of M_{LDO}. A better performance can be achieved using a clock boost circuit to increase the effectiveness of the switches. For the temperature range between 0 and 120 °C the temperature coefficient is 43 ppm/°C. At room temperature the supply current is 490 nA.

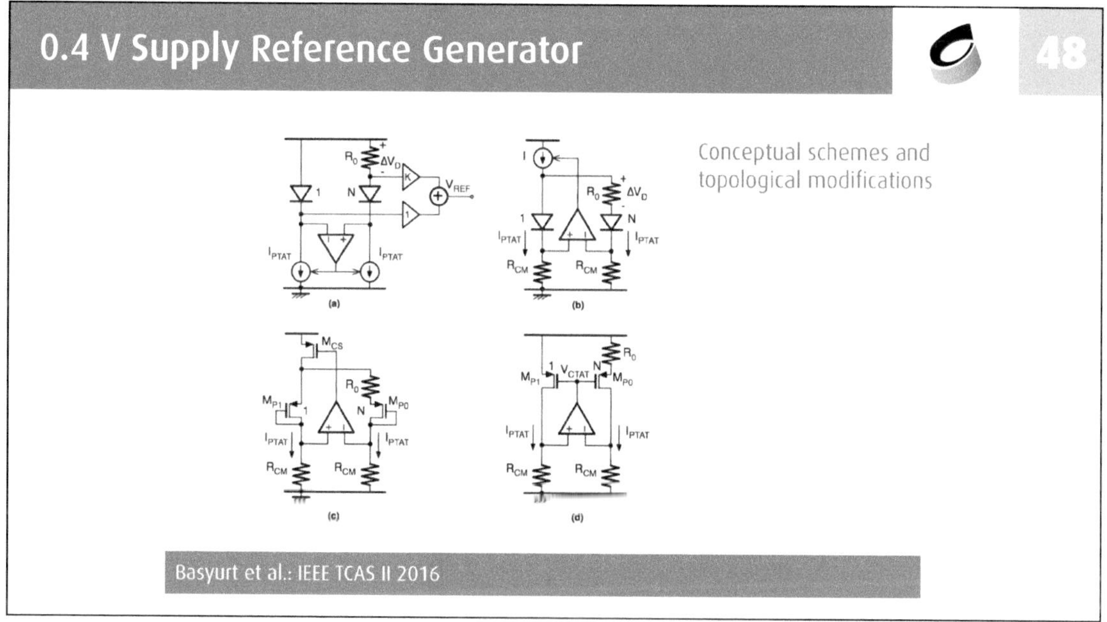

0.4 V Supply Reference Generator
48
(a)
(b)
(c)
(d)
Conceptual schemes and
topological modifications
Basyurt et al.: IEEE TCAS II 2016

The figure shows the measured results of the reference generator. Without the clock boost the circuit is not operating below a given temperature because of the increased threshold of transistors. The reference voltage is in the 200 mV range with 0.65 V supply.

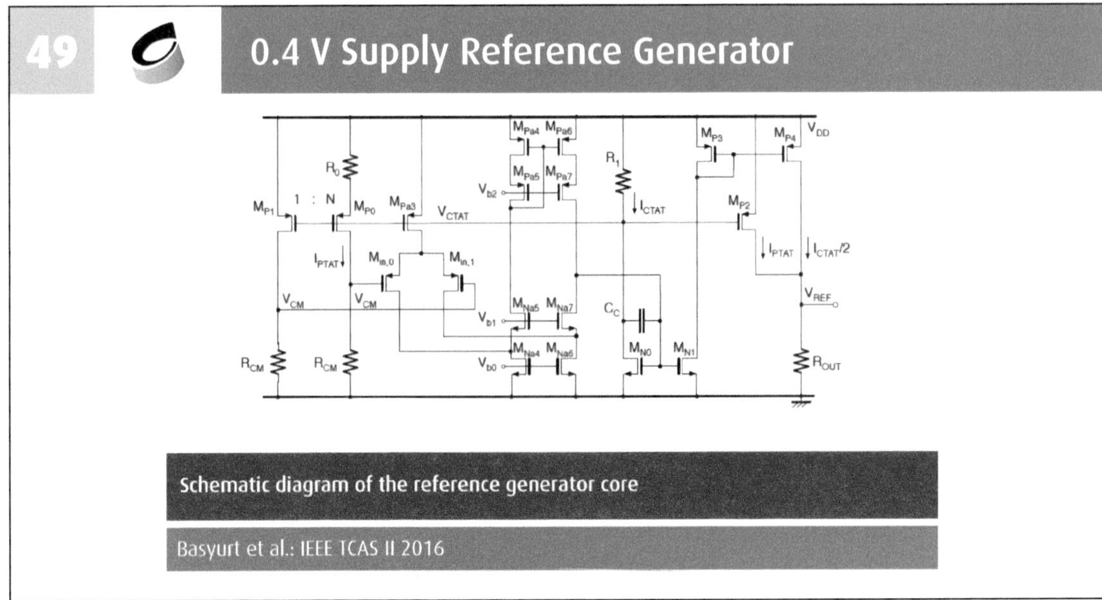

49
0.4 V Supply Reference Generator
Schematic diagram of the reference generator core
Basyurt et al.: IEEE TCAS II 2016

The figure shows different conceptual schemes of reference generators leading up to the circuit shown in the next slide. Circuit (a) shows the basic band-gap circuit, where the two equal currents generate a proportional to absolute temperature (PTAT) and a complementary to absolute temperature (CTAT) voltage. The reference voltage results from adding both these terms with the appropriate weights, equalizing the positive and negative temperature dependencies. Circuit (b) shows a variation of the basic circuit using current-mode operation to generate the PTAT current and circuit (c) its transistor level implementation. Circuit (d) generates the CTAT voltage through the VGS of transistor M_{P1}, while the PTAT current flows in the two branches. To generate the temperature compensated current, the V_{GS} of transistor M_{P1} needs to be transformed into a current.

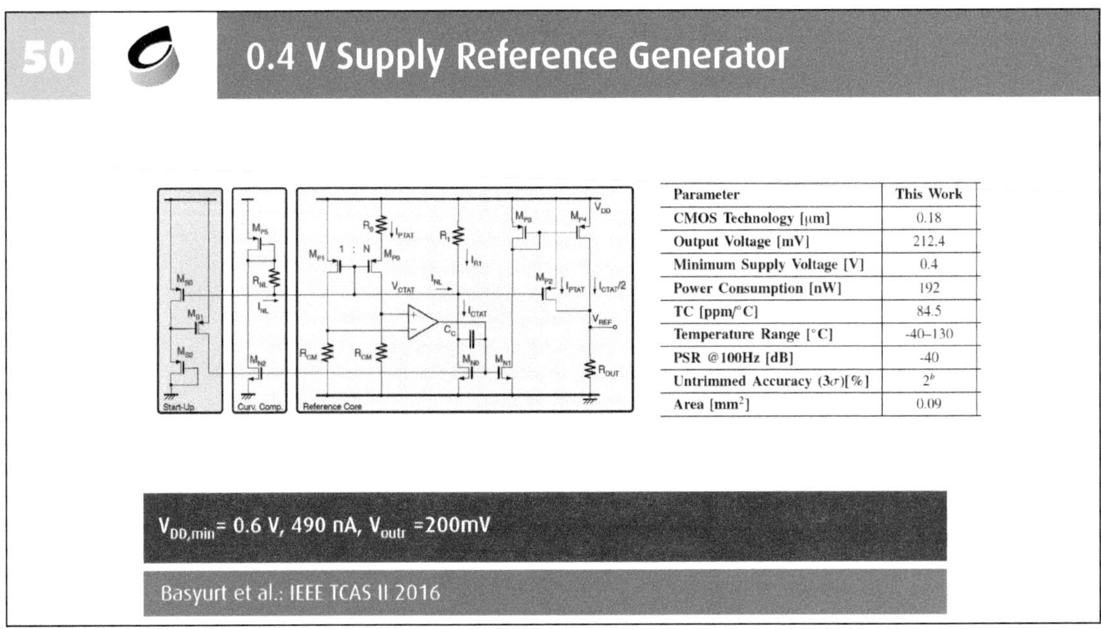

Parameter	This Work
CMOS Technology [μm]	0.18
Output Voltage [mV]	212.4
Minimum Supply Voltage [V]	0.4
Power Consumption [nW]	192
TC [ppm/°C]	84.5
Temperature Range [°C]	-40–130
PSR @100Hz [dB]	-40
Untrimmed Accuracy (3σ)[%]	2^b
Area [mm²]	0.09

The topological evolution of the reference scheme of the previous slide leads to the scheme shown here. The amplifier block achieves a large gain by using a folded-cascode scheme, biased by a scaled replica of the PTAT current and using PMOS transistors at the input and a resistive load in the second stage.

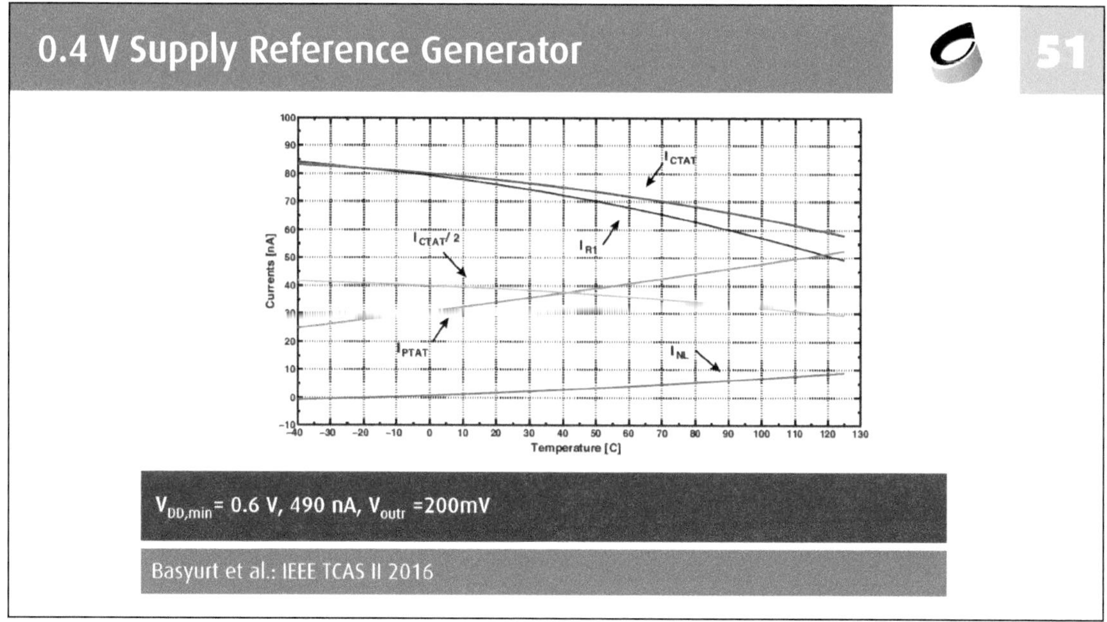

The circuit requires start-up and to increase the linearity over a wide temperature range the scheme includes a curvature correction circuitry. Curvature correction is needed due to the non-linear behavior of the V_{GS} voltage (M_{P0} and M_{P1}) when operating in the sub-threshold region.

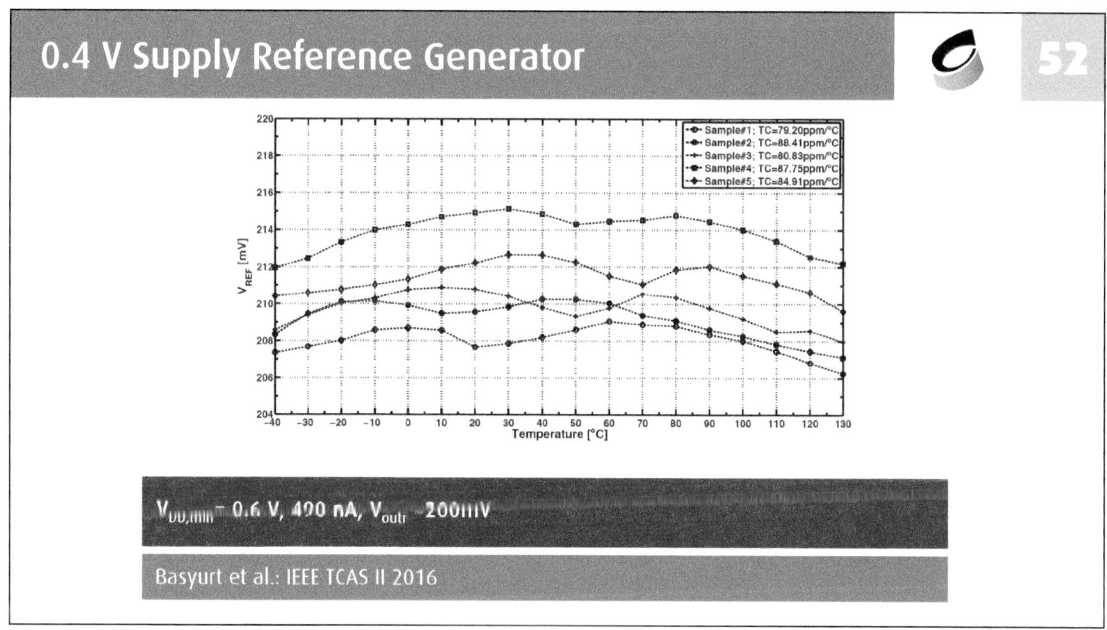

The graph shows the simulated results of the currents flowing in the different branches of the circuit as a function of the temperature. The difference between currents I_{R1} and I_{CTAT} is due to the non-linear temperature dependence of the gate-source voltage of transistor M_{P1}.

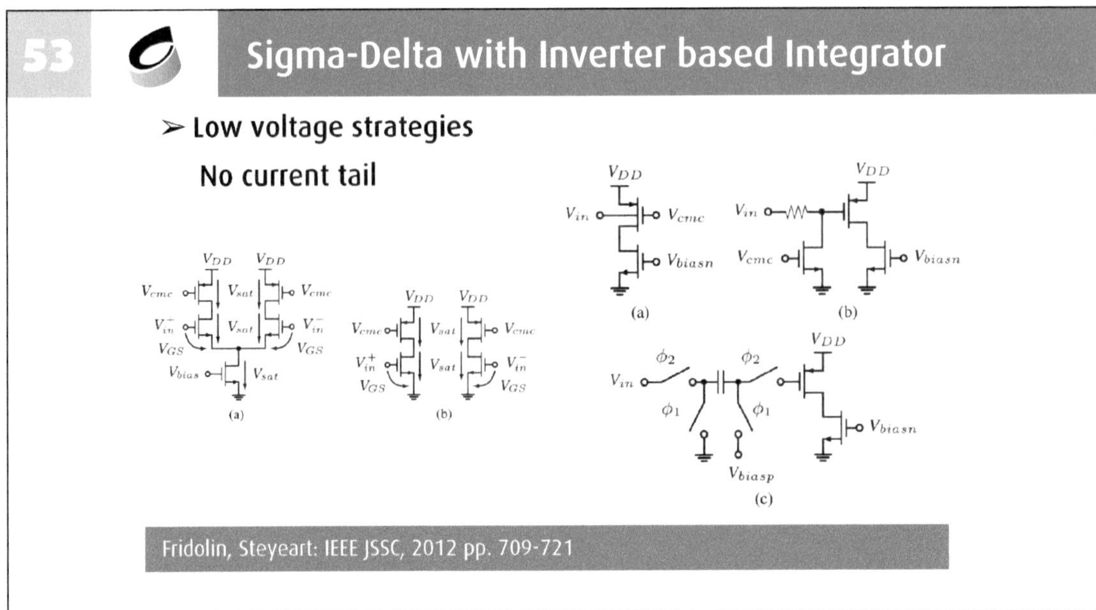

53
Sigma-Delta with Inverter based Integrator
➢ Low voltage strategies
No current tail
(a)
(b)
(a)
(b)
(c)
Fridolin, Steyeart: IEEE JSSC, 2012 pp. 709-721

The graph shows the measured reference voltage variation as a function of the temperature for 5 difference samples. In order to achieve the best possible temperature coefficient, resistor R1 was trimmed. The average reference voltage is 212.4 mV with a mean temperature coefficient of 84.5 ppm/°C over a temperature range between -40 and 130 °C. At room temperature the supply current is 480 nA and the power consumption is 192 nW.

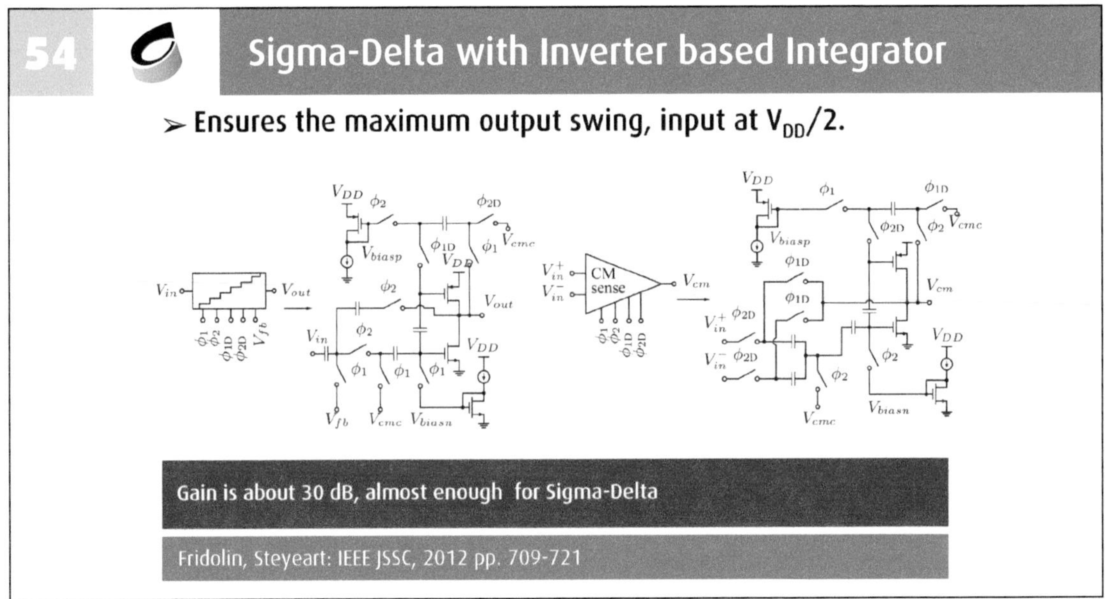

54
Sigma-Delta with Inverter based Integrator
➢ Ensures the maximum output swing, input at V_{DD}/2.
CM sense
Gain is about 30 dB, almost enough for Sigma-Delta
Fridolin, Steyeart: IEEE JSSC, 2012 pp. 709-721

In a typical differential pair the minimum supply voltage is limited by the saturation voltage and by the threshold voltage variation, while the input common-mode range is mainly limited by threshold voltage. To enable operation below 340 mV, the tail current source has to be removed, allowing the minimum supply voltage to drop to two times the saturation voltage (≈ 180 mV), although the common-mode rejection ratio decreases. In inverter based designs, the minimum swing that the analog circuit can still operate on also needs to be considered. To achieve large overdrive voltages with low supply voltages different techniques can be used, such as bulk input, resistive level shift, and capacitive level shift.

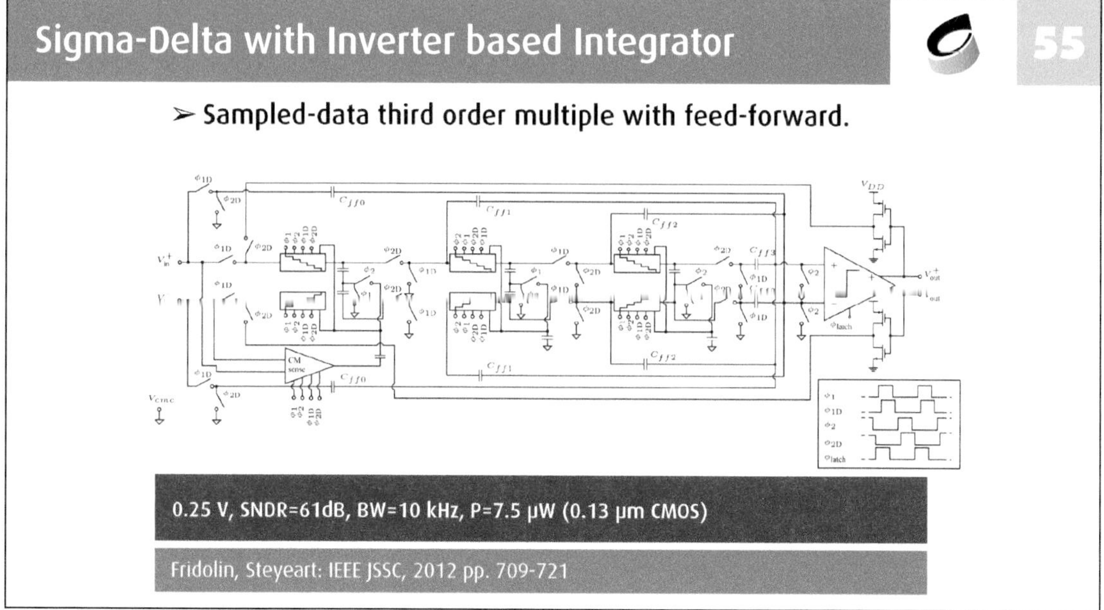

Sigma-Delta with Inverter based Integrator
55
➢ Sampled-data third order multiple with feed-forward.
0.25 V, SNDR=61dB, BW=10 kHz, P=7.5 µW (0.13 µm CMOS)
Fridolin, Steyeart: IEEE JSSC, 2012 pp. 709-721

The integrators are implemented using the SC technique due to its superiority in realizing large overdrive and common-mode cancellation. Using the pseudo differential inverter based topology, it is possible to double the gain, however, in order to achieve better control in the bias current and larger overdrives, the gates of the NMOS and PMOS devices in the inverter are capacitive biased independently. Phase Φ_1 is used as a biasing and offset compensation phase and phase Φ_2 as an amplification phase. The input common-mode sense amplifier uses the same biased inverter amplifier structure but with opposite phases since the common-mode needs to be available for the integrator during his biasing phase Φ_1.

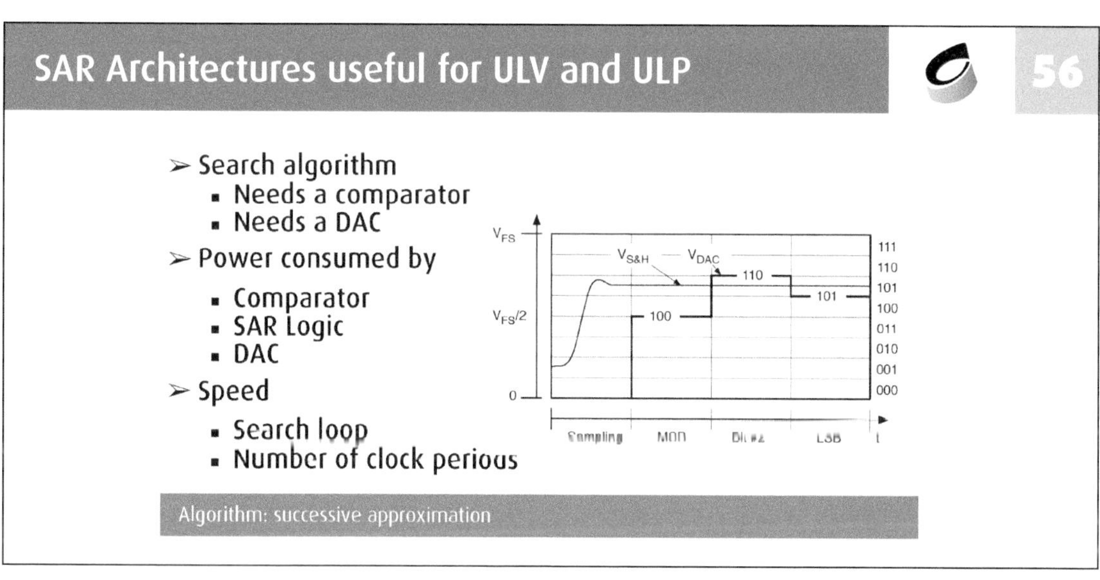

SAR Architectures useful for ULV and ULP
56
➢ Search algorithm
▪ Needs a comparator
▪ Needs a DAC
➢ Power consumed by
▪ Comparator
▪ SAR Logic
▪ DAC
➢ Speed
▪ Search loop
▪ Number of clock perious
V_FS
V_FS/2
0
V_S&H
V_DAC
110
100
101
111
110
101
100
011
010
001
000
LSB
t
Algorithm: successive approximation

This circuit uses the above described techniques to realize a sigma-delta modulator. Due to the ultra-low supply voltage, a pseudo differential design is reccommanded. A discrete-time common-mode loop is used in all integrators to reduce their common-mode gain to 1, in order to prevent accumulation by the integrators. Since large common-modes can still saturate the output of the integrators, common-mode cancellation is used in the first integrator and in the comparator.

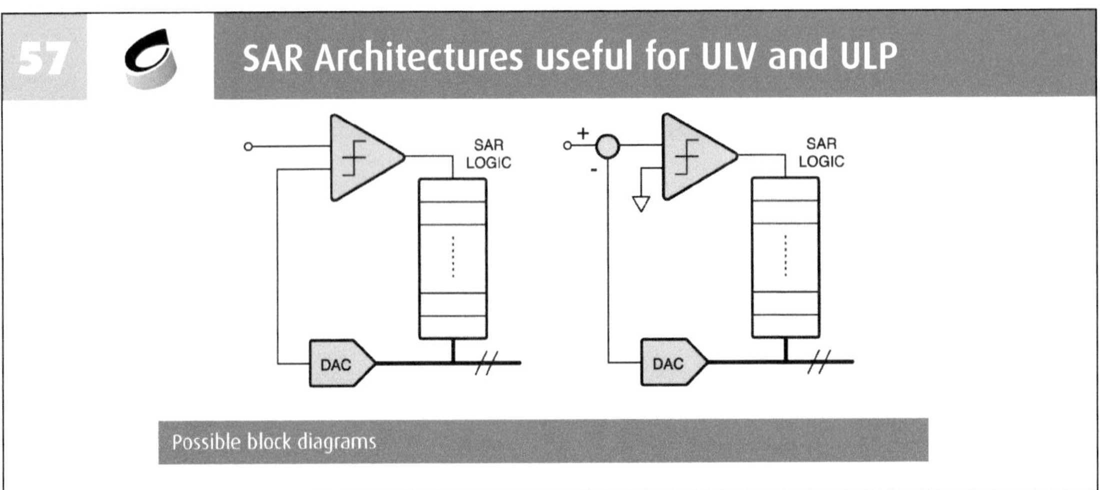

Possible block diagrams

SAR architectures are suitable for low power. They use a sample-and-hold circuit, a comparator, a DAC, and a SAR logic block. Depending on the architecture chosen the sample-and-hold circuit might not be needed, if the DAC used provides an inherent sample-and-hold operation. One example of the sequency of operations of a SAR ADC is shown in the graph where, after the sampling phase, the output of the DAC is successively compared with the common-mode voltage (the second input of the comparator) until finding the respective digital word for the current sampled signal.

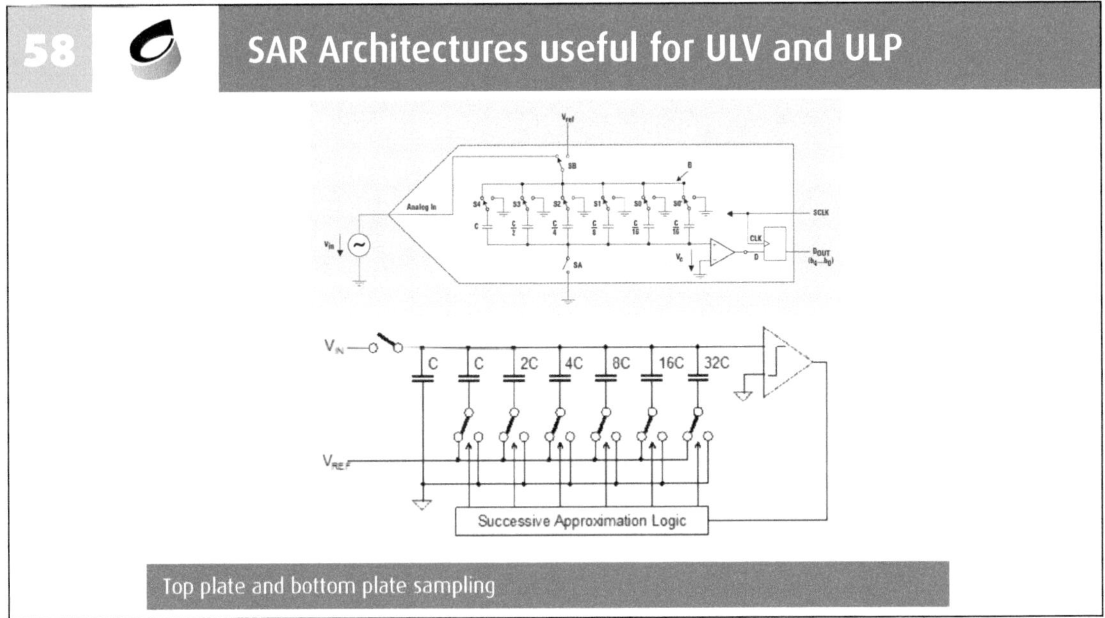

Top plate and bottom plate sampling

Two different architectures can be used to implement a SAR ADC. In the first (left), the input comes from a sample-and-hold circuit since the signal must remain constant during the conversion cycle. The drawback of this approach is the variable common mode input voltage that often requires additional and high power consumption to ensure linearity of the needed implement this sample-and-hold circuit.

The second architecture (charge redistribution SAR ADC) relies on a DAC operating as a sample-and-hold circuit to first sample the signal into the DACs capacitor array and then compare it to the common-mode voltage during n clock cycles. The main advantage of this architecture is the power saved in the sample-and-hold circuit due to the inherent sample-and-hold operation of the capacitive DAC.

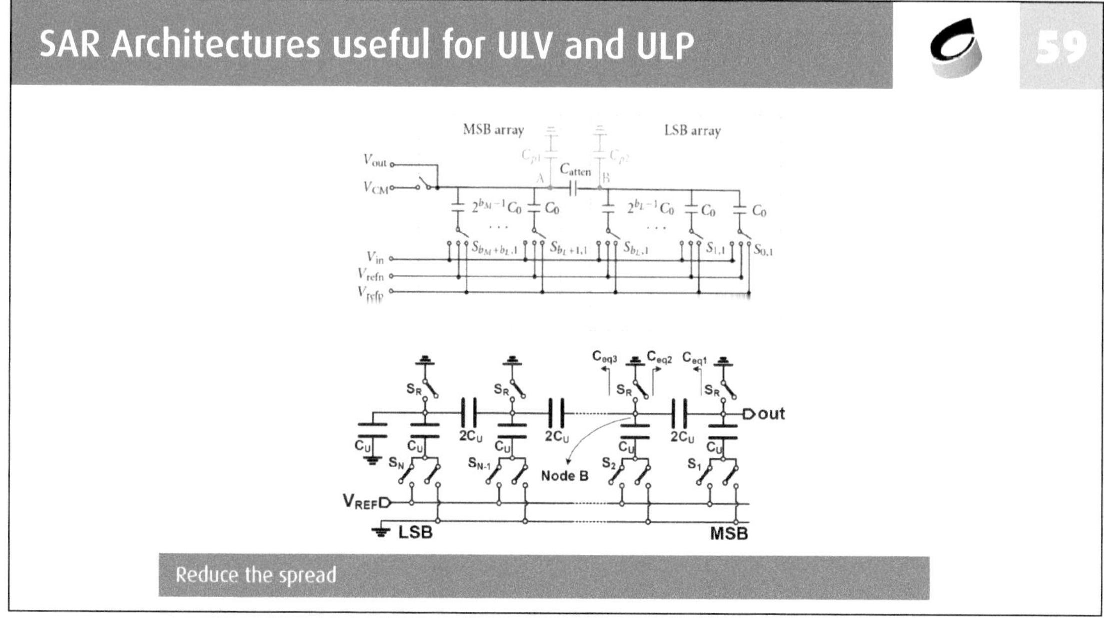

This DAC consists of a binary weighted capacitive array and a dummy LSB capacitor, resulting in a total capacitance value of $2^N \times C$. This array is both used for sampling and for conversion without requiring an additional phase to compare. One of the problems of this structure is the capacitor spread when the number of bits increases and the fact that the power consumption increases exponentially with the number of bits.

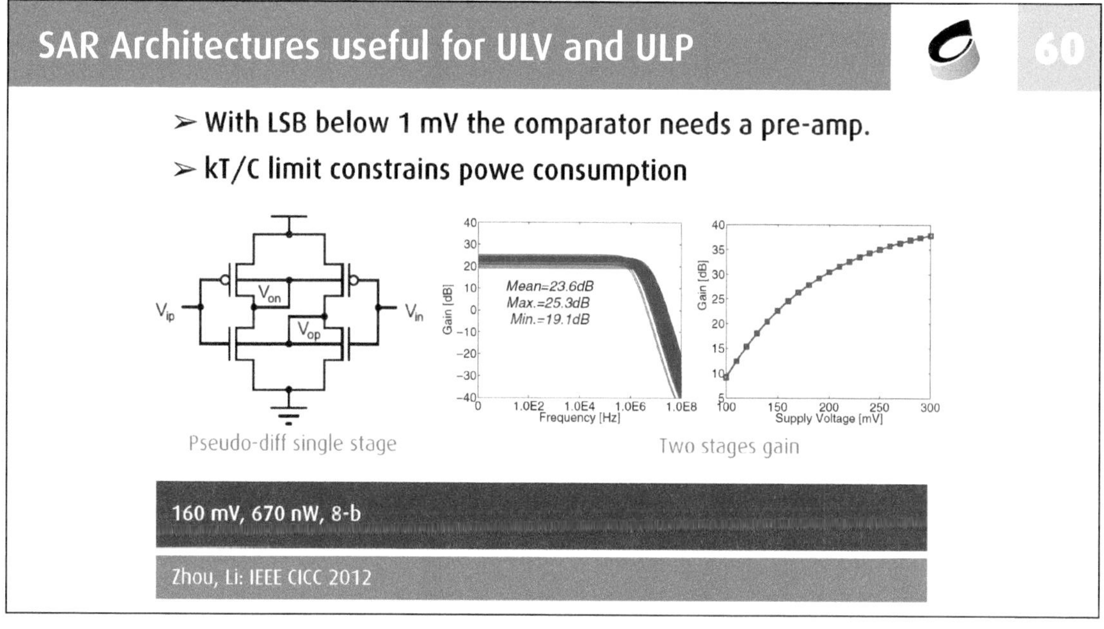

SAR Architectures useful for ULV and ULP
60
➢ With LSB below 1 mV the comparator needs a pre-amp.
➢ kT/C limit constrains powe consumption
Mean=23.6dB
Max.=25.3dB
Min.=19.1dB
Gain [dB]
Frequency [Hz]
Supply Voltage [mV]
Pseudo-diff single stage
Two stages gain
160 mV, 670 nW, 8-b
Zhou, Li: IEEE CICC 2012

To reduce the problem of the capacitor spread in the binary-weighted capacitor array from the previous slide, a split capacitor array or a C-2C structure can be employed. Using these capacitor configurations, the capacitance values are significantly reduced and for the C-2C scheme the power consumption increases linearly with the number of bits. Due to the smaller size of the capacitors, it is possible to achieve higher speeds while dissipating less power. The drawback of this approach is the degradation of the SARs linearity due to the effect of the parasitic capacitances.

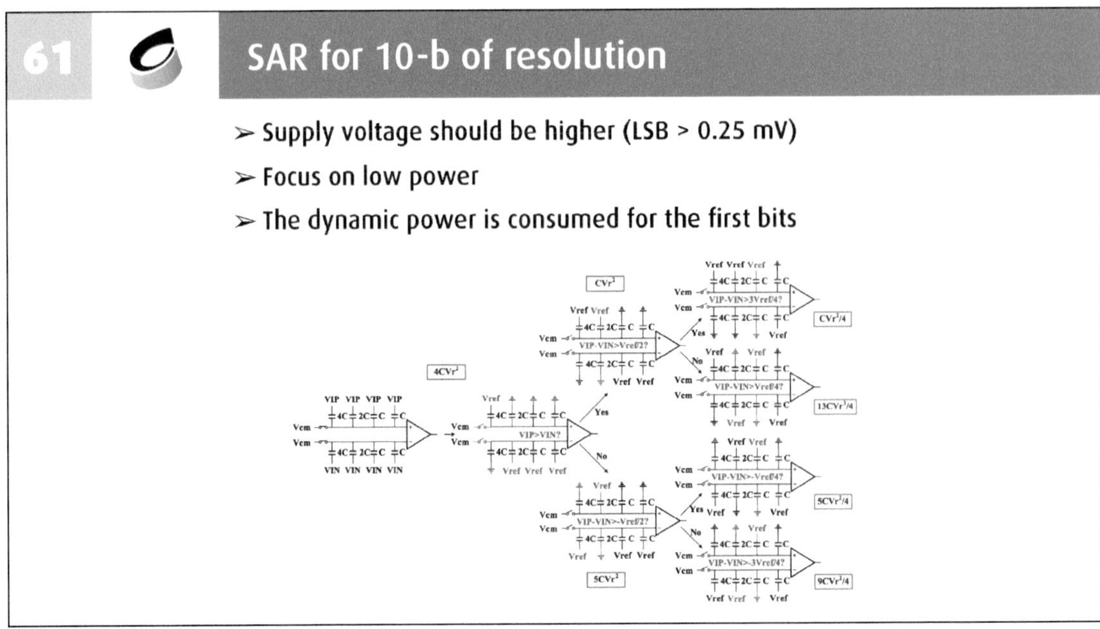

To achieve comparisons below the 1 mV range, the subthreshold SAR uses an inverter-based amplifier architecture, relying on the body-input technique to reduce the voltage constraints while achieving a reasonable gain for a supply voltage of 160 mV. For an 8-bit SAR, it is possible to achieve a SNDR of 45.7 dB with a sampling frequency of 40 kHz and power consumption of 670nW.

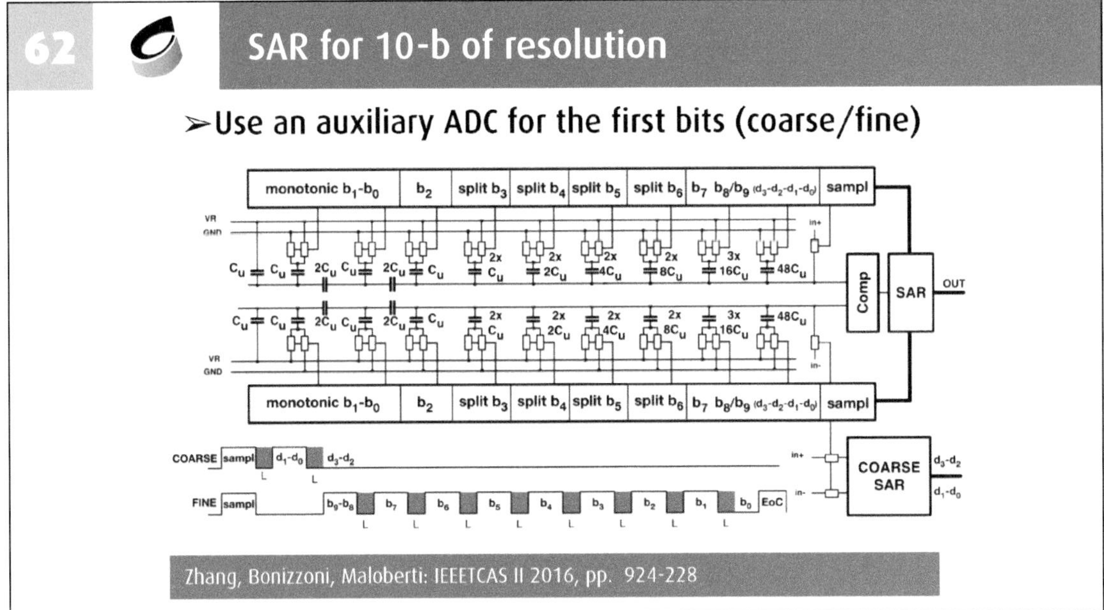

When designing an ultra low power SAR converter it is necessary to reduce the power consumption in the comparator, in the SAR logic, and in the DAC, because when the connections in the capacitors change position, charge is moving between capacitors, consuming power. The output of the comparator will determine which switches change position, so the power consumption will be dependent on the code that is generated. The diagram shows the consumed power as a function of the generated code for the first three bits of conversion.

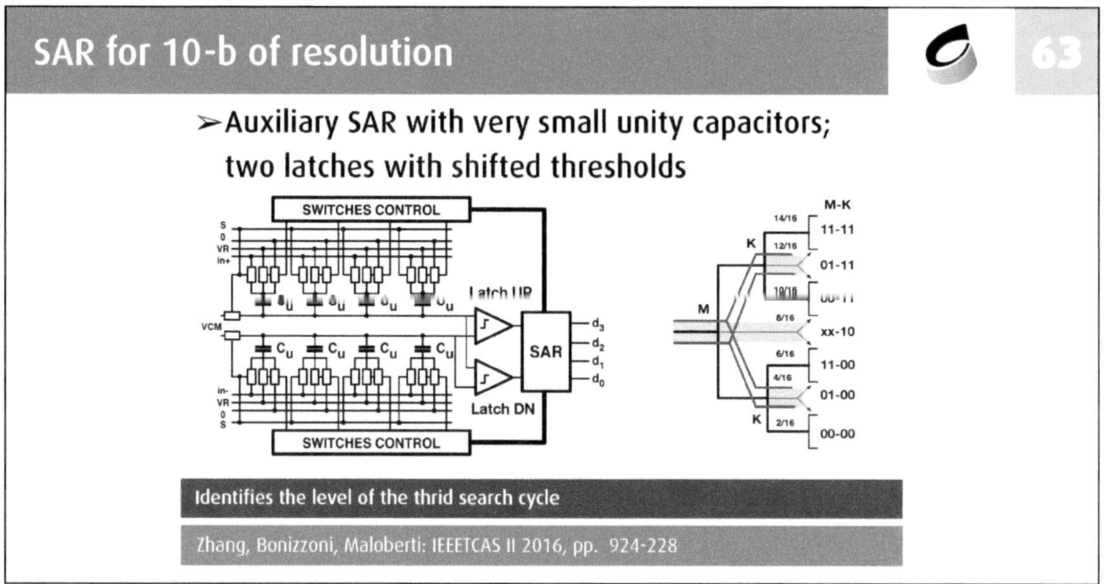

If the goal is to achieve 10-bit the LSB becomes too low with deep sub Volt supply. In this case the focus is mainly on low power. The design issues are architectural and on how obtaining comparison with large sensitivity. This architecture consists of a coarse and a fine SAR ADC. The 3 bit coarse SAR presets the two MSB capacitive arrays of the fine SAR, avoiding the largest source of dynamic power consumption. The comparator used in the fine SAR ADC has high sensitivity and very low power due to the use of a gain enhanced dynamic preamplifier. In the clocked mode, for a sampling rate of 200 kHz, the ADC achieves 9.05 effective number of bits with a power consumption of 200 nW.

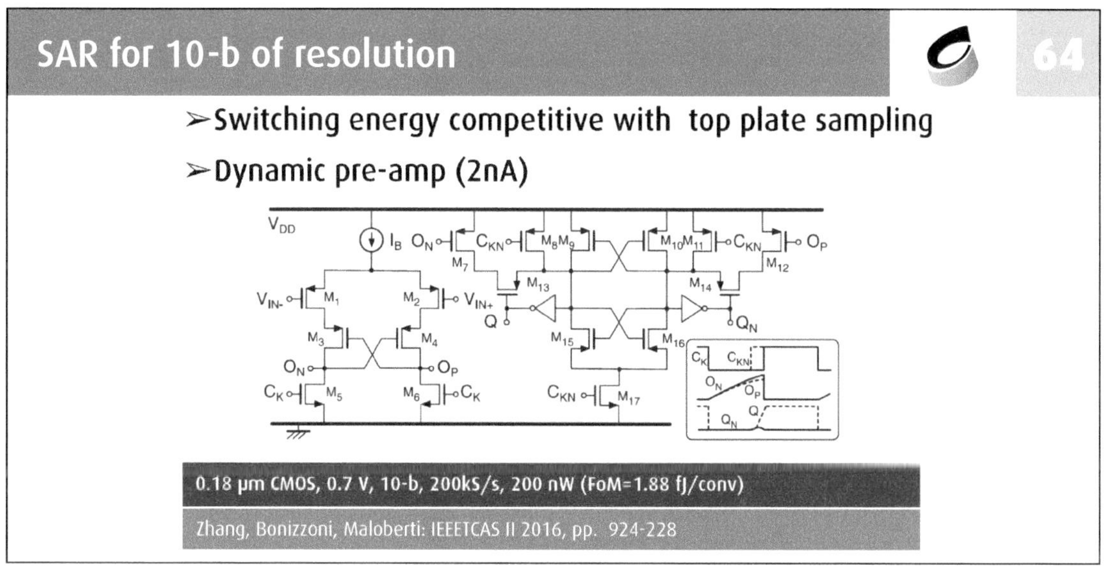

The coarse SAR ADC uses two comparators (one with the threshold shifted $V_{ref}/8$ up and the other shifted down by the same amount) and a 2 bit DAC (can be a simple latch). After the sampling phase, in the first estimation (M), the latches generate output d3 and d2, indicating if the input is well above $V_{ref}/2$, well below $V_{ref}/2$, or in the gray region. The second estimation (K) works in the similar manner to generate the other two outputs (d_1 and d_0) but around $3V_{ref}/4$ and $V_{ref}/4$. Due to the small size of the unity capacitors (3.1 fF) and the use of two simple latches the power consumption of the coarse SAR ADC is negligible.

65 

How obtaining lower power and voltage?

- ➢Use of non conventional circuits
- ➢Use of non conventional algorithm
- ➢Use of nanometer technologies
- ➢Accept low dynamic range
- ➢Transmit only what is essential
- ➢On board processing
- ➢Minimize the digital processing

The comparator used in the fine SAR ADC is implemented using a two stage approach. The first stage performs a dynamic pre-amplification of the LSB voltage, where cross-coupled transistors M3 and M4 are used to enhance the gain and transistors M_5 and M_6 operate as switches and as capacitive loads depending on whether the switch is ON or OFF. The second stage is a conventional regenerative latch using two inverters to avoid static current, achieving rail-to-rail digital outputs.

SAR with supply lower than 0.3 V?

- ➢ On going project
- ➢ Resolution of 5-bit

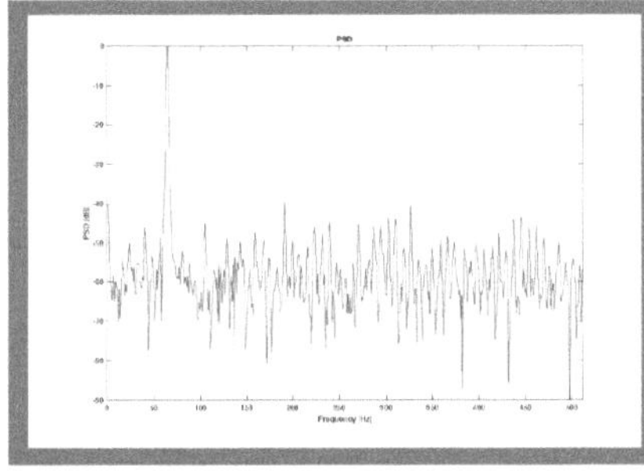

28 nm CMOS, 0.3 V, 5-b, 100kS/s, 30 nW

Basyurt, Bonizzoni, Maloberti

To obtain low power and low voltage, conventional circuits can be used, although very often having a nonconventional solution can be very interesting. Nanometer technologies are also very important since due to the switching between transmission and sleep modes, the parasitic capacitances are charged and discharged consuming additional power, and since these capacitances are scaled with the technology node, the switches consume less power in smaller technologies. To achieve low power it is also necessary to accept low dynamic range, or implement a dynamic solution capable of using more power to achieve higher resolution only when necessary. Only the essential information should be transmitted and with the minimum use of digital circuits. This is an example of an ongoing project capable of achieving low resolution with a supply voltage of 0.3 V, using 28 nm technology, consuming only 30 nW.

Conclusions

67

- IoT will be a very important breakthrough:
  - Impact on human life (safety and privacy)
  - Increase efficiency of things
- To become IoE (everything and not something)
  - Link wired, refueled and autonomous things
  - Harvesting and storing power
  - Ultra low voltage and micro or better nano-power

Voltage in the few 100mV range
Power in the few 10 nW range

In conclusion, the Internet of Things is very important for affecting the human life, to improve the efficiency of things, but if we want to have the Internet of everything and not the Internet of some-thing it is necessary to have activities in the study of new architectures, harvesting and storing power, but also to achieve ultra low voltage (100 mV) and micro or nano-power (10 nW) operation.

CHAPTER 2

SAR ADCs for Internet of Things: Basics and Innovations

Pieter Harpe
Eindhoven University of Technology
Eindhoven, The Netherlands

This chapter will cover the basics and recent innovations in the field of SAR ADCs for IoT. Even though SAR ADCs have been known for a very long time, they receive a lot of attention recently thanks to their power-efficiency and beneficial scaling with technology. This lecture starts with a basic overview of SAR ADC design, and continues with discussing several recent innovations aiming for better power efficiency, improved performance, and more versatility.

1 

Outline

1. **Introduction to ADCs for IoT**
2. Basic SAR ADC Architecture
3. Circuit Design Considerations
4. Advanced Examples for IoT
5. Summary and Conclusions

2 

ADCs for IoT

- Applications in IoT
  - Sensor interfaces
  - Radio front-ends
- Specifications
  - Very low power
  - Moderate precision
  - Relatively low speed
- Features of interest
  - Duty-cycled operation (on-demand sensing)
  - Versatility (speed, precision)

[1]

[2]

In IoT, the applications are mainly sensor interfaces and radio front-ends. Depending on the application, some specifications can be more relevant. In general, low power is crucial due to limited energy sources, and often moderate precision and speed are sufficient for applications like sensors and low data-rate radios. There are additional features of interest such as duty-cycled operation, where the hardware is idle (low power consumption) for a certain period of time and is only active when required. Another feature could be versatility, to accommodate different applications with a single design.

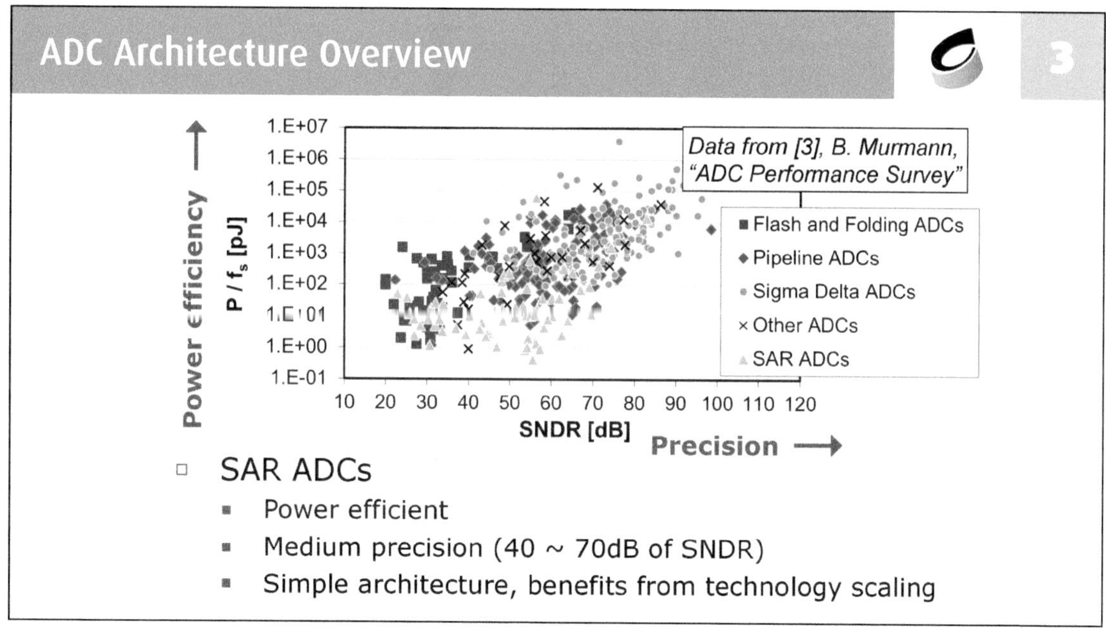

For low-power medium-precision (6-12b) applications, SAR ADCs pose themselves as the most viable architecture, due to their scaling-friendly nature. Since SAR ADCs use a switched-capacitor structure, they directly benefit from faster transistor speeds, while gradually lower intrinsic transistor gains (g_m/g_{ds}) are not as critical as for other architectures, since here precision amplifiers are not required. For higher resolutions, Sigma-Delta ($\Sigma\Delta$) ADCs are still the architecture of choice.

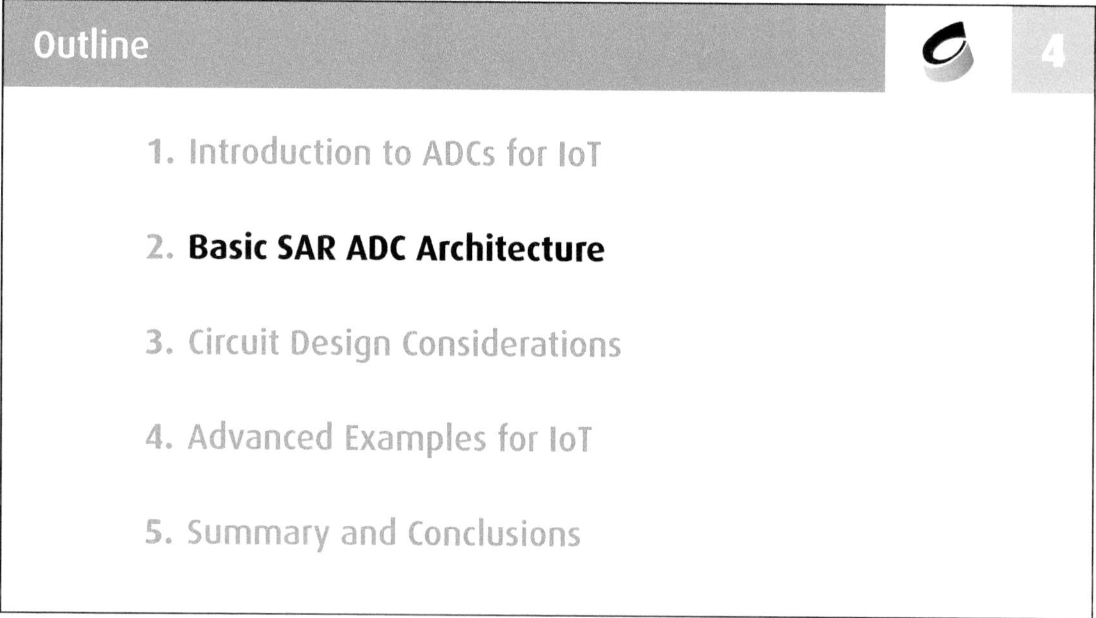

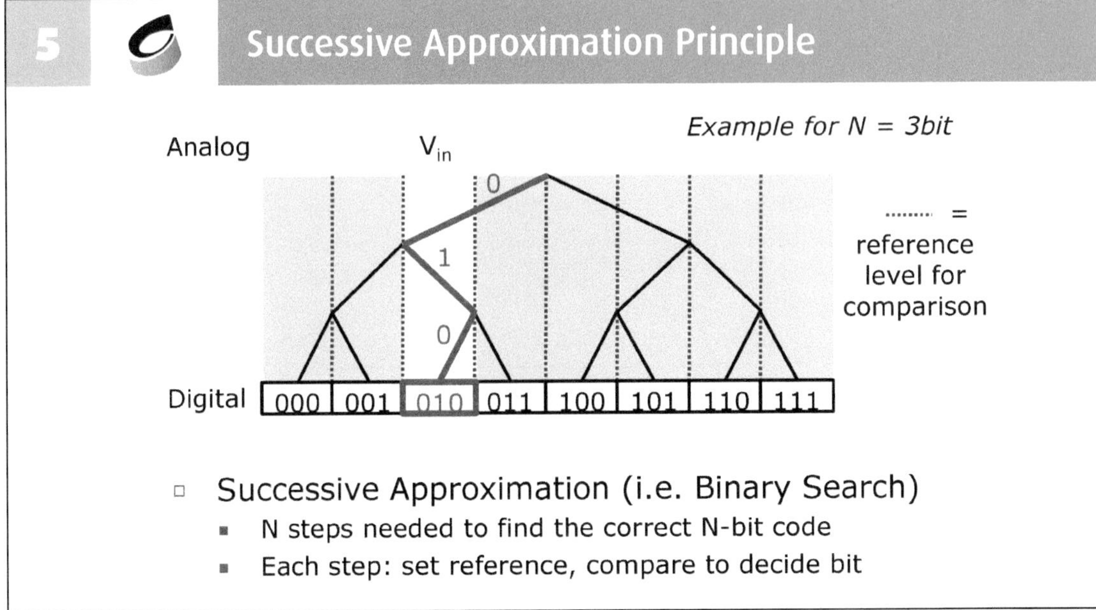

Example for N = 3bit

N steps needed to find the correct N-bit code
Each step: set reference, compare to decide bit

Successive approximation is based on a binary search algorithm, trading speed for accuracy and efficiency. Initially, the input signal is compared to a reference voltage, where the most significant bit (MSB) is decided. Afterwards, the reference voltage is updated (i.e., the search range is reduced by 2X) and the second bit is determined. This process continues until the digital N-bit word is achieved. As a result, this principle requires N cycles to generate the final output.

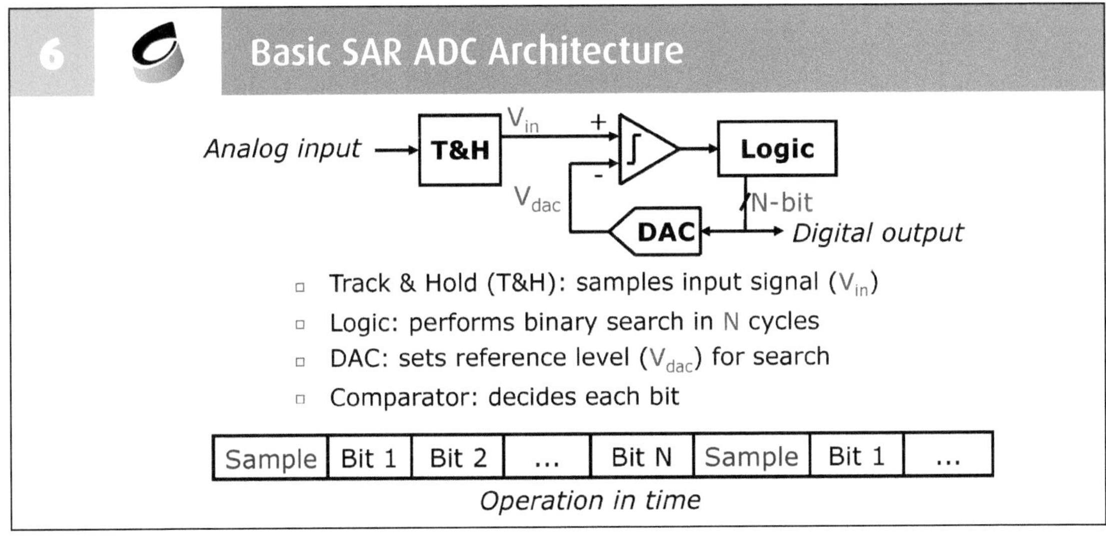

6
Basic SAR ADC Architecture
Analog input
T&H
Vin
+
-
Logic
Vdac
N-bit
DAC
Digital output
Track & Hold (T&H): samples input signal (Vin)
Logic: performs binary search in N cycles
DAC: sets reference level (Vdac) for search
Comparator: decides each bit
Sample Bit 1 Bit 2 ... Bit N Sample Bit 1 ...
Operation in time

The most basic SAR ADC architecture has four main blocks:

- The Track and Hold (T&H): responsible for sampling the analog input signal into a sampling capacitor Cs, enabling a steady signal during the quantization phase;
- The Control Logic: performs the binary search in N cycles and feeds the digital-to-analog converter (DAC) with the corresponding digital code;
- The DAC: sets its output, the reference level, in accordance with the digital word received from the control logic, narrowing the search range, i.e. bringing the difference between the sampled input and the reference level towards zero;
- The comparator: decides each bit, based on the difference between the sampled input signal and the reference voltage.

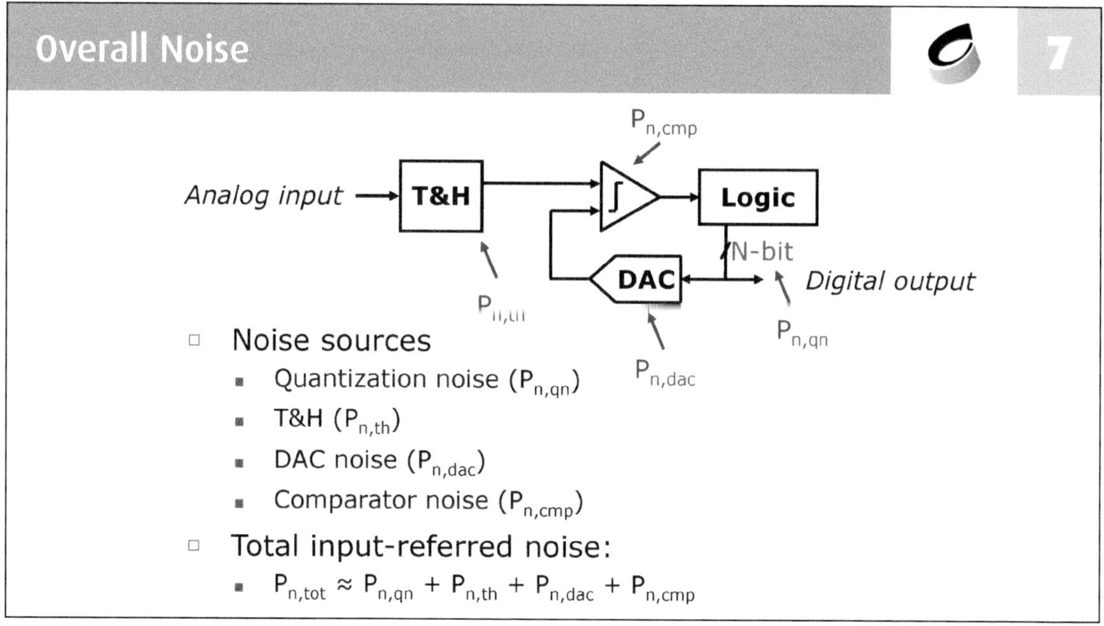

Overall Noise
7
$P_{n,cmp}$
Overall Linearity, Gain, Offset
Analog input
T&H
Logic
N-bit
DAC
Digital output
R_{on}, Charge injection,
Mismatch, Settling errors
$P_{n,th}$
$P_{n,qn}$
$P_{n,dac}$
Noise sources
Quantization noise ($P_{n,qn}$)
T&H ($P_{n,th}$)
DAC noise ($P_{n,dac}$)
Comparator noise ($P_{n,cmp}$)
Total input-referred noise:
$P_{n,tot} \approx P_{n,qn} + P_{n,th} + P_{n,dac} + P_{n,cmp}$

Among the noise sources in this architecture, the ones generated by both the comparator and the DAC are the most critical in terms of performance. However, during the circuit design, quantization (which is inherently present and can only be reduced by increasing the converter's resolution) and thermal noise should both be accounted for in the total input-referred noise.

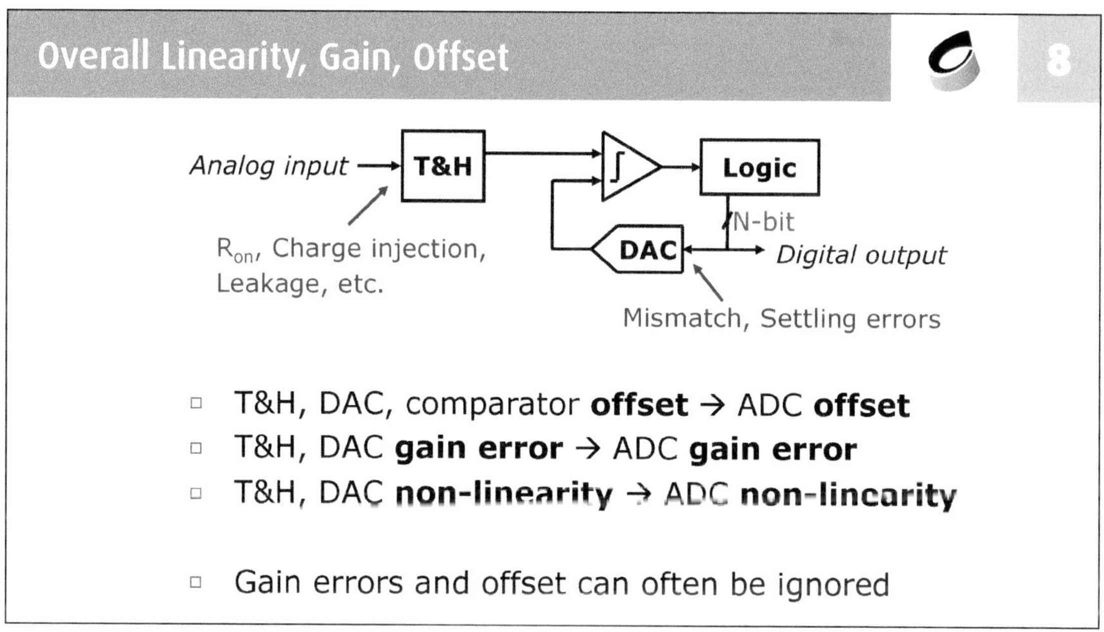

T&H, DAC **gain error** → ADC **gain error**

Beyond the aforementioned noise sources, there are static and dynamic errors such as gain and offset errors and non-linearities that can compromise the performance. T&H, DAC and the comparator all contribute with offset error. Gain errors and non-linearities are caused by the T&H and DAC especially. Depending on the application, gain and offset errors can sometimes be ignored.

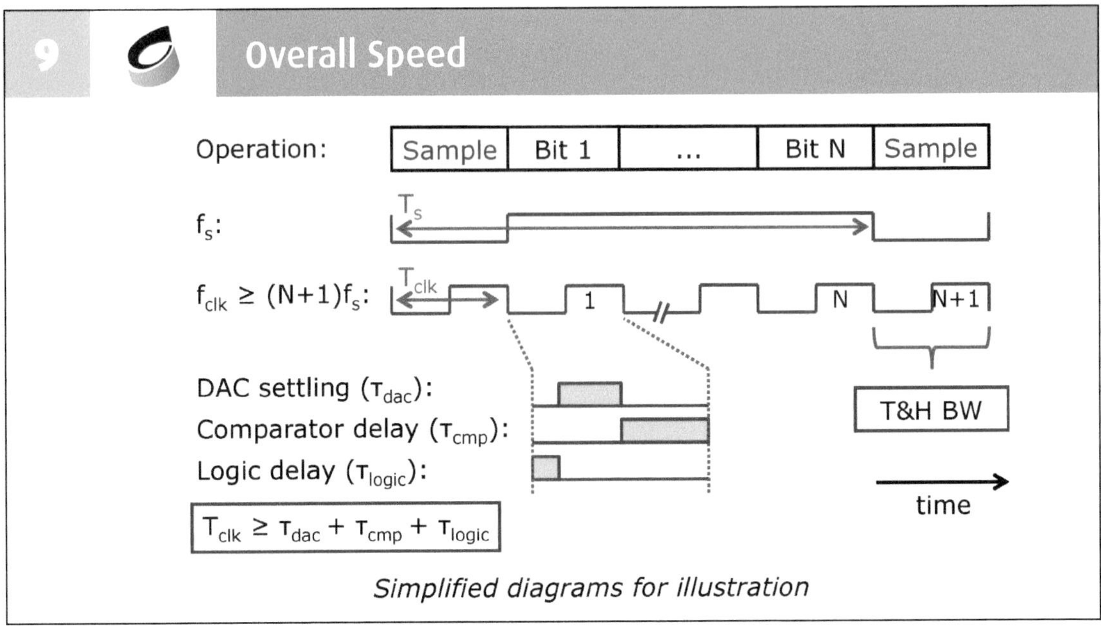

Taking into account that the bits are decided one-by-one, the clock frequency (F_{clk}) within the SAR is, at least, N+1 times higher than the sampling frequency (F_s). The duration of each bit decision cycle has to accommodate for three distinct operations:

- The reference has to be settled (T_{DAC});
- The comparator has to make a decision (T_{cmp});
- The binary search has to be performed by the logic circuitry (T_{logic}).

The additional (+1) cycle accounts for the time it takes for the T&H circuit to sample the input signal.

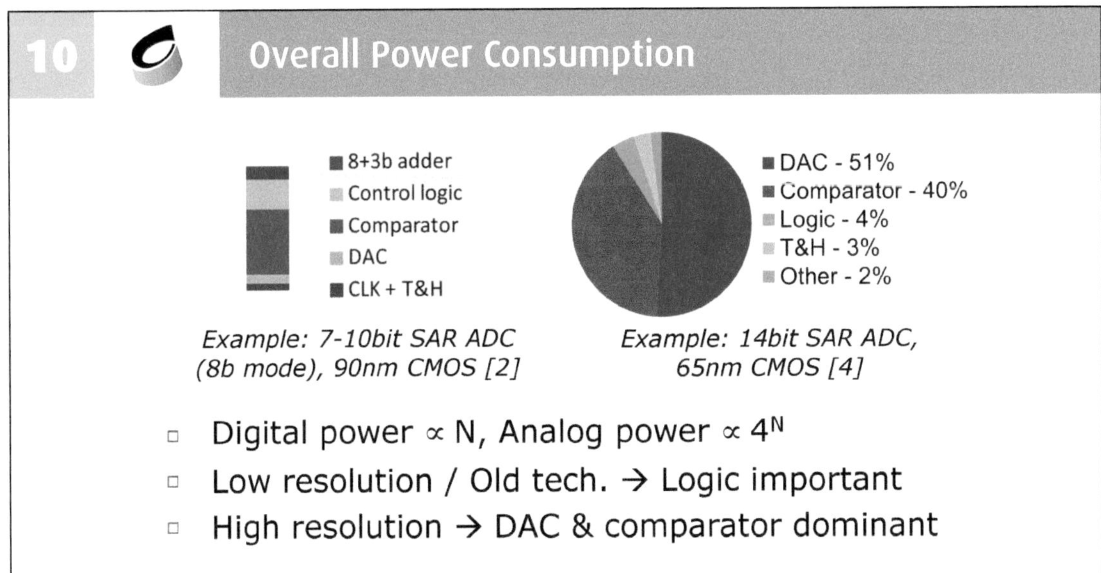

In SAR ADCs, the analog blocks contribution to power consumption is proportional to 4^N, while it is proportional to N for the digital logic. Due to its exponential growth, analog blocks such as the DAC and comparator are often dominating, although for low resolution implementations (< 8b) the contribution of the digital blocks in the power consumption can reach significant proportions (> 30%).

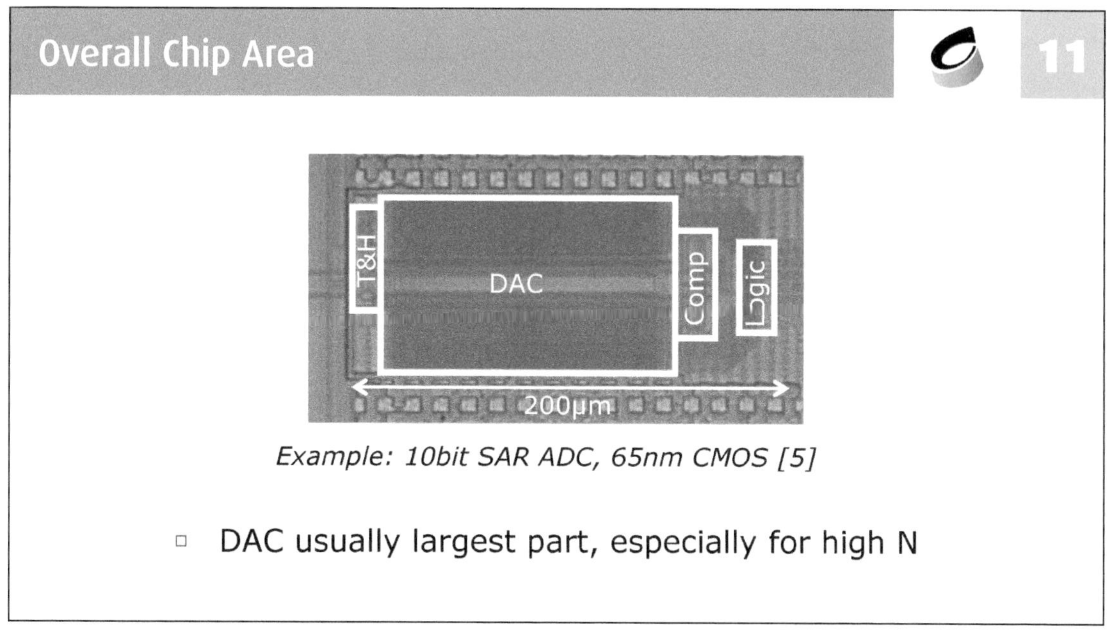

In regards to the overall chip area, the most dominant block is usually the DAC, due to its high number of elements/capacitors. This becomes even more relevant for a differential implementation where two DACs are required (in most switching schemes) or for high resolution, increasing the total area.

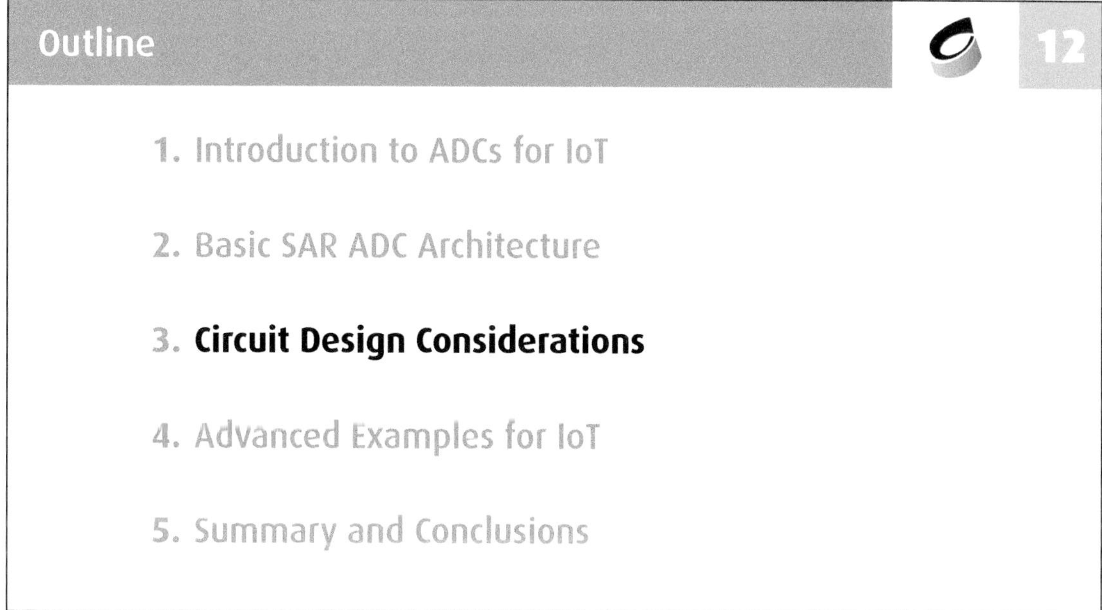

In the following section, circuit design considerations for each individual block are detailed.

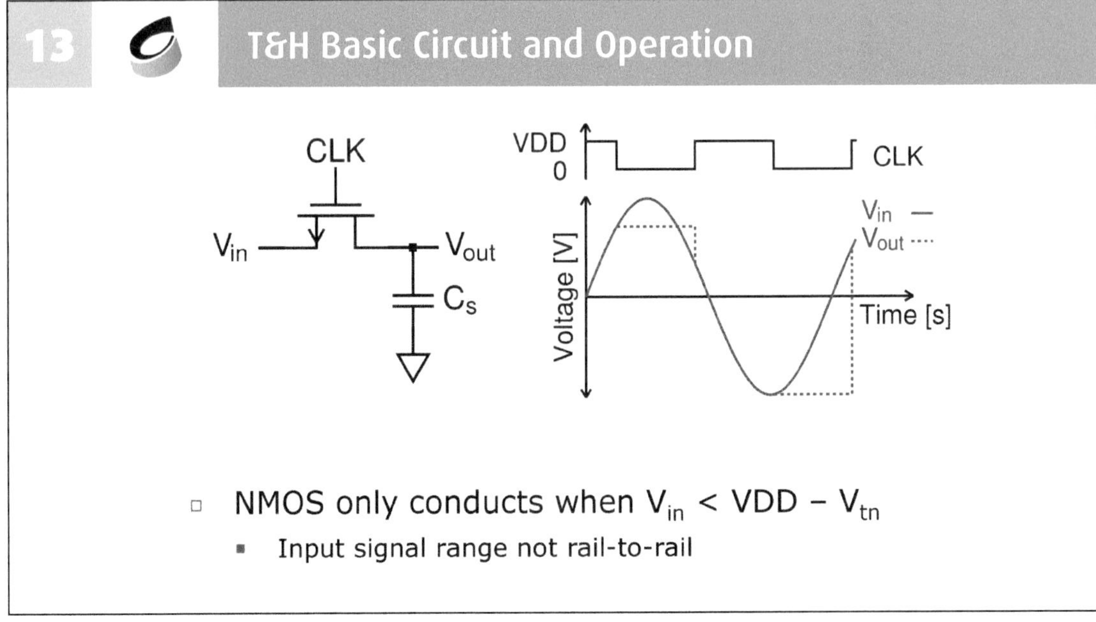

The most basic T&H implementation consists of a single NMOS switch coupled with a sampling capacitor. On the falling edge of the clock signal the switch is turned off and the input signal is sampled onto capacitor C_S. However, this only occurs for input signals that are at least a threshold voltage (V_{tn}) lower than the supply (V_{DD}), limiting the input signal range.

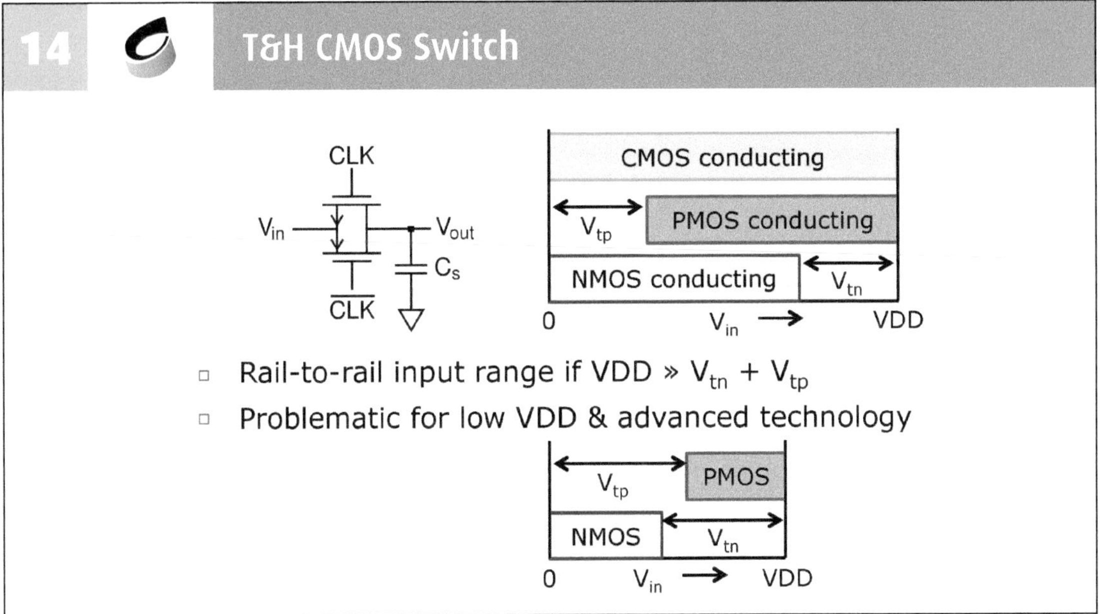

A common T&H implementation uses a CMOS switch (transmission gate) in order to ensure rail-to-rail operation as long as the V_{DD} is higher than the sum of the thresholds of the NMOS and PMOS switch (V_{tn} + V_{tp}). For advanced technology nodes and/or low V_{DD}, this becomes critical since the overdrive voltage is smaller (as the threshold voltages do not scale down as fast as the V_{DD}). As a consequence, the CMOS switch does not longer operate as expected over the entire signal range.

T&H Boosting Techniques

- CLK boosted >VDD
  - Swing: $0 \le V_{in} \le VDD$
  - Large overdrive V_{ov}
  - V_{gs} can be >VDD
- CLK shifted to VDD + V_{in}
  - Swing: $0 \le V_{in} \le VDD$
  - Constant overdrive V_{ov}
  - V_{gs} equal to VDD

To ensure that the switch operates for the entire input range, two boosting techniques can be employed: clock boosting and bootstrapping. In the former, through some sort of charge-pump method, the clock amplitude will extend from zero to (for instance) two times the V_{DD} (being completely independent from the input signal), significantly increasing the overdrive voltage. This approach can compromise the circuit operation, as V_{gs} can be higher than V_{DD}. On the other hand, bootstrapping implementations let the clock amplitude swing from zero to a maximum of $V_{DD} + V_{in}$. Being dependent on the input signal, a constant overdrive voltage can be provided and V_{gs} is always kept equal do V_{DD} during tracking.

T&H Imperfections

- **ON** state:
  - On-resistance
- Sampling moment (**ON → OFF**):
  - Charge injection
  - KT/C sampling noise
- **OFF** state:
  - Leakage
  - Capacitive coupling

Next, T&H imperfections that span across both operating modes (ON & OFF) will be described in detail.

17 

T&H On-Resistance

R_{on}

V_{in} V_{out} C_s

NMOS:

$$R_{on} = \frac{1}{\mu_n C_{ox} \frac{W}{L} (V_{CLK} - V_{in} - V_{tn})}$$

- Issues:
  - Limited bandwidth, $f_{3dB} = 1 / (2\pi R_{on} \cdot C_s)$
  - Non-linearity (R_{on} depends on V_{in})
  - Frequency dependent distortion (worse for higher f)
- Solutions:
  - W/L, low-V_t device, CMOS, boosting / bootstrapping

During its ON state, the device is operating in the triode region. However, the on-resistance of the switch, which ideally should be constant, can compromise the circuit performance if otherwise. The R_{on} in conjunction with C_s forms a low-pass filter and therefore the bandwidth of the T&H is determined by it. Also, since the on-resistance shows a dependence on the input signal, it will not be constant and non-linearities will arise. Furthermore, this distortion is frequency dependent and becomes more critical at higher input frequencies, since the sampling capacitor's impedance is no longer infinitely high when compared to the R_{on} (as it is at lower frequencies). These issues can be reduced through proper sizing of the devices (W/L), the usage of low-V_t devices and/or CMOS switches and recurring to boosting/bootstrapping techniques, etc.

18 

T&H Charge Injection

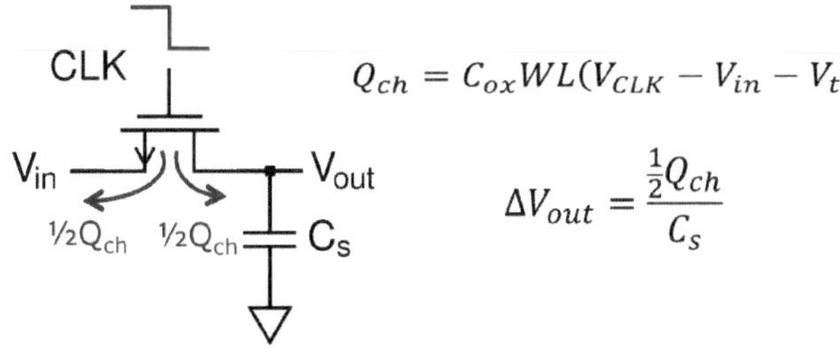

$$Q_{ch} = C_{ox} WL (V_{CLK} - V_{in} - V_{tn})$$

$$\Delta V_{out} = \frac{\frac{1}{2} Q_{ch}}{C_s}$$

- Issues:
  - Non-linearity (frequency independent)
- Solutions:
  - WL, bottom-plate sampling, boosting / bootstrapping

During the sampling instant (when the switched is turned from ON to OFF) the output voltage can exhibit a small variation due to charge injection[1]. Again, since this accumulated charge is input signal dependent, it can cause non-linearities. This phenomenon is more relevant for larger device sizes as they store more charge. To reduce this problem, techniques such as bottom-plate sampling or boosting/bootstrapping can be used.

1. Where a small charge from the switch device is injected in the sampling capacitor.

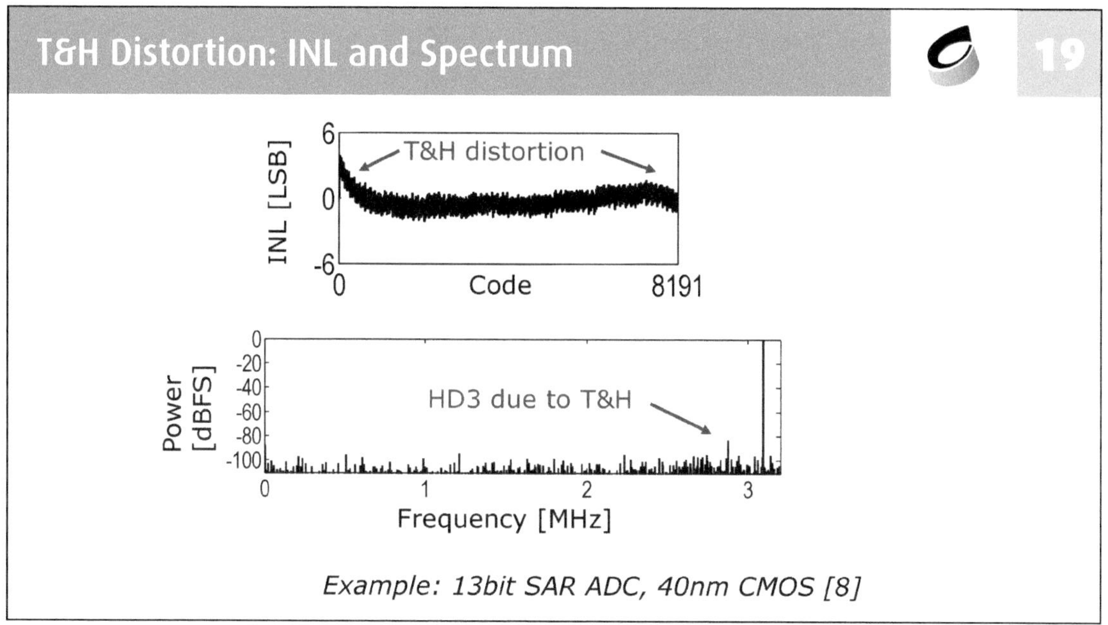

As an example, these figures show how T&H distortion can compromise the linearity of a 13b SAR ADC in 40nm CMOS in both the INL and output spectrum graphs. In this typical example, the T&H produces mostly third order harmonic distortion.

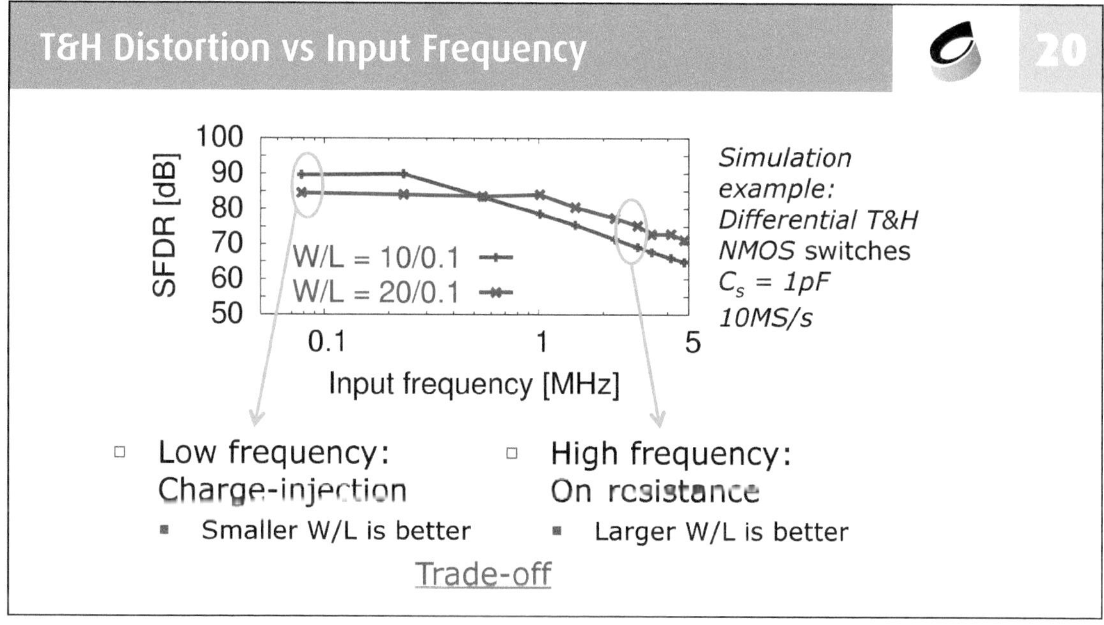

This simulation shows T&H linearity versus input frequency for different sizes of the sampling switch. At lower frequencies, where charge-injection dominates, a smaller aspect ratio is preferable, whilst at higher frequencies the more pressing factor is the R_{on}, at which a larger W/L is better. As such, there is an aspect ratio trade-off that must be considered depending on the intended application and its frequency range of operation.

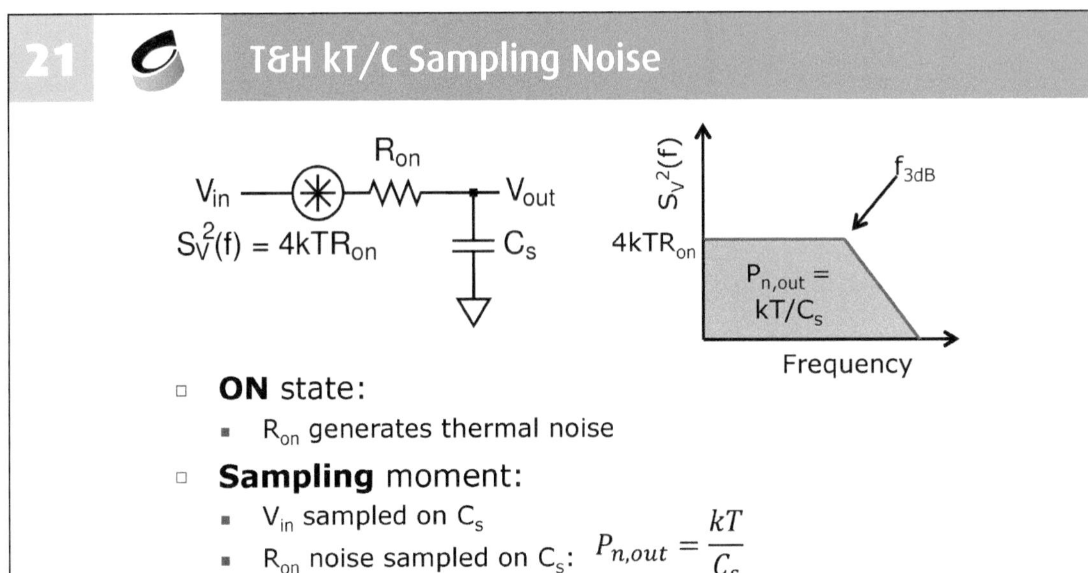

21
T&H kT/C Sampling Noise
Ron
Vin
Vout
Cs
$S_V^2(f) = 4kTR_{on}$
$S_V^2(f)$
4kTRon
f3dB
$P_{n,out} = kT/C_s$
Frequency
ON state:
R_{on} generates thermal noise
Sampling moment:
V_{in} sampled on C_s
R_{on} noise sampled on C_s: $P_{n,out} = \frac{kT}{C_s}$

Inevitably, the switch resistance will generate thermal noise whenever it is turned ON. This noise will end up being sampled onto C_s during the sampling phase. Its power can be reduced by increasing C_s, with subsequent area and power consumption penalties.

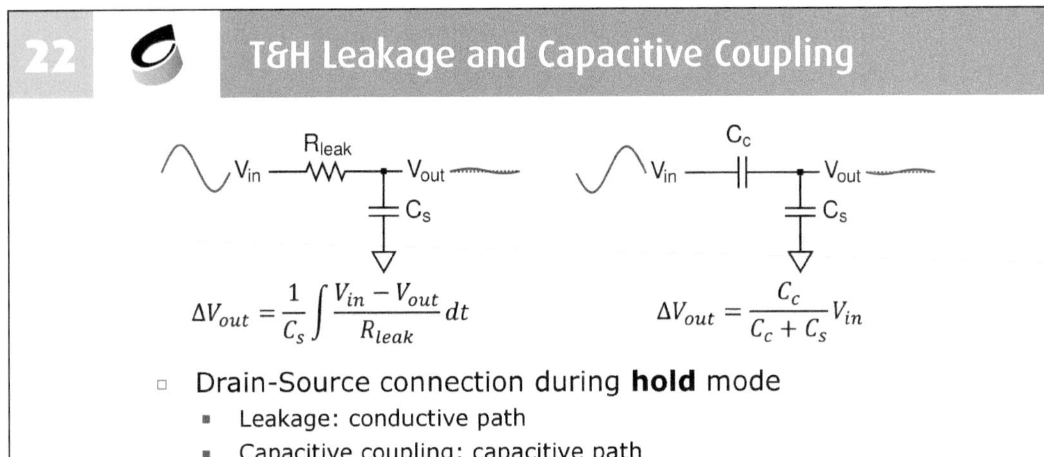

22
T&H Leakage and Capacitive Coupling
Rleak
Vin
Vout
Cs
Cc
$\Delta V_{out} = \frac{1}{C_s}\int \frac{V_{in} - V_{out}}{R_{leak}} dt$
$\Delta V_{out} = \frac{C_c}{C_c + C_s} V_{in}$
Drain-Source connection during **hold** mode
Leakage: conductive path
Capacitive coupling: capacitive path
Result: ΔV_{out} during AD conversion → Distortion
Worse in scaled CMOS (leakage ↑, dimensions ↓)
Solutions: layout, compensating dummies, high-V_t

After the switch is turned OFF both leakage and capacitive coupling can cause slight variations in the output voltage, as there is an unwanted small but existing path between both the drain and the source of the device. As a result, during the quantization phase the sampled signal will still exhibit a small variation, thus causing distortion. These effects become more pronounced in scaled CMOS, and can be compensated through several methods, such as the use of dummies, higher V_t transistors and by careful layout.

T&H Overview

23

Problem	Solution
On resistance	**Maximize W/L**; Low-V_t; Boosting
Charge injection	**Minimize WL**; Boosting; Bottom-plate sampling
kT/C sampling noise	Sizing of C_s
Leakage	**Minimize W/L**; High-V_t
Capacitive coupling	**Minimize W/L**; Layout; Dummies

- Various trade-offs for determining size (W, L)
- Careful selection of switch implementation, size optimization, and layout needed

In summary, the T&H block places several design challenges described in the adjacent table, together with some appropriate solutions. According to the application and its most important aspects/specifications, various trade-offs must be considered. One can also observe that the sizing of the transistors ends up being one of, if not the most critical design parameter.

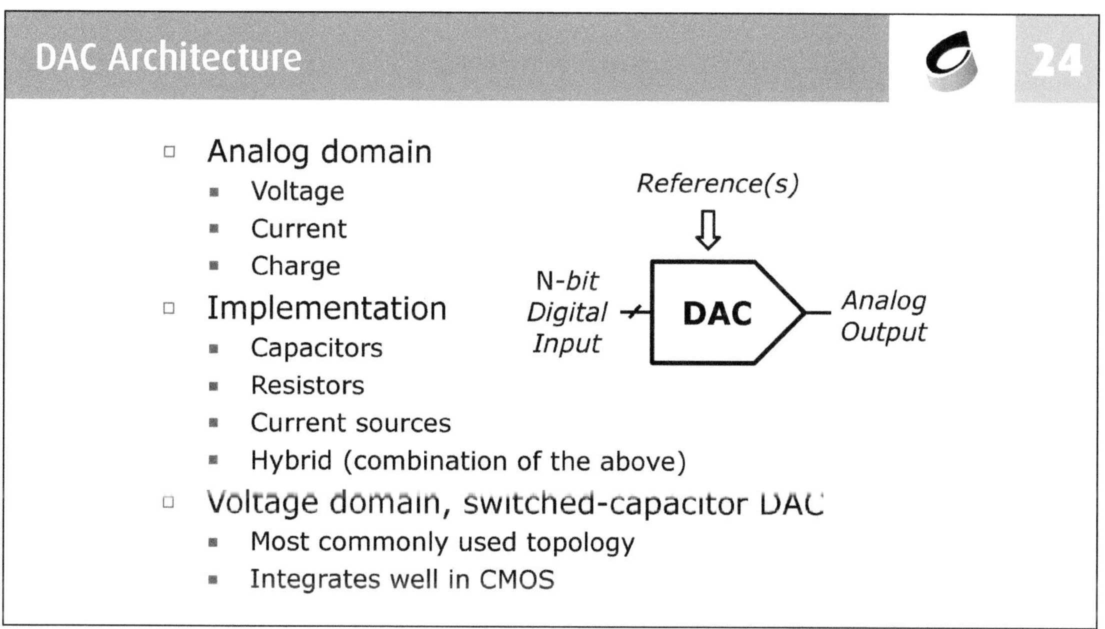

DAC Architecture
24
Analog domain
Voltage
Current
Charge
Implementation
Capacitors
Resistors
Current sources
Hybrid (combination of the above)
Voltage domain, switched-capacitor DAC
Most commonly used topology
Integrates well in CMOS
Reference(s)
N-bit Digital Input
DAC
Analog Output

The following section focuses on the DAC architecture. The DAC itself can be implemented in several ways, either in "voltage", "current" or "charge-mode" and rely on capacitors, resistors, current-sources or a hybrid combination of all of them. Here, the focus is on "voltage-mode" switched-capacitor DACs, as it is the most commonly used topology inside SAR ADCs.

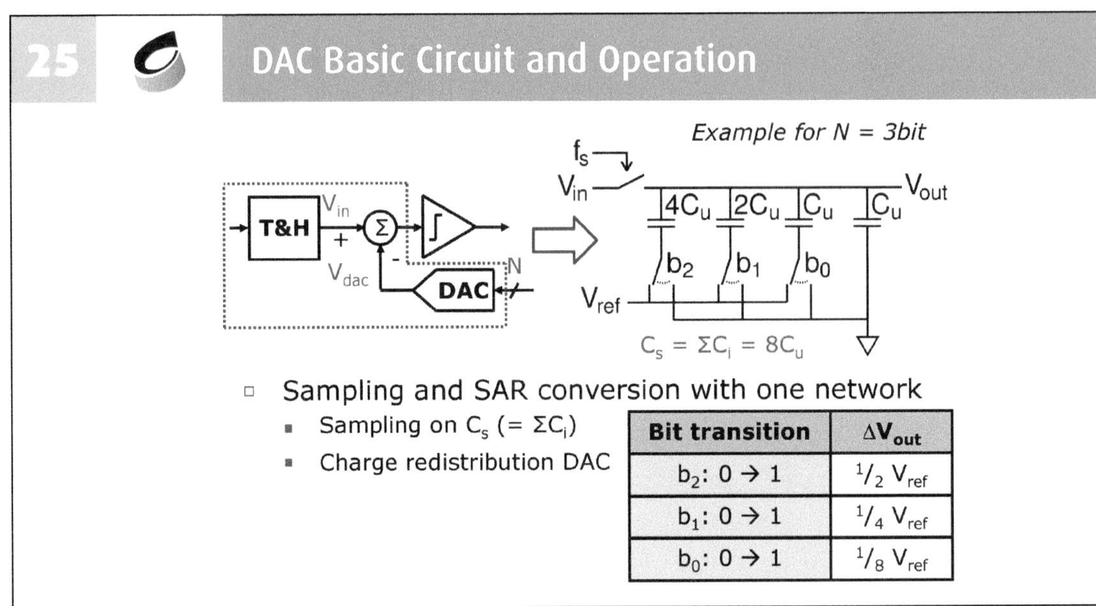

Bit transition	ΔV_{out}
b_2: 0 → 1	$^1/_2\ V_{ref}$
b_1: 0 → 1	$^1/_4\ V_{ref}$
b_0: 0 → 1	$^1/_8\ V_{ref}$

Most common SAR architectures embed the T&H and subtracting functions in the DAC capacitive-array, with the entire array functioning as the sampling capacitor. Since most DAC implementations follow a charge-redistribution (CR) switching scheme, their total capacitance is kept unchanged throughout the conversion and is usually formed by a set of binary-scaled capacitances. Since the capacitances are binary-scaled, whenever one of the switches in the array is turned ON, the output voltage will also experience a binary-scaled shift.

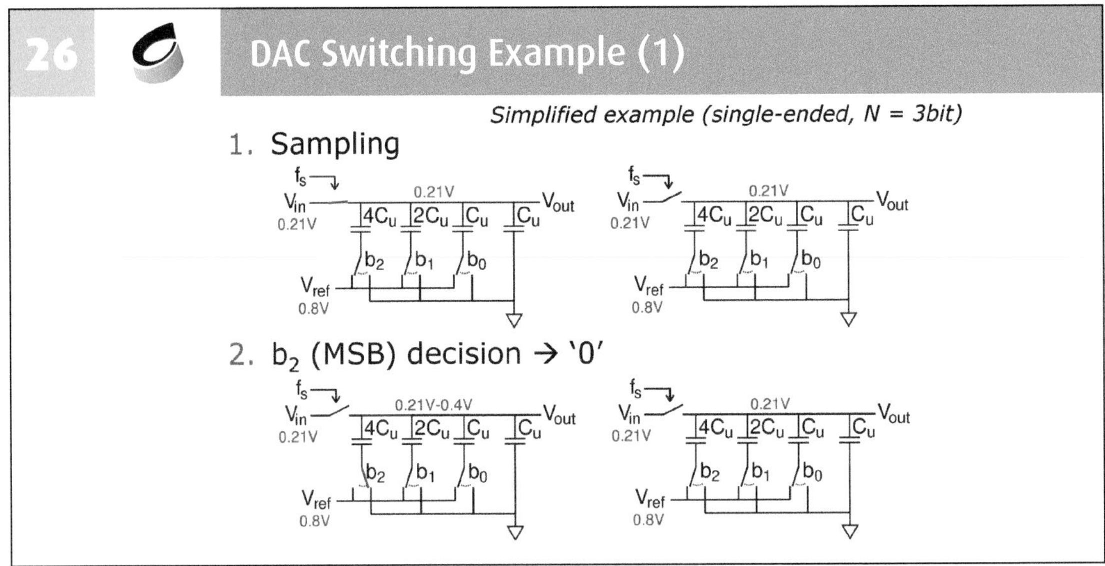

A 3-bit single-ended SAR ADC operation is described as an example, for an input signal of 0.21 V and a V_{Ref} of 0.8 V. During the sampling phase, the input signal is sampled onto the entire capacitive array. Afterwards, in the quantization phase, the sampling switch is turned OFF and in order to determine the most significant bit (MSB), b_2 is switched from V_{Ref} to ground/zero (shifting the reference voltage to 0.4 V). As a result, the voltage potential across the MSB capacitor is evaluated and determined to be negative/below zero (= -0.19 V). This implies that the MSB is logic 'zero', i.e., the input signal is (at least) on the lower half of the ADC's signal range, and b_2 is switched back to V_{Ref}.

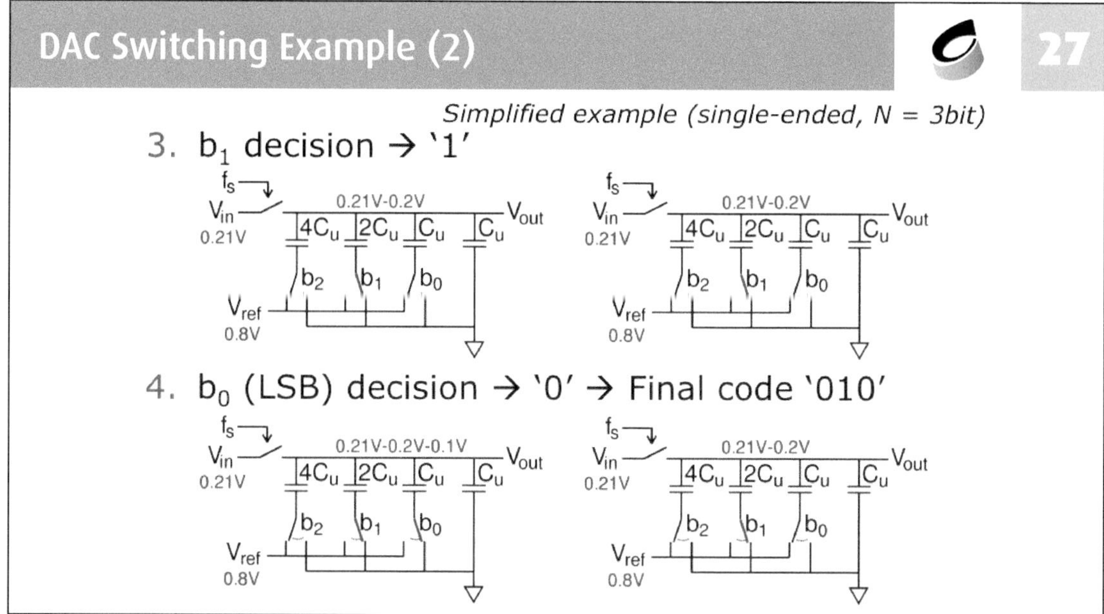

The following bits are estimated/determined, following the same thought as before with the search range being continuously halved for each bit. As such, for the 2nd bit the search range is within 0 to 0.4 V and b_1 is switched to ground/zero, bringing the reference voltage down to 0.2 V. Since the result of the comparison between the input and 0.2 V is positive, a logic 'one' is determined for b_1 thereby keeping its corresponding switch connected to ground/zero. Finally, the search interval is reduced to 0.2 V to 0.4 V and since the input signal is below 0.3 V, the final code is evaluated as '010'.

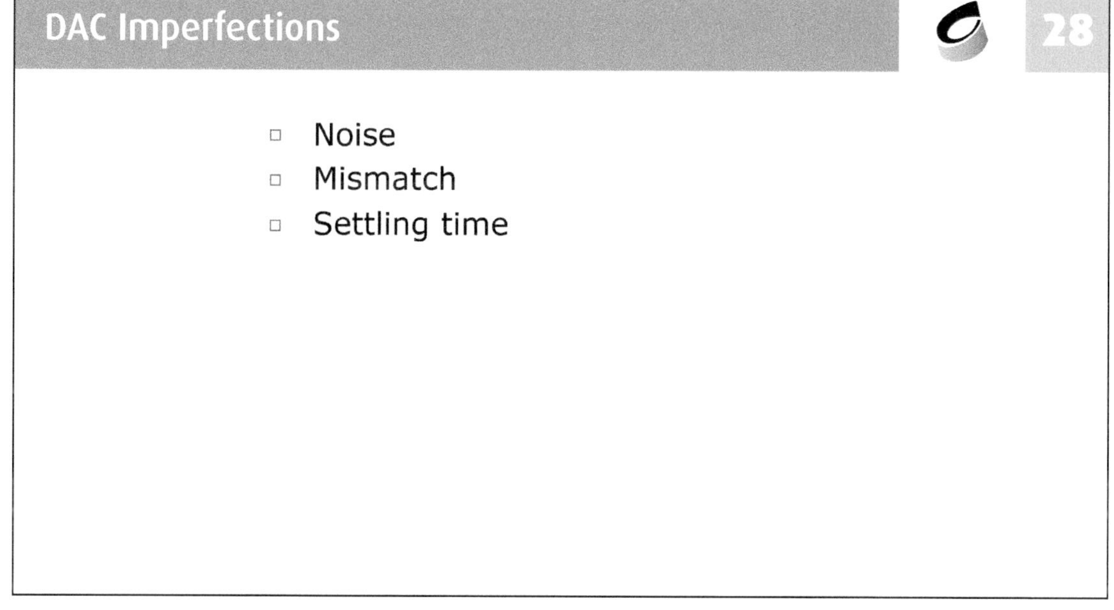

The following slides will focus on the main sources of DAC imperfections such as noise, mismatch and settling time.

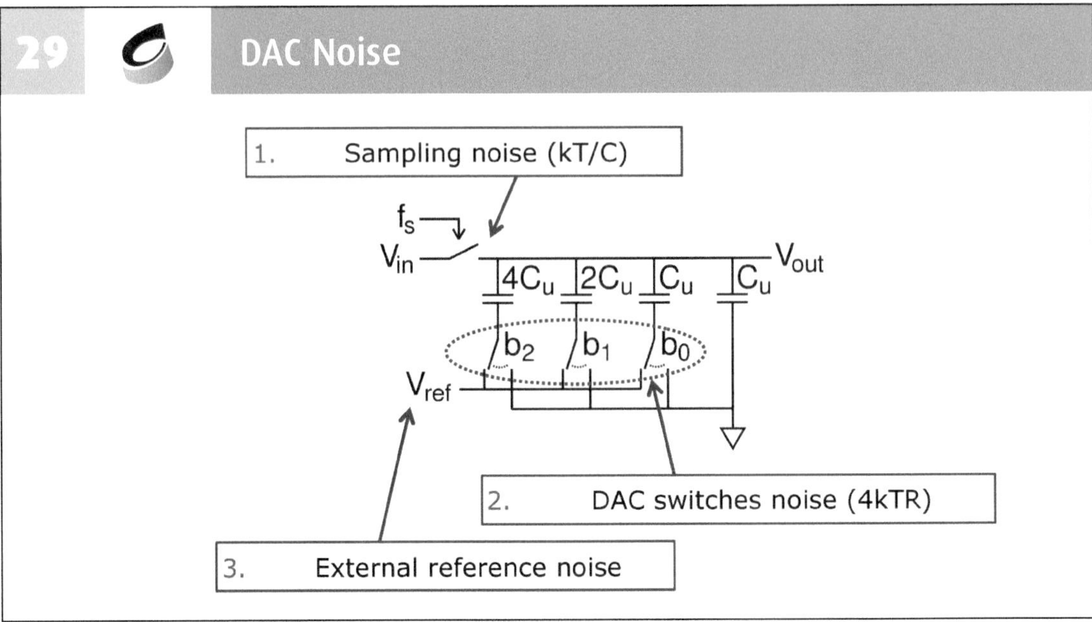

Within the DAC block there are several noise sources. Among them are the following:

- Sampling noise: originated in the T&H section of the DAC and given by kT/C, i.e., it is inversely proportional to the total size of the capacitive-array;
- DAC switches noise: as these switches are implemented by transistors, which have R_{on}, the noise generated by them is given by $4kTR_{on}$;
- External reference noise: as the reference is not ideal, it will also exhibit noise.

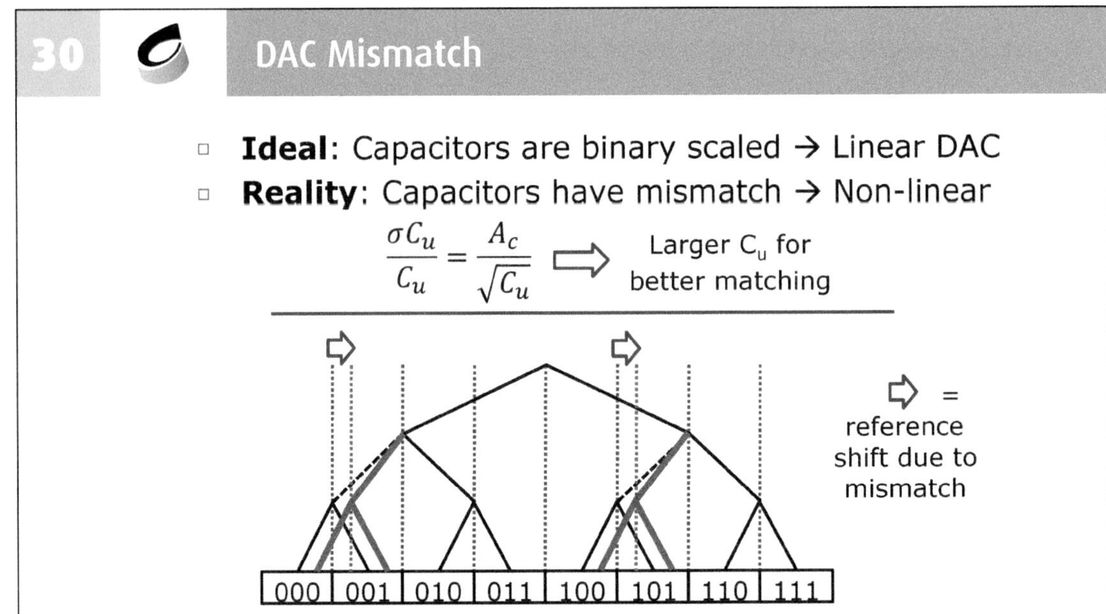

30
DAC Mismatch
Ideal: Capacitors are binary scaled → Linear DAC
Reality: Capacitors have mismatch → Non-linear
$\frac{\sigma C_u}{C_u} = \frac{A_c}{\sqrt{C_u}}$
Larger Cu for better matching
= reference shift due to mismatch
000
001
010
011
100
101
110
111

Another source of imperfection in the DAC is the mismatch between elements. Ideally, the capacitors in the array should be perfectly binary-scaled, resulting in a linear DAC. However, slight mismatches between them lead to non-linearities and the reference levels will no longer be precise. The dimensions of the capacitors can be increased to improve matching, but that deteriorates power efficiency, area and speed.

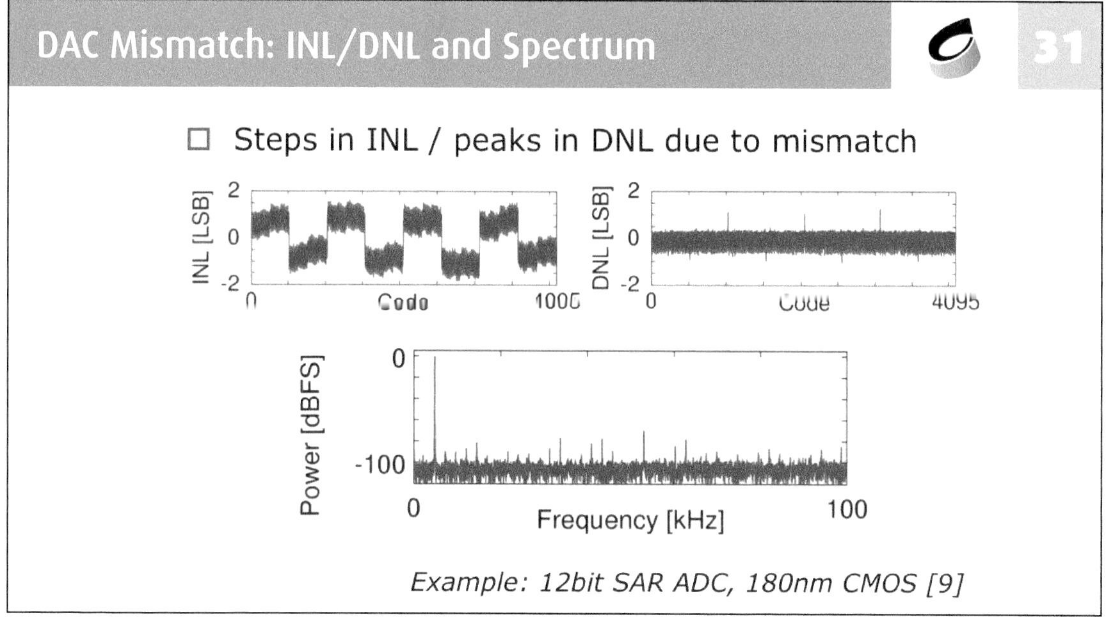

Again as an example, these figures show how DAC mismatch impact the linearity of a 12b SAR ADC in 180nm CMOS, as observed in the steps in the INL graph and the corresponding transition peaks in the DNL plot. Note that this type of distortion has a very different signature as compared to the T&H distortion discussed earlier.

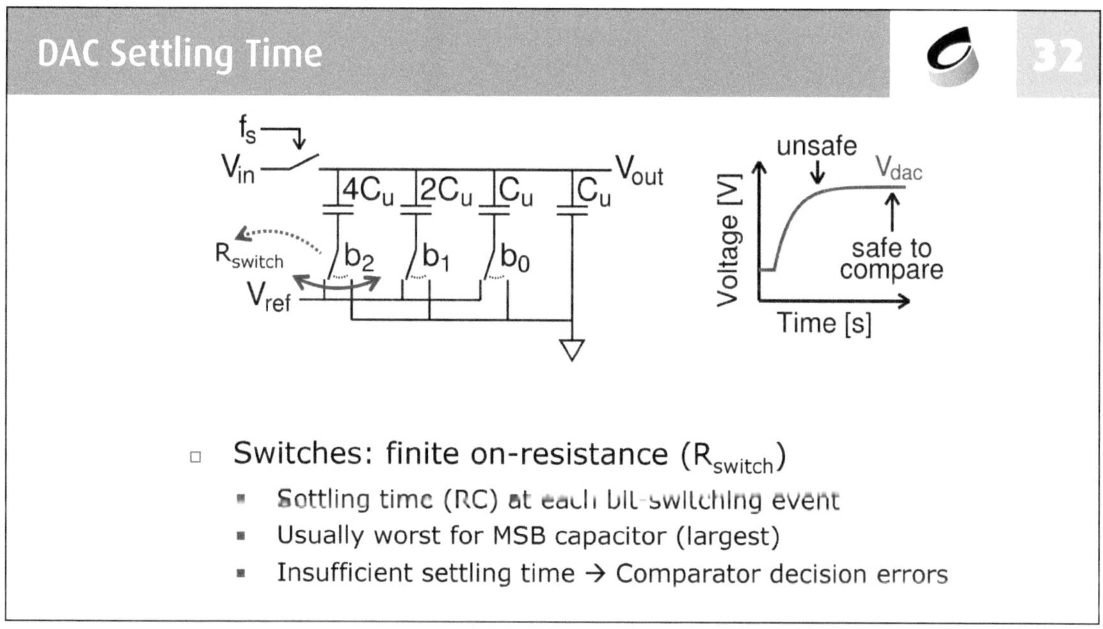

Lastly, the settling time within the DAC is also crucial. Each bit-switching event will present a different settling time, as each switch has its own finite R_{on}. As each switch is connected to each binary-scaled capacitor in the array (forming an RC circuit), the settling time for the MSB capacitor is usually the worst-case scenario. Regardless, comparisons can only be performed after proper settling, at the risk of decision errors.

33 

DAC Design Strategy

- Key choice: Value of C_s ($\approx 2^N \cdot C_u$)
- Small C_s is good for:
  - Power consumption
  - Chip area
  - Speed
- Large C_s is good for:
  - Noise
  - Matching / Linearity
- To overcome limitations:
  - Switching schemes
  - Calibration techniques

In summary, the proper choice of the value of C_s is key when designing the DAC. Smaller dimensions yield lower power consumption, smaller area and higher speed at the cost of noise, matching and linearity. To cope with this trade-off, different switching schemes and calibration techniques can be employed, improving the power consumption and lowering the impact of mismatch, respectively.

34 

DAC Switching Schemes (1)

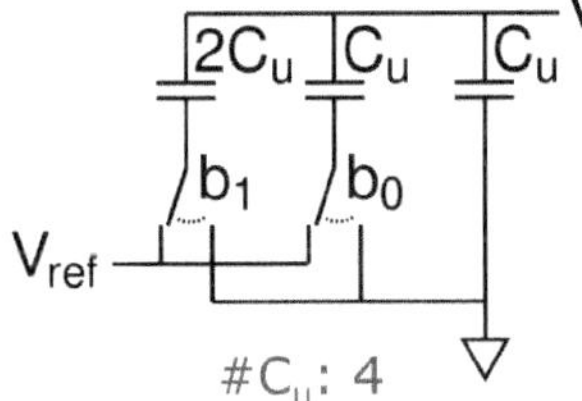

#C_u: 4

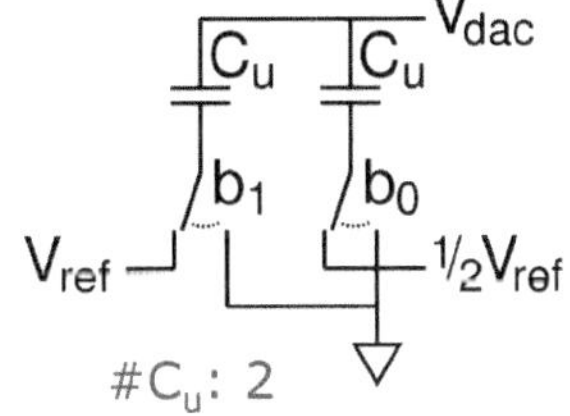

#C_u: 2

Bit transition	ΔV_{dac}
b_1: 0 → 1	$^1/_2\ V_{ref}$
b_0: 0 → 1	$^1/_4\ V_{ref}$

- Different circuits can realize same functionality
  - Power can be saved with smarter switching schemes
  - Number of C_u's be reduced and save chip area

Switching schemes are particularly useful when aiming to reduce power consumption. Next, ttwo 2-bit DACs are presented next to each other. Although structurally different, they perform the same functionality. The one on the right does that so with half the number of elements, saving chip area. However, it uses two different references, which may increase the power consumption or complexity.

DAC Switching Schemes (2)

35

- Many examples in recent literature, like:
  - Split capacitor switching method [10]
  - Monotonic capacitor switching [11]
  - Merged capacitor switching [12]
  - Charge average switching [13]
  - Detect and skip, aligned switching [14]
  - ...
- Overall: power can be saved, but be aware of:
  - Noise
  - Matching
  - Circuit overhead
  - Reset and conversion power
  - Number of V_{ref}'s

In literature many switching schemes have been reported, all with the goal of saving power, but with different applicability. These schemes can save significant amounts of power. However, when choosing the switching scheme, one must be aware of all consequences (such as noise, matching, circuit overhead, etc.) that this particular scheme may have and how they affect the overall circuit performance. Also, for very-low power applications it might be more worth it to try and reduce the power consumption in other key blocks such as digital circuitry.

DAC Mismatch Calibration (1)

36

1. Error detection
   - Background: during normal ADC operation
   - Foreground: while ADC is off line (e.g. at start-up)
2. Error correction
   - Analog: component tuning / compensation for errors
   - Digital: post-processing to cancel errors

- Be aware of:
  - Complexity
  - Overhead in circuits, power consumption, chip area
  - Calibration precision
  - Restrictions on input signal during background cal.

In regards to mismatch calibration, the goal is to detect and correct mismatches between the DAC capacitors. The detection might occur during the ADCs normal operation, in background, which might be a somewhat complex operation, or during an idle phase, either during the ADCs start-up or during a specific instant where the ADC is stopped. These errors are then corrected either via component tuning on the analog side, or via post-processing to cancel said errors on the digital side. This can yield significant results in terms of linearity but at the cost of several factors such as complexity, power and area.

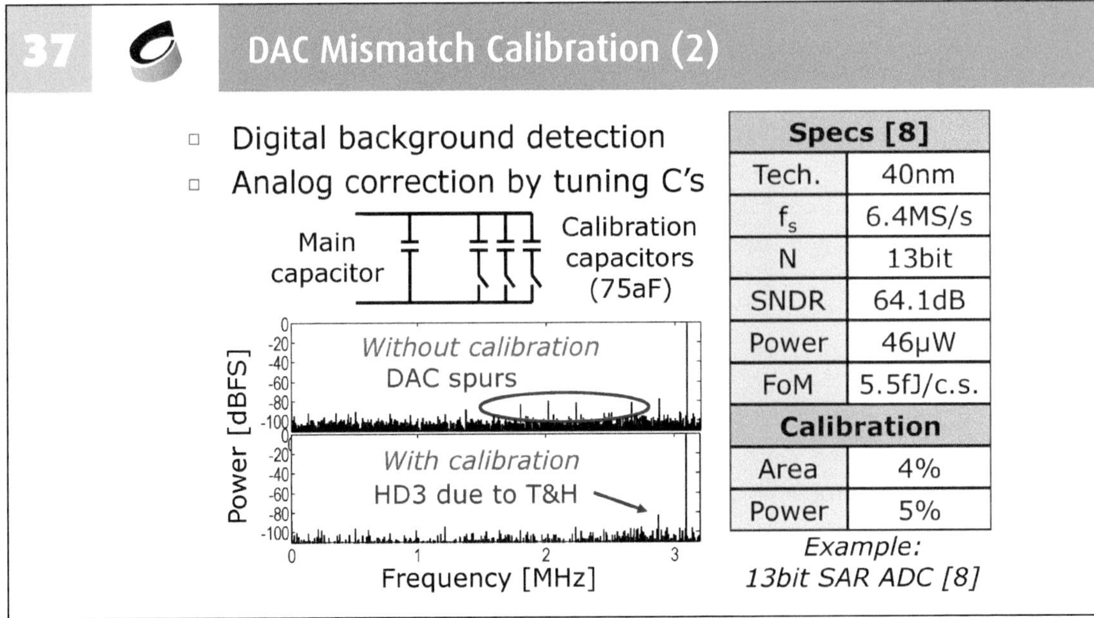

37
DAC Mismatch Calibration (2)
Digital background detection
Analog correction by tuning C's
Main capacitor
Calibration capacitors (75aF)
Without calibration
DAC spurs
With calibration
HD3 due to T&H
Power [dBFS]
Frequency [MHz]
Specs [8]
Tech. 40nm
fs 6.4MS/s
N 13bit
SNDR 64.1dB
Power 46μW
FoM 5.5fJ/c.s.
Calibration
Area 4%
Power 5%
Example:
13bit SAR ADC [8]

As an example, for the same 13b SAR ADC shown before, a digital background calibration technique is employed, where analog correction is used. Very small capacitors (in the order of aF) can optionally be connected to the capacitor(s) to be calibrated. The calibration logic and the bank of capacitors only take up 4% of the overall chip area and account for 5% of the total power, a trade-off deemed acceptable when one observes the improvements in terms of overall distortion (as the DAC spurs are greatly mitigated). The distortion caused by the T&H is not corrected and as such, its impact still shows in the HD3.

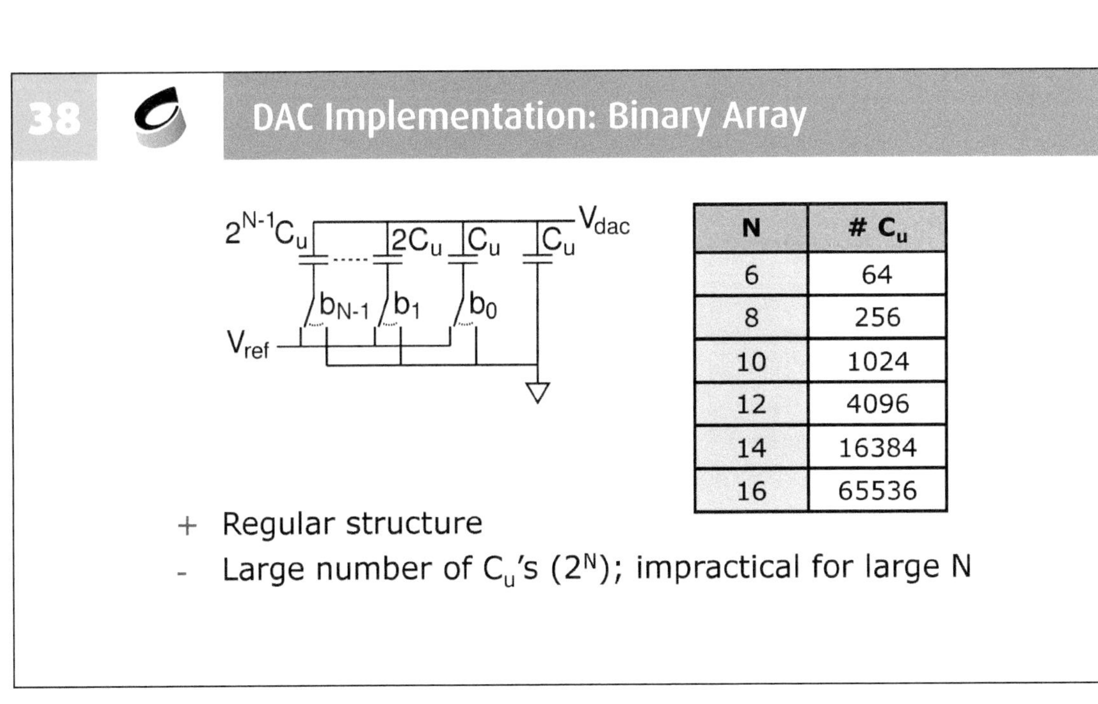

38
DAC Implementation: Binary Array
2^(N-1)Cu
2Cu
Cu
Cu
Vdac
bN-1
b1
b0
Vref
N | # Cu
6 | 64
8 | 256
10 | 1024
12 | 4096
14 | 16384
16 | 65536
+ Regular structure
- Large number of Cu's (2^N); impractical for large N

Next, we look at the implementation of the capacitive array. Although the structure is quite regular when the array is designed in a conventional binary-scaled fashion, it exponentially grows with the desired resolution N, making it impractical for high N. Several alternative solutions exist, with the goal of saving chip area.

DAC Implementation: Split Array

39

- Split capacitor array with bridge capacitor (C_b)
  - Main DAC (N-k-bit) connected to output as usual
  - Sub DAC (k-bit) is attenuated by bridge capacitor

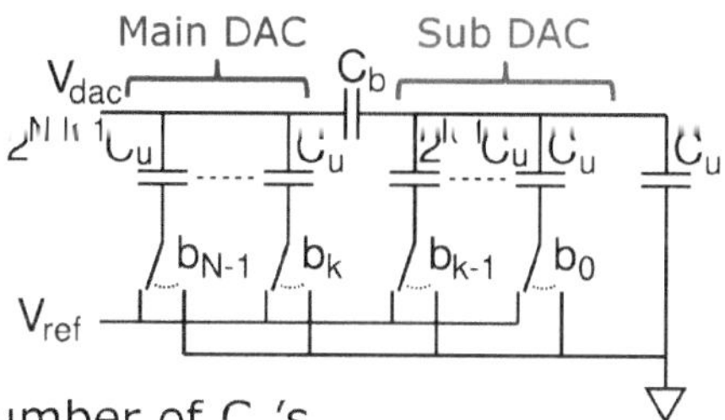

\+ Small number of C_u's

\- Structure can have systematic mismatch
  - 'Gain' mismatch between sub and main DAC [15]

One alternative relies on the usage of a split capacitor array with a bridge capacitor C_b (in literature, this capacitor is also known as attenuating capacitor). Thus, the DAC is split into two sections (although more than a single split can be used), one for k LSB capacitors and one for N-k MSB capacitors, with the MSB capacitors size being decreased by a factor of 2^{N-k}. With this decrease in overall size, area and power consumption can be reduced. However, there is a risk for systematic mismatch between both sections, causing distortion.

DAC Implementation: MIMCAPs

40

Area (A)

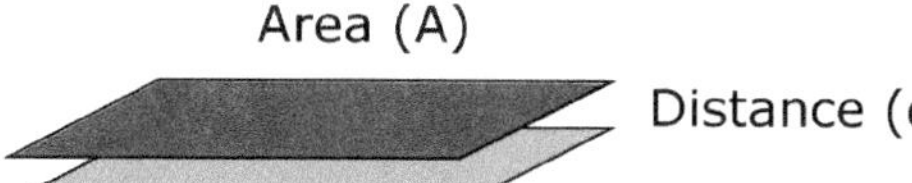

Distance (d)

2 metal layers on top of each other

$$C_u = \varepsilon_0 \varepsilon_r \frac{A}{d}$$

$$\frac{\sigma C_u}{C_u} = \frac{A_c}{\sqrt{C_u}}$$

- Only one design parameter: Area (A)
  - Determines capacitance (C_u)
  - Determines matching ($\sigma C_u / C_u$)
- Smallest value of C_u relatively high
  - >1fF up to >30fF, dependent on technology
  - For small C_u, relatively large area overhead

When it comes to the implementation of each element, it might be fabricated as a metal-insulator-metal (MIM) capacitor or a metal-oxide-metal (MOM) capacitor. (Other implementations, such as MOSCAPs, are not detailed here). The former consists of a parallel-plate capacitor formed by two planes of metal with area *A* on top of each other, separated by a dielectric, by a distance *d*. While the *d* is defined by the technology process, A is the only design parameter available, which is directly proportional to capacitance and has influence on matching (a smaller value of capacitance will produce a relatively worse matching).

41  DAC Implementation: MOMCAPs

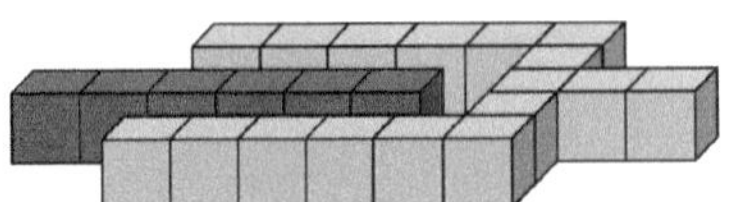

l = length of finger
w = width of metal
d = distance between fingers
n = number of stacked metals

- Various design parameters: l, w, d, n
  - Optimize capacitance and matching
- C_u can be as small as desired

	N [bit]	C_u [fF]	SFDR [dB]	FoM [fJ/c.s.]
[16] 2010	8	1	59	30
[17] 2011	8	0.5	62	12
[18] 2013	12	0.25	69	2.2
[4] 2014	14	0.55	79	4.4

MOM capacitors are interdigitated multi-finger capacitors formed by single or multiple metal layers (optionally connected by vias) in the vertical BEOL (back-end-of-line) stack separated by inter-metal dielectrics. As opposed to MIMs, this technology offers various design parameters allowing to optimize the devices in terms of capacitance and matching. Much smaller capacitor values can be implemented in this way as compared to MIM capacitors. Also, the chip area occupancy is more efficient.

42  DAC Layout

- Matching
  - Common-centroid layout
  - Dummies

D	B_2	B_1	B_2	B_0	B_2	B_1	B_2	D

D	D	D	D	D
D	B_1	B_2	D	D
D	B_2	B_0	B_2	D
D	D	B_2	B_1	D
D	D	D	D	D

- Parasitics (R and C)
  - Speed loss
  - Increase in power consumption
  - Mismatch
- RCX verification

Finally, the layout of the DAC block in particular should follow the common-centroid technique and dummies should be used in order to improve matching. Also, the designer should be aware of the impact of parasitics (both capacitive and resistive) on the overall speed and power consumption of the circuit and their relative contributions to component mismatch, and therefore perform an RCX (parasitic-extraction) verification.

Comparator Basic Operation

43

- Task: Compare inputs → i.e.: $V_1 > V_2$?

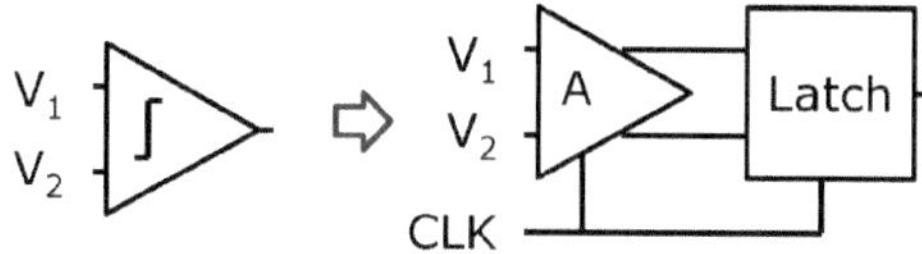

- Common approach:
  - Topology: preamplifier and latch
    - Preamplifier: gain, avoid kickback/noise from latch
    - Latch: decision element, positive feedback
  - Discrete-time operation, dynamic biasing

The main task of the comparator is to compare two inputs and output a (digital) signal indicating which one is larger. The large swing at the comparator's output (0 to V_{DD}) within a small time frame (high slew-rate) may cause glitches at its inputs, an effect known as kickback noise. As a result, common topologies rely on a preamplifier and latch combo, where the former's main objective is to provide gain and isolate the inputs from the latch output. Very often these comparators are dynamic, meaning that they only work in discrete-time thereby consuming power only during specific instants.

Comparator Basic Circuit Example

44

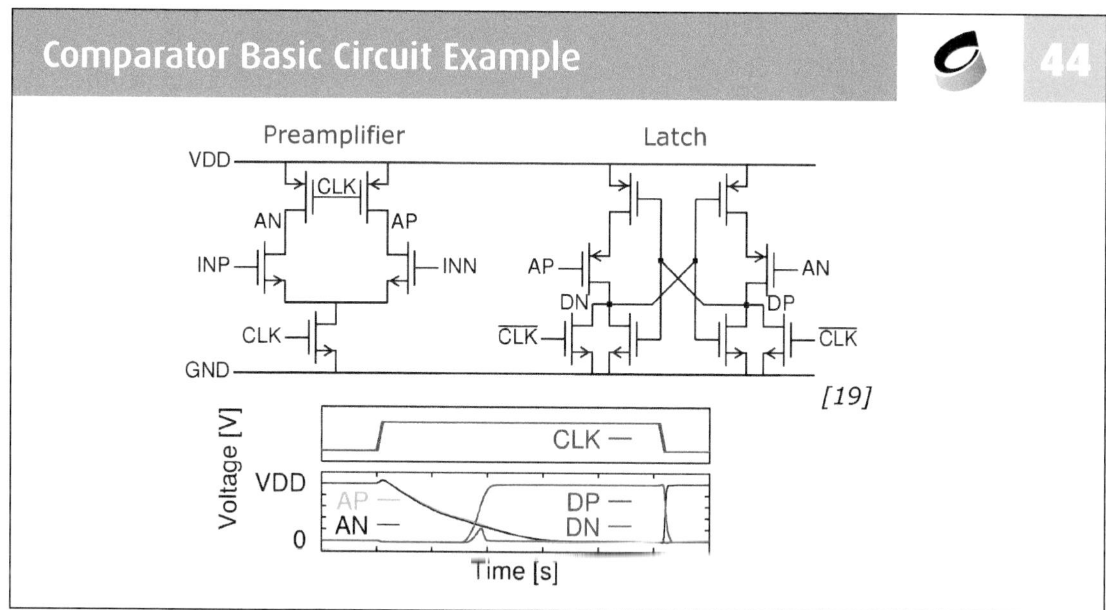

A dynamic comparator is shown in the left, as an example. The circuit works as follows: When the clock signal is low, the PMOS devices on the preamplifier are conducting and $A_{N,P}$ are connected to the supply, since the tail current is switched off. As soon as the clock signal goes high, the PMOS devices are switched off and the differential pair starts to work, with the only load being the parasitic capacitances at the output nodes $A_{N,P}$. If a differential input is applied, the differential pair will be slightly imbalanced and one output will drop faster than the other, thus performing dynamic amplification. As soon as the output goes down far enough, the latch will be triggered and make a decision to evaluate the final output voltages $D_{N,P}$.

45  Comparator Imperfections

- Noise
- Offset
- Speed & Meta-stability

A few problems when designing the comparator concern noise, offset and speed. These will be described next.

46  Comparator Noise

- Comparator thermal noise modeling:
  - Input: analog noise source
  - Output: digital Bit-Error-Rate

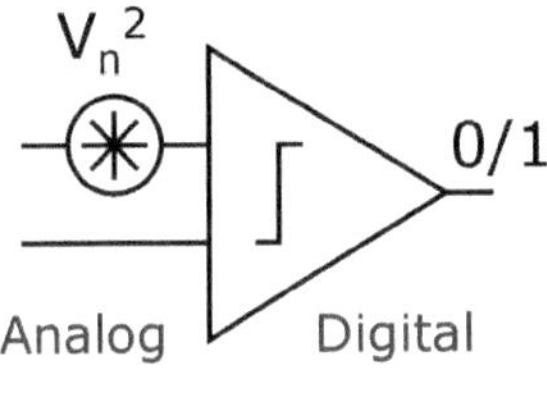

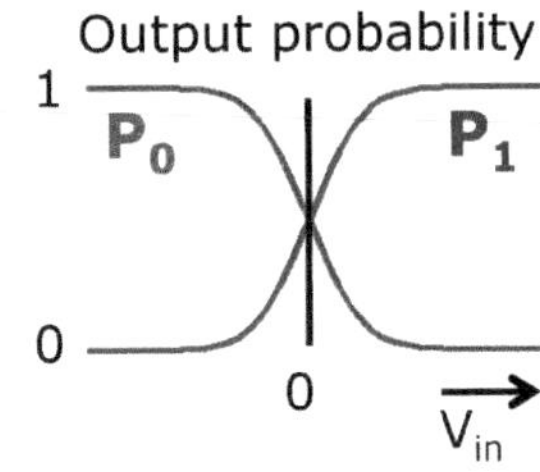

$$CDF = \frac{1}{2}\left[1 + erf\left(\frac{V_{in}}{\sqrt{2V_n^2}}\right)\right]$$

$$P_1 = CDF$$
$$P_0 + P_1 = 1$$

Since the comparator itself takes analog inputs and evaluates in digital outputs, it is not trivial to model its thermal noise. If one desires to determine how noisy the comparator is, one can observe the output error probability. For shorted inputs (differential input equal to zero), there is a theoretical 50-50 chance of either getting a logic 'zero' (P_0) or 'one' (P_1) at the output, due to noise. As the input increases, so does the chance of obtaining a logic 'one'. The relation between the equivalent input-referred noise (V_n^2) and the output probability is given by the CDF function shown.

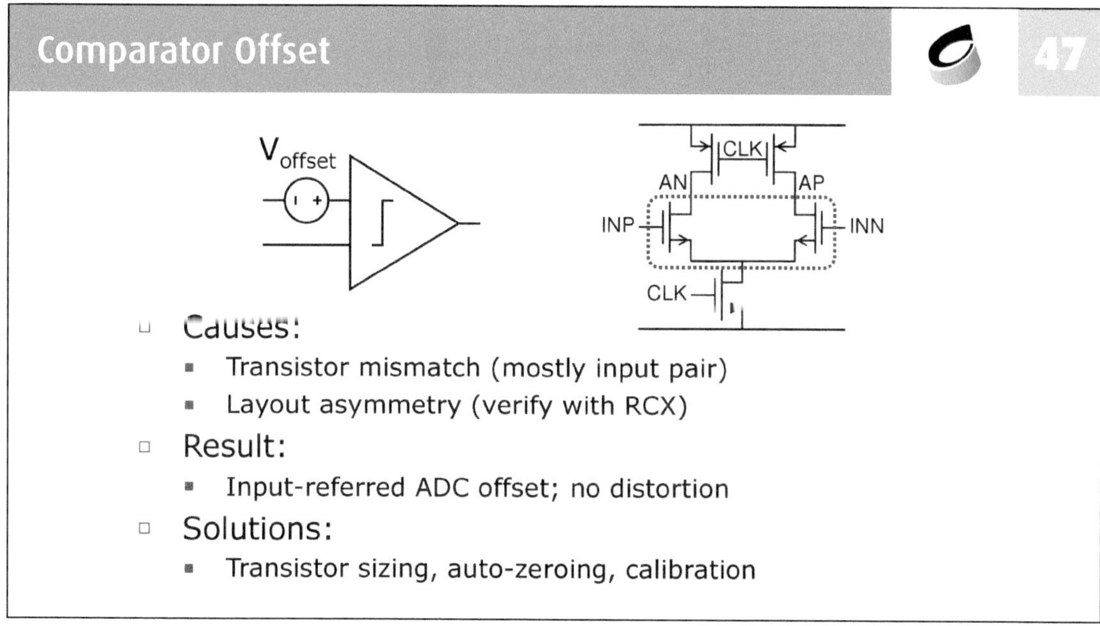

Comparator Offset
47
Voffset
CLK
AN
AP
INP
INN
CLK
Causes:
Transistor mismatch (mostly input pair)
Layout asymmetry (verify with RCX)
Result:
Input-referred ADC offset; no distortion
Solutions:
Transistor sizing, auto-zeroing, calibration

Existing mismatches between transistors (mainly the input pair) as well as layout asymmetry may generate an offset that is input-referred. Fortunately, since this offset does not cause distortion, it is often not a critical design factor, as mentioned before. However, if needed, it can be mitigated through proper transistor sizing or by recurring to auto-zeroing and calibration techniques.

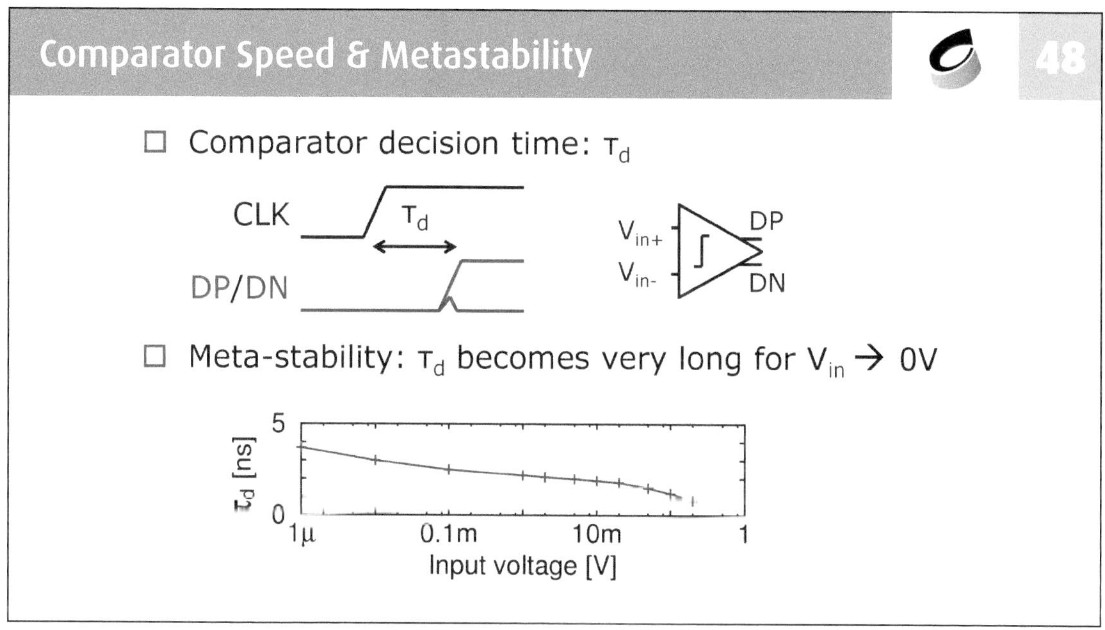

Meta-stability: τd becomes very long for Vin → 0V

When evaluating the comparator's speed, it should be taken into account that the time that the latch takes to make a decision (T_d) is not constant and becomes larger for smaller signals. Therefore, it is a good design choice to simulate the comparator for both low and high amplitude input signals to get a good estimate of the speed it can actually achieve.

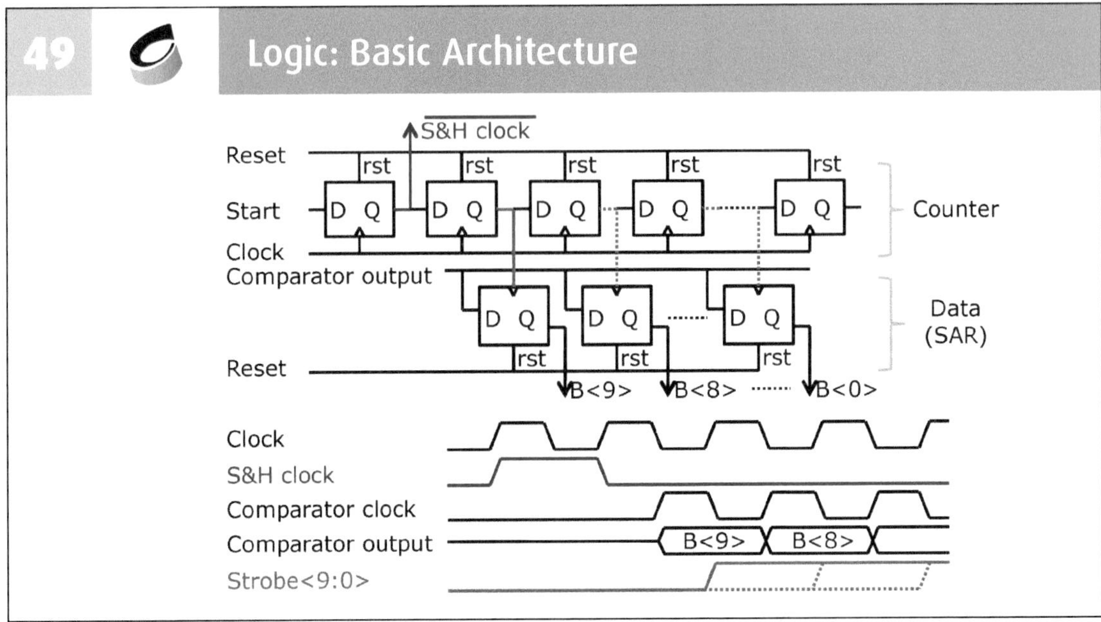

The last block to be discussed is the Logic. A basic implementation is shown where two rows of D-type flip-flops are used for distinct tasks. The top row's main function is to work as a counter, to keep track of the bit decision cycle. The bottom row works as a storage unit, where each element stores the result output by the comparator for a given bit, i.e., for an N-bit ADC, N flip-flops are required. After each digital code has been decided, the entire chain is reset and prepared for the next sampled input. Beside the flip-flops, some additional logic gates (not shown) will be required to generate the appropriate control signals for the comparator, T&H and DAC.

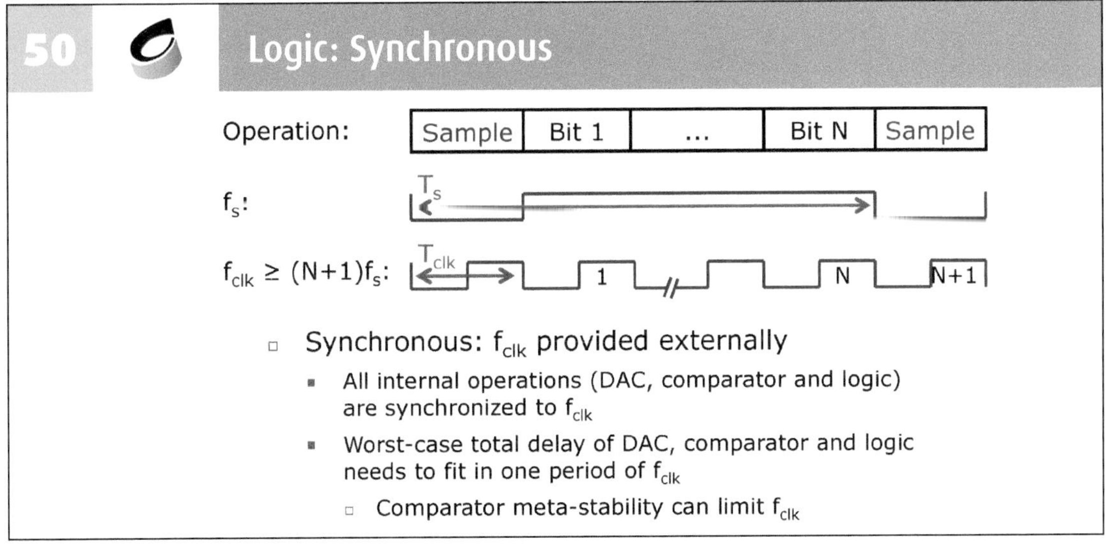

The Logic structure can operate either in a synchronous or asynchronous scheme. In synchronous implementations the clock should operate at a speed of at least (N+1) times the sampling frequency. As shown before, one period of this clock frequency needs to fit the sum of the delays of the DAC, comparator and logic. Since the comparator presents a non-constant delay time, this leaves the designer forced to generate an external clock signal where in some cases time, and therefore speed, will be lost, in order to ensure that the circuit can withstand the worst case scenario.

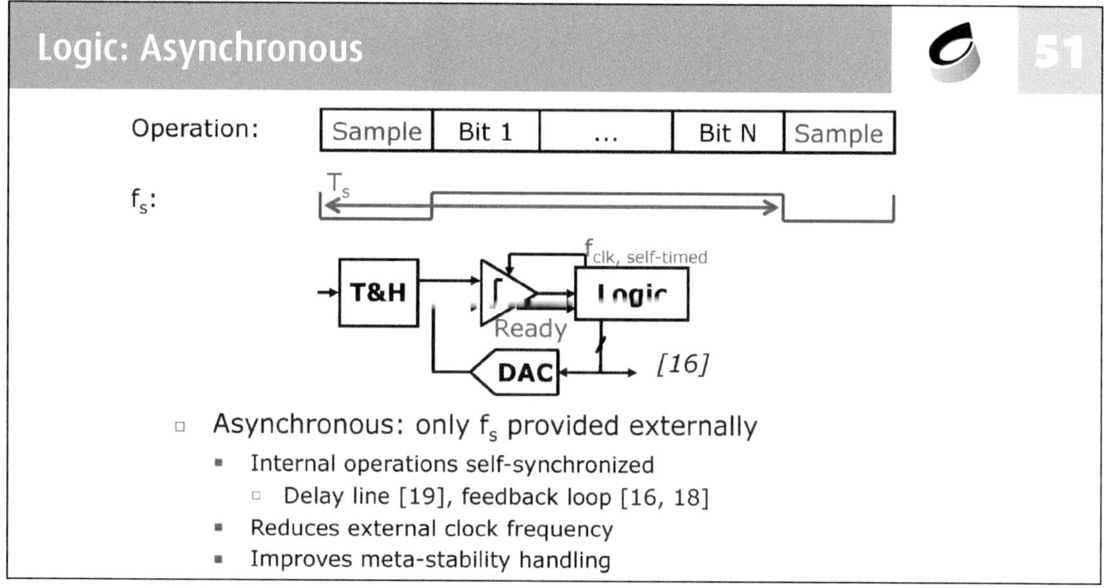

Otherwise, in asynchronous implementations only the sampling frequency needs to be provided, as internal operations can be self-synchronized through means of several techniques such as a delay line or a feedback loop. Thus, the comparator itself can operate close to its maximum speed for every single case and the external clock frequency can be reduced.

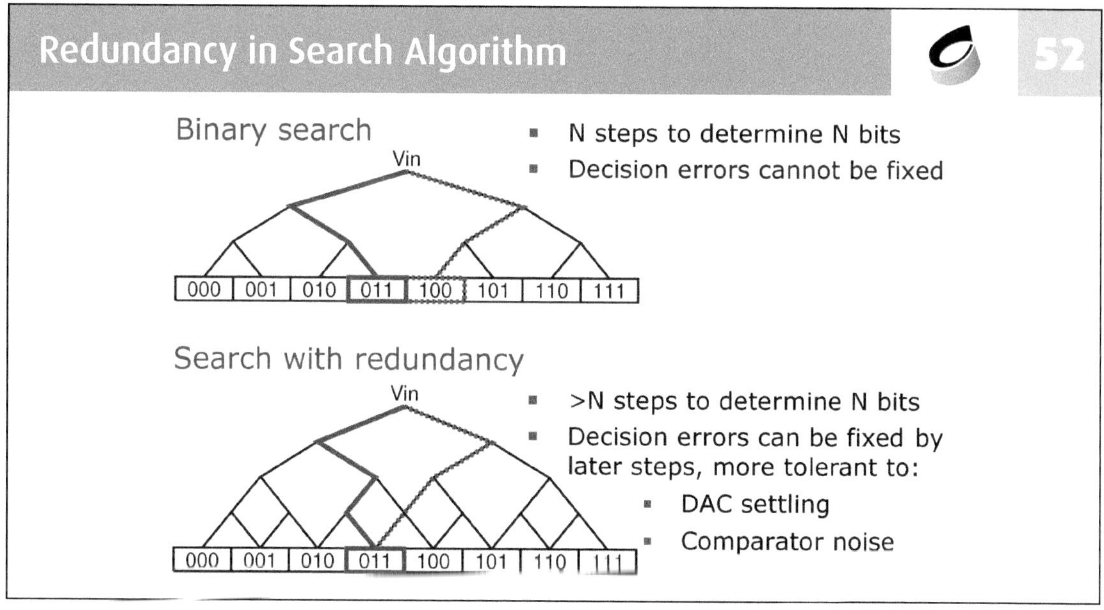

The last feature to be mentioned is the usage of redundancy in the search algorithm. In a simple binary search algorithm, once an error has been made by the circuit, the search range is inevitably reduced to a set of wrong possible outcomes and there is no possible way to correct the mistake. When redundancy is introduced, one can still reach the correct code, despite any error made (within the range of redundancy). In fact, it allows for relaxing the precision in early cycles of the SAR algorithm, knowing that errors that might occur here will be corrected further down the line (in this way, DAC settling and comparator noise can be more relaxed, thereby saving power). The major drawback here is the need for additional steps in the decision making and additional logic to generate a binary code out of a non-binary decision tree.

53  Outline

1. Introduction to ADCs for IoT
2. Basic SAR ADC Architecture
3. Circuit Design Considerations
4. **Advanced Examples for IoT**
5. Summary and Conclusions

The following slides will focus on several advanced examples for IoT applications together with brief explanations of their main features. Most of the circuits presented here are intended for low power applications, a key aspect of IoT devices.

54  10bit 0-100kS/s SAR ADC [5]

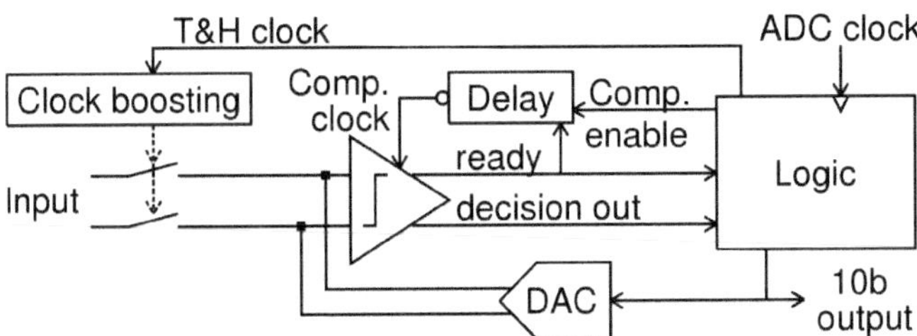

- Low VDD operation (down to 0.5V)
  - Clock boosting for T&H linearity
- Asynchronous clocking (f_s clock needed only)
- Dynamic circuits only → Power ∝ f_s
  - High-V_t devices & standby state for reduced leakage
- Custom design for logic

The first example consists of a 10b 100 kS/s SAR ADC particularly suited for sensing applications, which usually operate at low power. Hence, the circuit operates with a supply of 0.5 to 1 V and clock boosting is used to achieve sufficient T&H linearity. A self-synchronized architecture is used, relaxing the external sampling clock speed. Internal synchronization is achieved by a local oscillation loop, only active during quantization, created by the comparator and a delay element. Otherwise, the ADC is in a standby state. This dynamic operation, coupled with high-V_t devices, allows for reduced leakage power and consequently low power consumption.

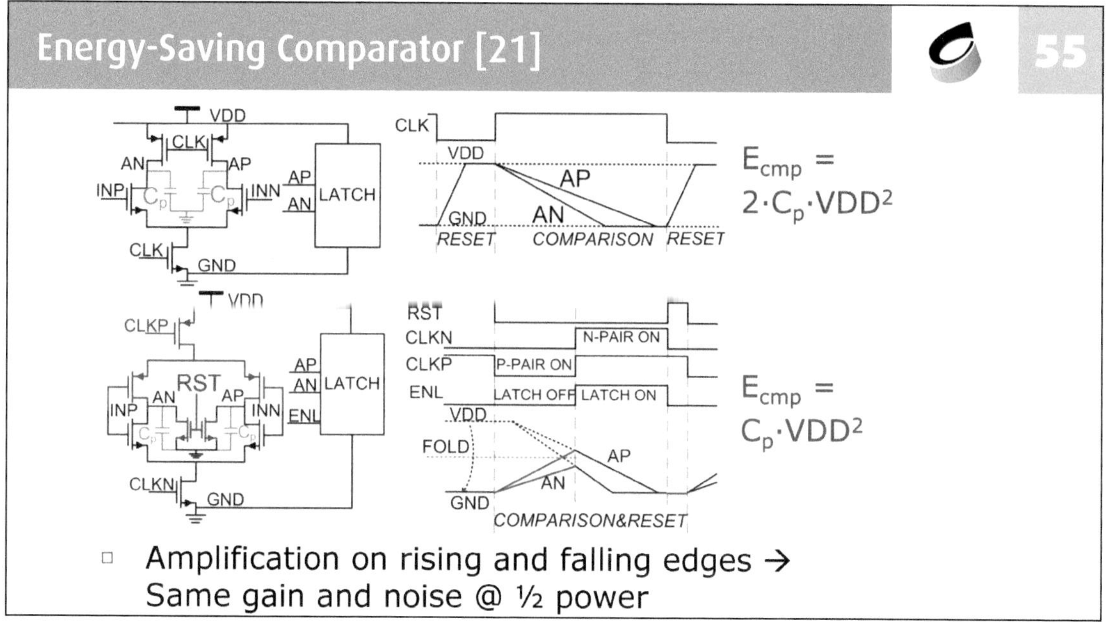

Inside the comparator, typical preamplifiers consist of a single entry pair, where amplification occurs during the discharging phase (falling edge) of the output capacitors C_p. The rising edge is only used for reset. Instead here the dynamic preamplifier is made with both an NMOS and PMOS entry pair, enabling the use of both rising and falling edges for amplification. Since the output capacitors only charge up to $V_{DD}/2$, the energy consumption is given by $C_p{}^*V_{DD}{}^2$, yielding a power saving in the order of 50% for the same gain and noise.

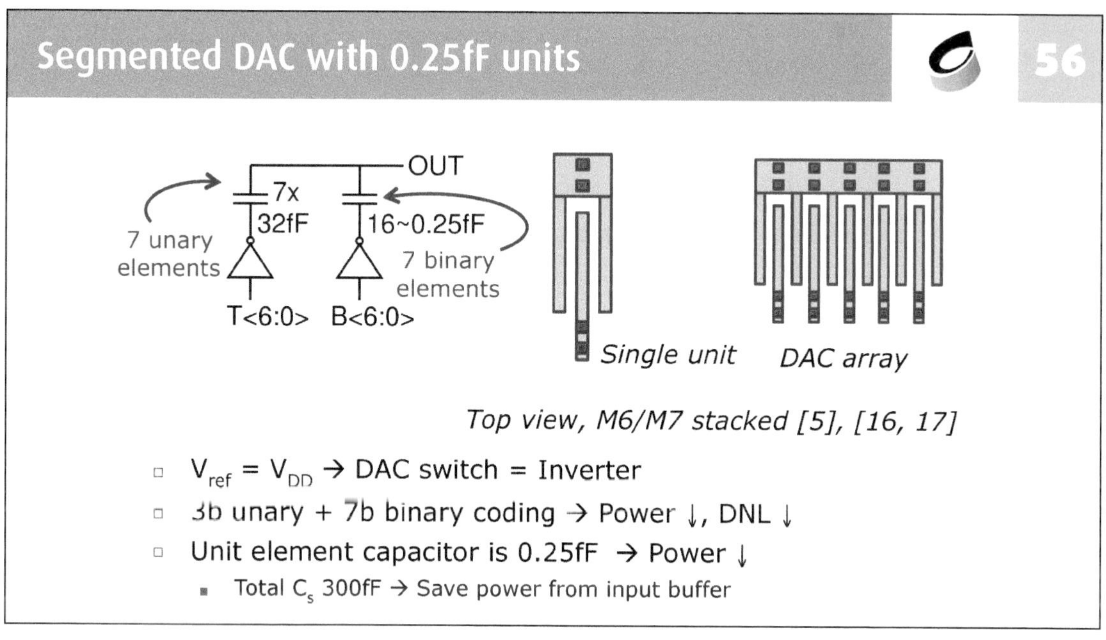

Segmented DAC with 0.25fF units
56
OUT
7x
32fF
16~0.25fF
7 unary
elements
7 binary
elements
T<6:0> B<6:0>
Single unit
DAC array
Top view, M6/M7 stacked [5], [16, 17]
Vref = VDD → DAC switch = Inverter
3b unary + 7b binary coding → Power ↓, DNL ↓
Unit element capacitor is 0.25fF → Power ↓
Total Cs 300fF → Save power from input buffer

A 10b segmented DAC is implemented with a unit element capacitor of 0.25 fF, yielding an overall capacitance of around 300 fF, including parasitics. This reduces the power of not only the ADC but also from its input buffer. The lower end of the array consists of a 7b binary-scaled structure, while the three MSBs are coded in a unary scheme by seven 32 fF capacitors. This feature allows for some distinct advantages, such as saving power and reducing distortion (by improving the DNL), as there is no scaling up of the capacitors.

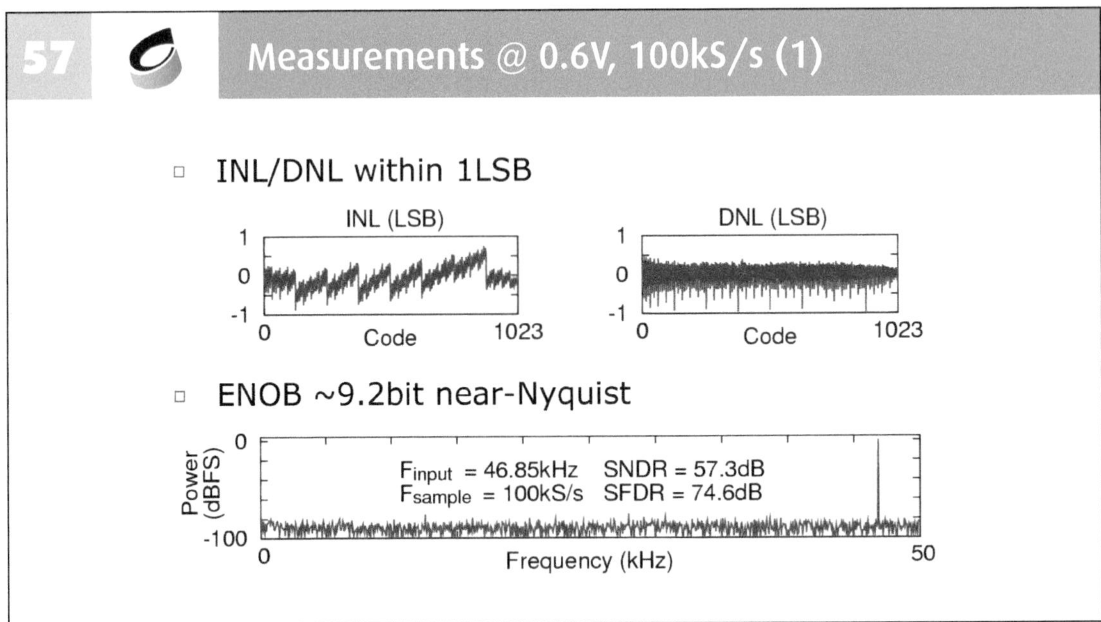

Measurements show that an INL of 0.87 LSB and a DNL of 0.96 LSB are achieved, both below 1 LSB, ensuring a monotonic transfer function with no missing codes. For an input signal with a frequency of 46.85 kHz the spectrum shows an SNDR of 57.3 dB and a SFDR of 74.6 dB, resulting in a 9.2b ENOB.

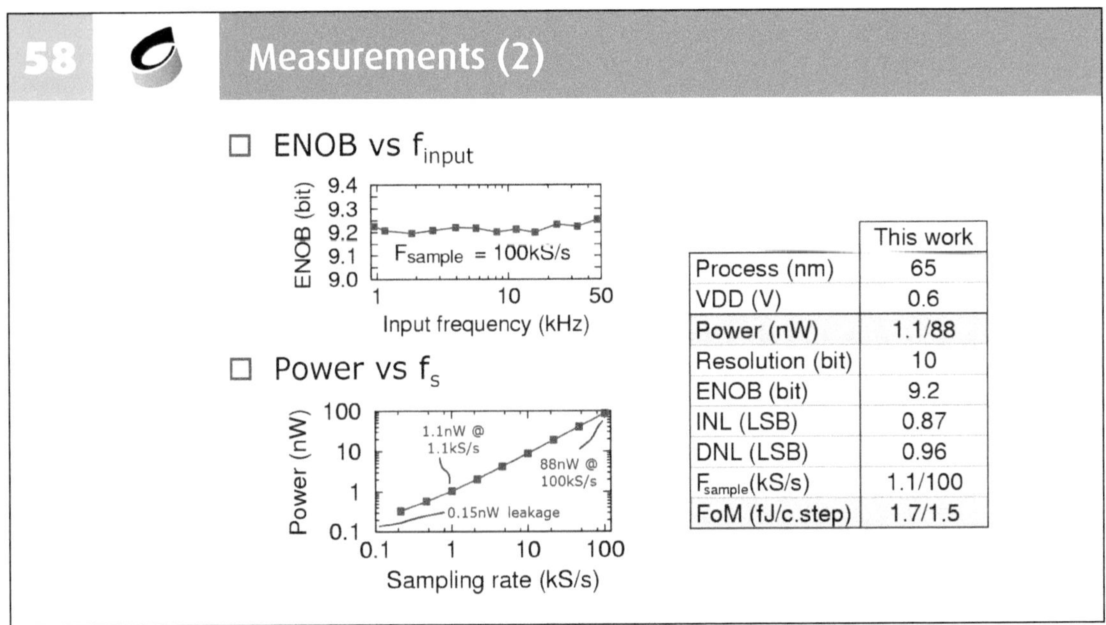

	This work
Process (nm)	65
VDD (V)	0.6
Power (nW)	1.1/88
Resolution (bit)	10
ENOB (bit)	9.2
INL (LSB)	0.87
DNL (LSB)	0.96
F_{sample}(kS/s)	1.1/100
FoM (fJ/c.step)	1.7/1.5

This ENOB is maintained throughout the entire bandwidth. The second graph confirms that the dynamic power consumption of the ADC scales with the sampling rate from roughly 88 nW at 100 kS/s down to a leakage level of 0.15 nW, when the clock is completely disabled. For lower sampling rates the impact of leakage power on the overall power consumption is more significant.

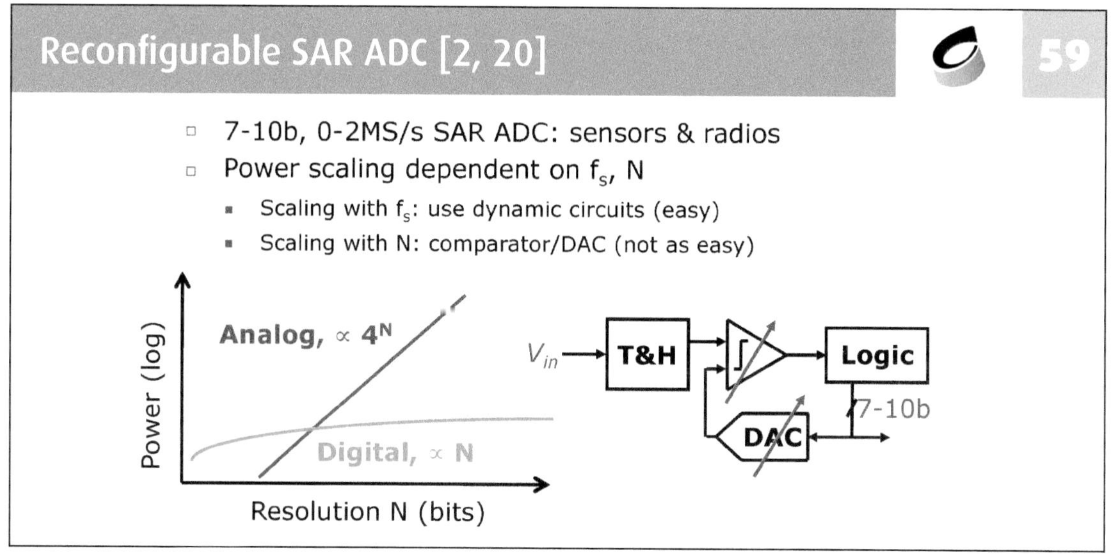

Next, a reconfigurable SAR ADC is presented. It can provide 7 to 10b of resolution and operate anywhere between 0 to 2 MS/s, covering a wide array of sensors and radios. For each possible configuration the power will scale accordingly with the required speed and resolution. Scaling power with speed is not as problematic as dynamic circuits can be used to cope with it. On the other hand, the power of analog blocks should in theory grow exponentially with resolution, as mentioned before. However, it is hard to implement adaptable analog circuits that can follow this exponential trend when scaling resolution. In SAR ADCs this specifically concerns the design of both the DAC and the comparator.

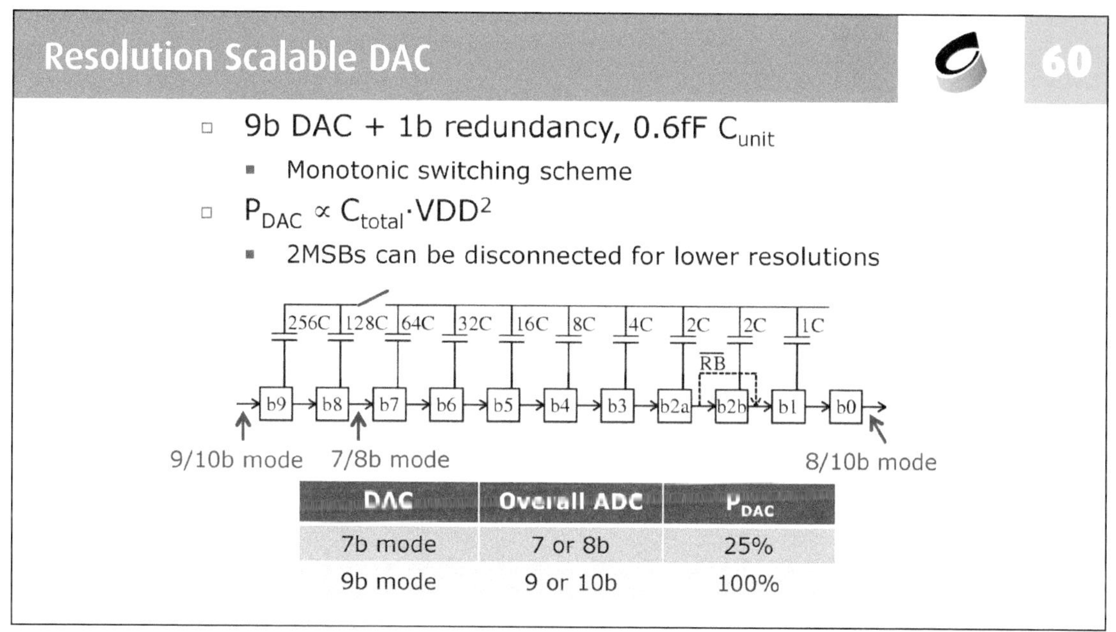

DAC	Overall ADC	P_{DAC}
7b mode	7 or 8b	25%
9b mode	9 or 10b	100%

For the former, a monotonic switching scheme is employed, where a 9b binary-scaled array of elements is enough to ensure the maximum possible resolution of 10b. 1 bit of redundancy is used at the second least significant bit for error correction, relaxing the initial cycles. Optionally, the 2 MSBs can be disconnected if the user only demands a maximum of 8b resolution. Since the overall power consumption is proportional to the total capacitance, this results in a 75% power saving. More resolutions could be enabled by adding more switches, at the cost of complexity.

61 Resolution Scalable Comparator

- Pre-amplifier: $E_{cmp} \propto C_a$; $V_{noise} \propto 1 / \sqrt{C_a}$

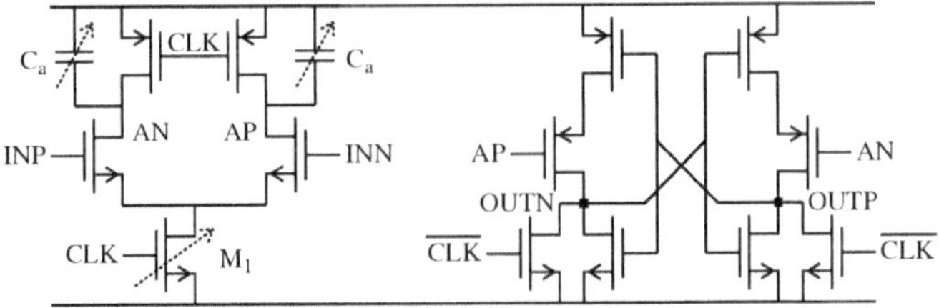

Accuracy setting	Energy per comparison	Input ref. noise
7b/8b	108fJ	0.24mVrms
9b	214fJ	0.18mVrms
10b	373fJ	0.12mVrms

Regarding the comparator, as with the previous circuit, the power consumption of the pre-amplifier will scale with the output capacitors. Furthermore, the input referred noise is also related with this capacitance. Hence, the load capacitor is made programmable to trade power consumption with input referred noise.

62 Two-Step Conversion with Redundancy

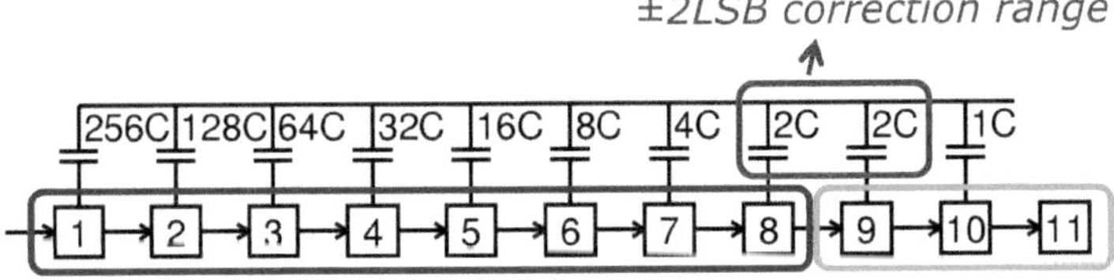

- Regular conversion:
  - 10 cycles full power: $E_{total} = 10E_{cmp}$
- 2-step conversion:
  - 8 cycles ~¼ power, 3 cycles full power: $E_{total} = 5E_{cmp}$

The proposed comparator, when coupled with the redundancy inserted by the DAC, allows for significant power savings during the conversion process. Since the precision requirements for the early cycles can be relaxed, the comparator can operate at a much lower power (i.e., smaller load capacitance). Only during the last three cycles does the comparator operate at full power, i.e., higher capacitances, with significantly less noise. When compared to a regular conversion, operating at full power continuously, this translates into power savings in the magnitude of 50%.

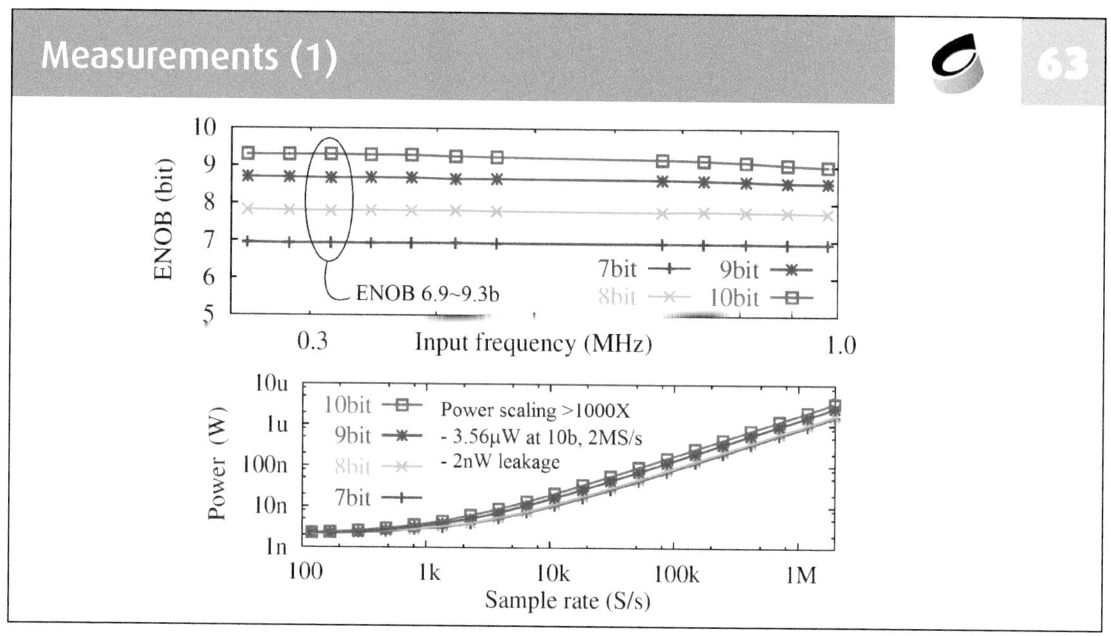

Results show how this reconfigurable ADC ensures a resolution ranging from 6.9 to 9.3 ENOB, kept relatively constant throughout the entire bandwidth. At maximum speed (2 MS/s) and resolution the ADC dissipates around 3.6 μW of power. Also, it can be observed how the power scales with the sample rate by more than a thousand times.

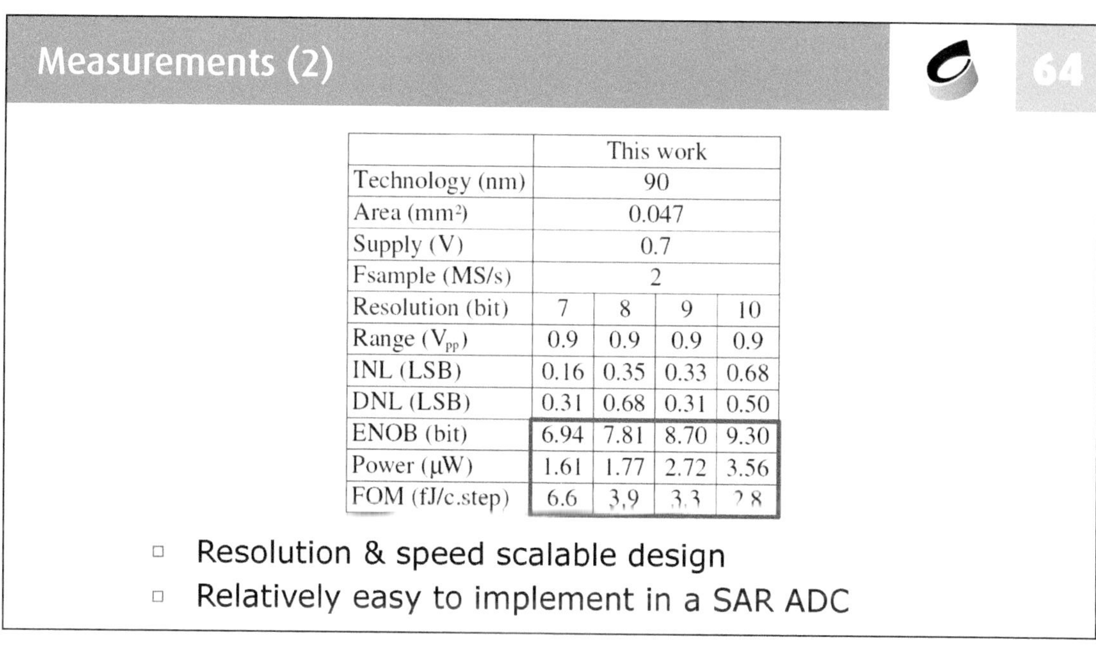

	This work			
Technology (nm)	90			
Area (mm²)	0.047			
Supply (V)	0.7			
Fsample (MS/s)	2			
Resolution (bit)	7	8	9	10
Range (V_{pp})	0.9	0.9	0.9	0.9
INL (LSB)	0.16	0.35	0.33	0.68
DNL (LSB)	0.31	0.68	0.31	0.50
ENOB (bit)	6.94	7.81	8.70	9.30
Power (μW)	1.61	1.77	2.72	3.56
FOM (fJ/c.step)	6.6	3.9	3.3	2.8

Resolution & speed scalable design
Relatively easy to implement in a SAR ADC

In short, SAR ADCs are an extremely viable solution when reconfigurability features such as speed and resolution come into play, particularly when low power is required. A monotonic transfer function is ensured in an ADC that is capable of offering moderate sampling rates (up to 2 MS/s) and resolution (up to 10b).

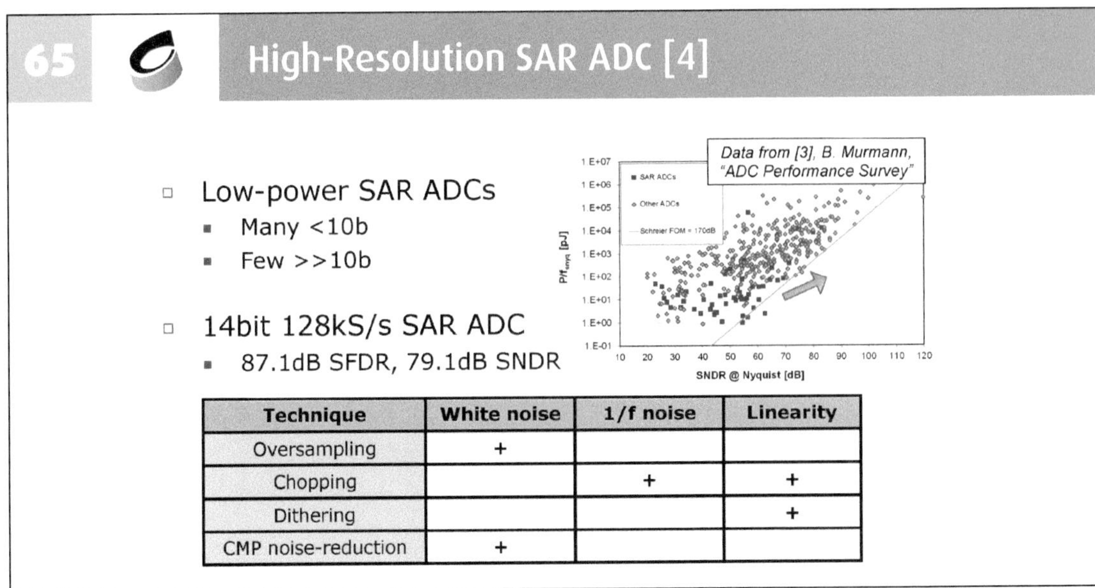

Technique	White noise	1/f noise	Linearity
Oversampling	+		
Chopping		+	+
Dithering			+
CMP noise-reduction	+		

Most reported power-efficient SAR ADCs fail to surpass the 10b resolution mark. The main contributor to this limitation in performance is the linearity, followed by noise. Several techniques can be used to cope with these issues, namely:

- Oversampling: which reduces the impact of white noise in the signal band;
- Chopping: which reduces both the impact of flicker noise and improves linearity;
- Dithering: which further helps reducing distortion in the bandwidth of interest, namely distortion due to INL errors;
- Comparator noise-reduction: where a majority voting method is used to improve noise performance in a power-efficient way.

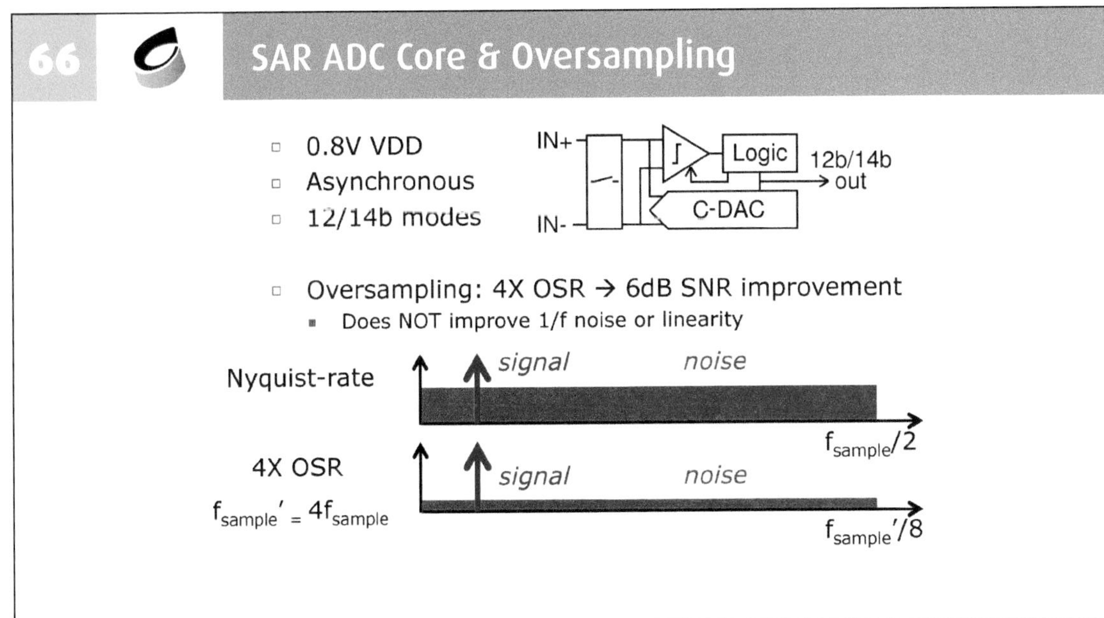

This SAR ADC operates at 0.8 V, in order to avoid the linearity issues caused by the T&H switches at lower supplies. Also it can be configured to provide either 12b or 14b of resolution. As example, an oversampling ratio of 4 could spread the noise over a 4x wider frequency range, improving the in-band SNR by around 6 dB. However, flicker noise and distortion still need to be addressed as they are not mitigated by oversampling.

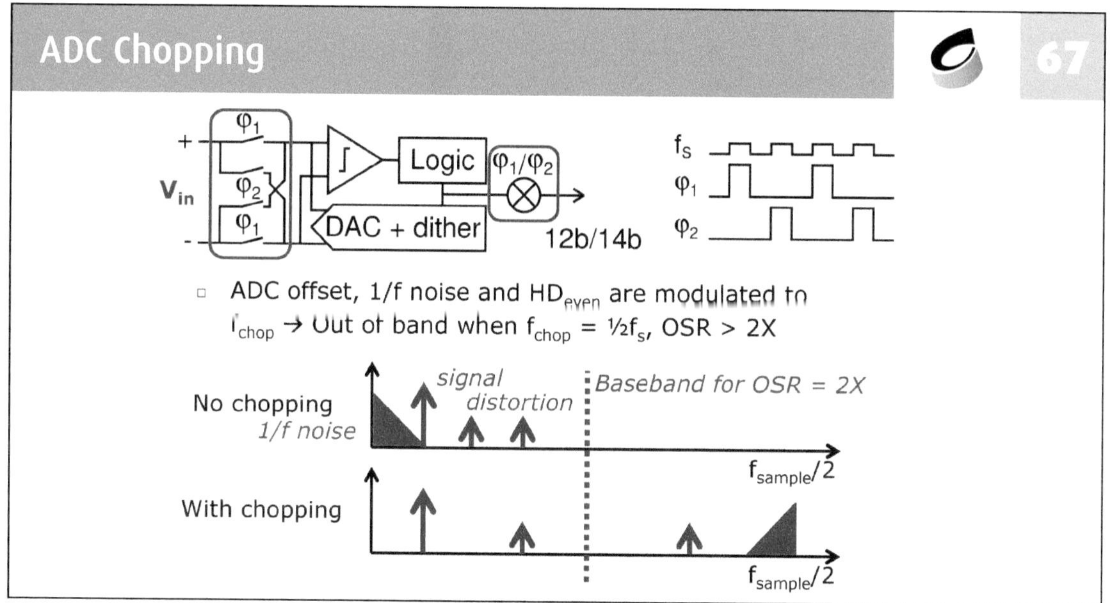

ADCs can benefit from using chopping much like amplifiers do, in terms of DC offset and flicker noise. Also, it makes it possible to modulate even-order distortion components and move them out of the signal bandwidth, when chopping at half the sampling rate. This scheme does not increase the overall power consumption, as despite the usage of two clock boosters, each operates at half the sample rate. The output chopping is performed in the digital domain, where a MUX selects the output data from either the non-inverted or the inverted output of the SAR.

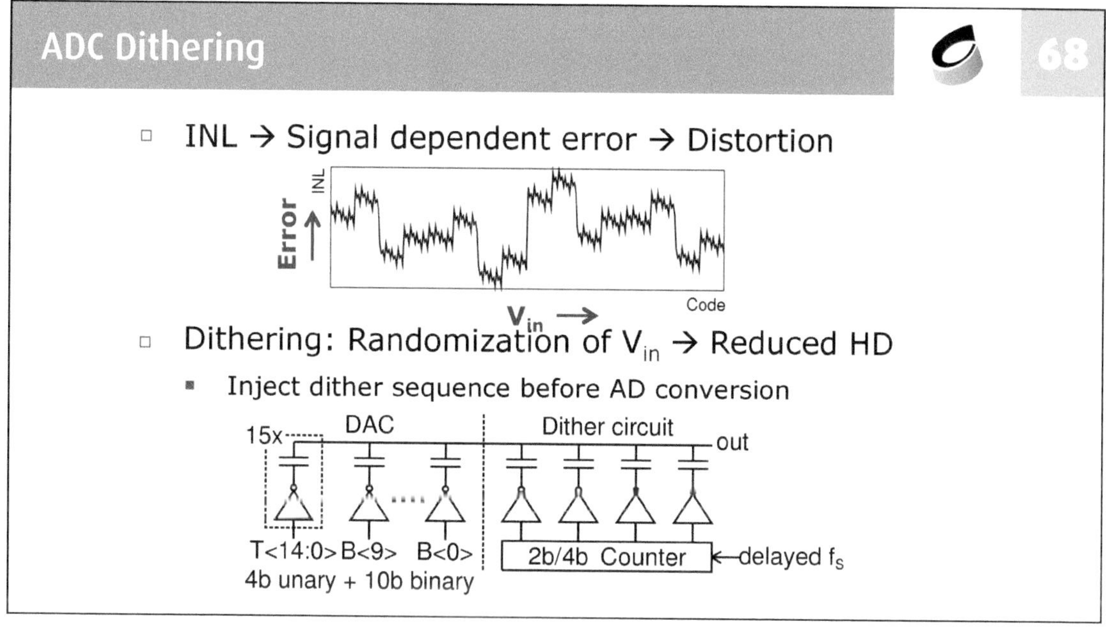

INL errors caused by capacitor mismatch generate a signal dependent error. In this sense, dithering can be used to randomize the input signal, which randomizes the error, thereby reducing distortion. Here, a dither circuit of 4 or 16 levels is connected in parallel to the DAC array, where the dither sequence is injected at the input of the ADC after sampling but prior to the actual A/D conversion. As the dither is a deterministic pattern, it results in spurious tones at multiples of fs/4 or fs/16, dependent on the selected length of the sequence.

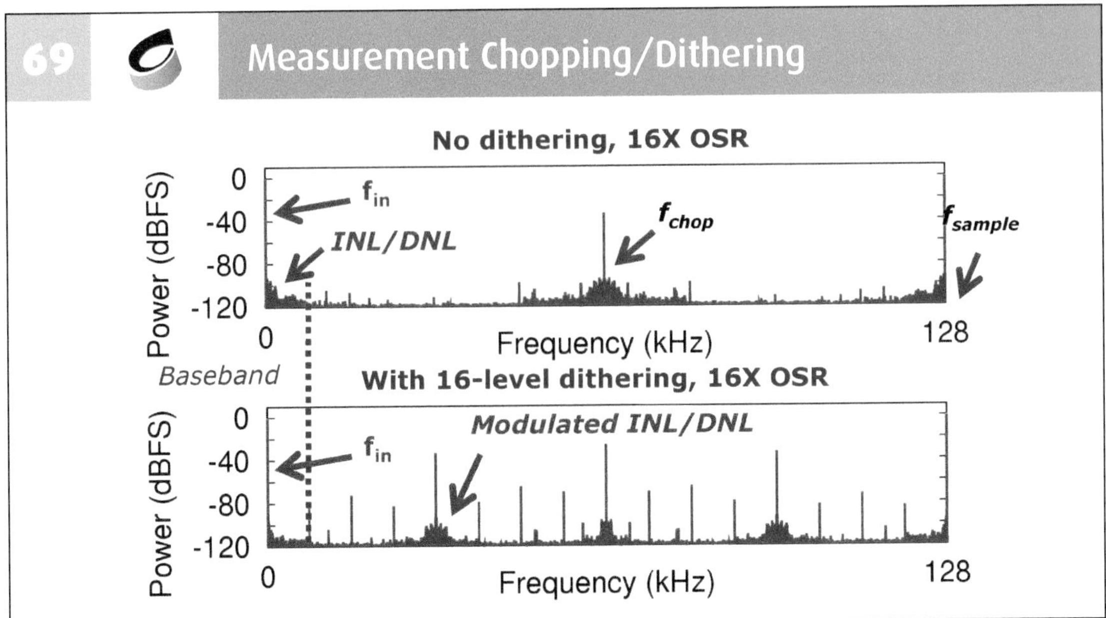

Measurement results show that when 16-level dithering is enabled, 16 tones will appear across the spectrum, but in-band distortion is reduced, leading to linearity improvements. It can also be observed how 1/f noise is modulated out-of-band by chopping.

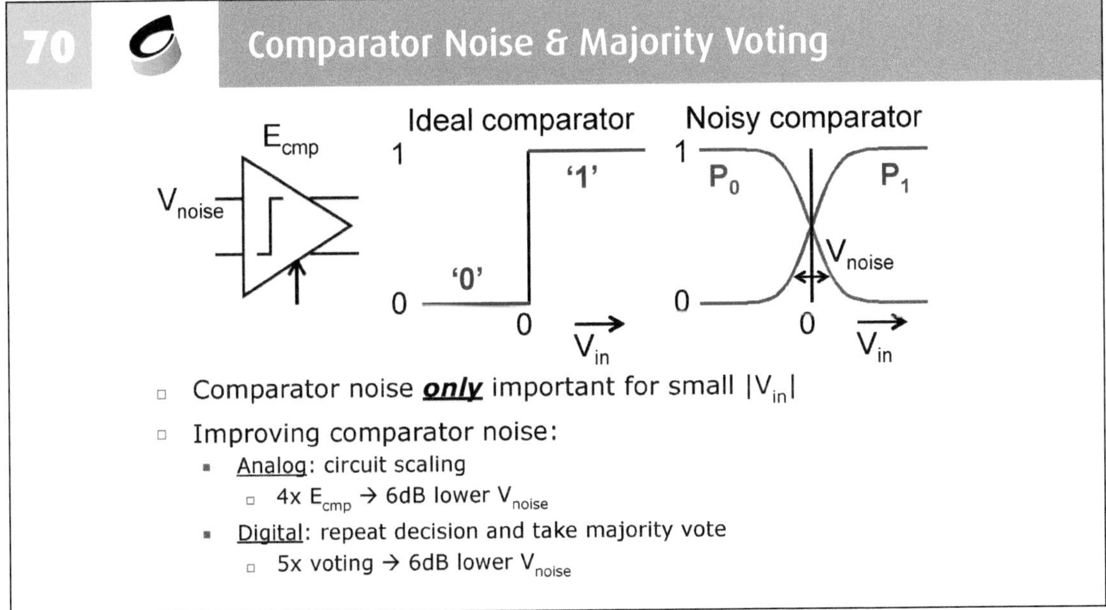

70
Comparator Noise & Majority Voting
E_{cmp}
V_{noise}
Ideal comparator
Noisy comparator
1
'1'
'0'
0
V_{in}
P_0
P_1
V_{noise}
Comparator noise only important for small $|V_{in}|$
Improving comparator noise:
Analog: circuit scaling
4x E_{cmp} → 6dB lower V_{noise}
Digital: repeat decision and take majority vote
5x voting → 6dB lower V_{noise}

Finally, the performance of a noisy 1-bit comparator is only critical for small input signals, as detailed before. Possible improvements rely on increasing the dimensions of the circuit and, consequently, power consumption, or instead keep the same noisy comparator but make it repeat the same decision an N_v number of times for N_c noise-critical bit-cycles, through a data-driven noise reduction (DDNR) technique. In other words, the comparator will sometimes output a logic 'one' and sometimes a logic 'zero'. Taking the majority of occurrences of either '1' or '0' makes this noisy comparator more accurate than if a single decision was made for the same low amplitude input.

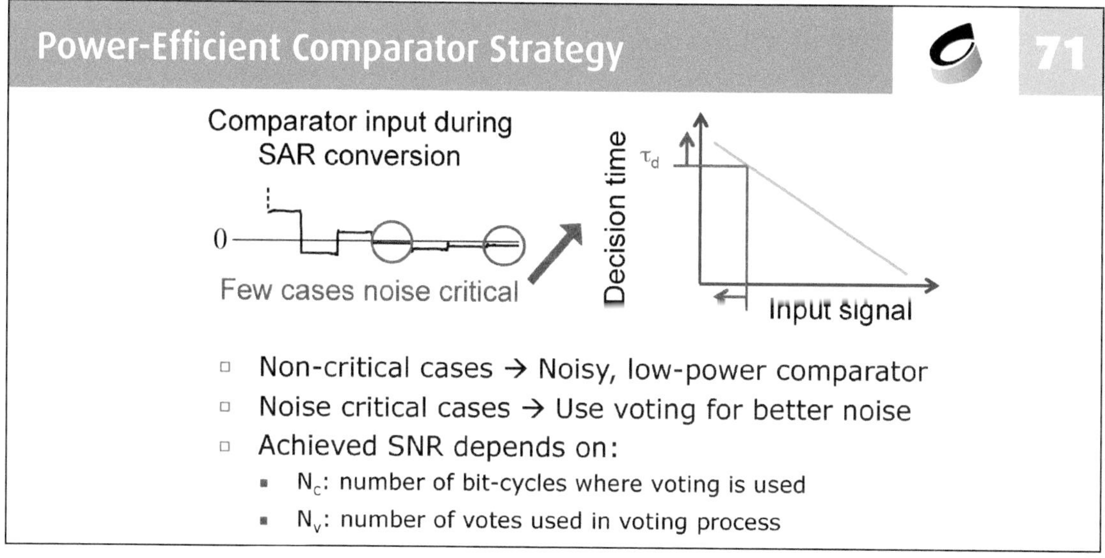

Power-Efficient Comparator Strategy
71
Comparator input during SAR conversion
0
Few cases noise critical
Decision time
τd
Input signal
Non-critical cases → Noisy, low-power comparator
Noise critical cases → Use voting for better noise
Achieved SNR depends on:
Nc: number of bit-cycles where voting is used
Nv: number of votes used in voting process

The amount of noise reduction depends on two parameters: N_v (a higher number results in better noise averaging) and N_c. The detection of which cycles are the noise-critical bit-cycles is done by triggering a reference delay cell together with the comparator. For large input signals that are not noise-critical, the comparator generates an output before the reference delay, and the decision is immediately provided to the logic. Small input levels that can be corrupted by noise are detected whenever the comparator is running slow and the voting logic produces the majority value as the final output.

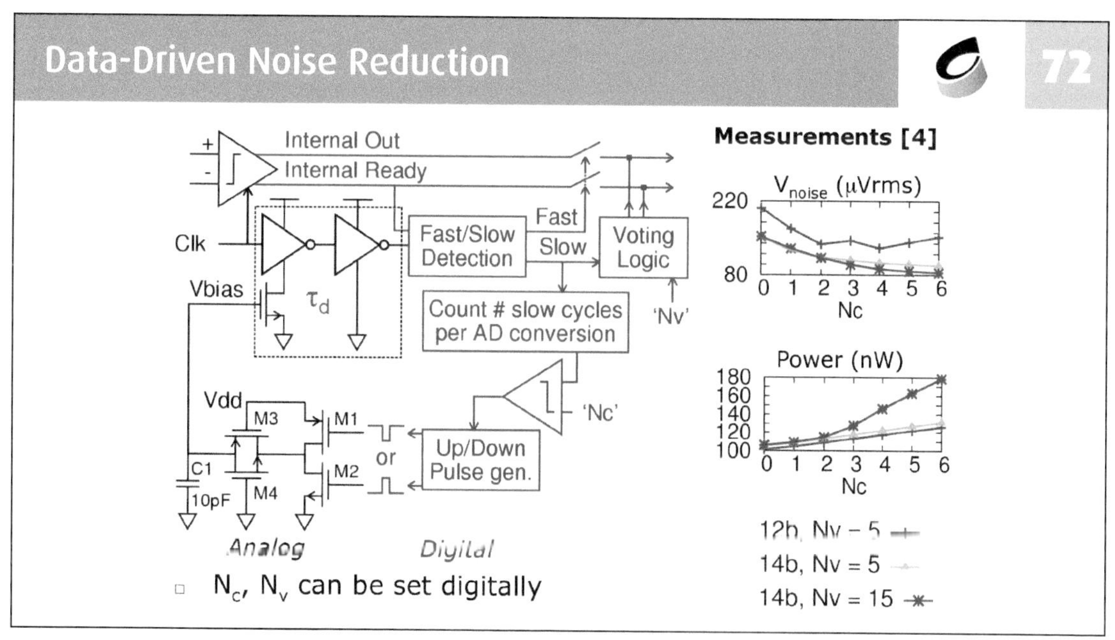

Here, parameters N_v and N_c are set digitally. N_v is simply used in the digital voting logic to count the number of repetitive decisions. N_c is used to control the reference delay by means of a feedback loop that drives V_{bias}. The actual number of voting cycles is determined by counting the number of times a slow, noise-critical decision is detected during a conversion. This number is compared against the desired N_c. Dependent on the comparison result, V_{bias} is either increased or decreased by a charge or discharge pulse on C_1 through $M_{1,2}$. To achieve a slow time-constant in the loop without needing an excessively large capacitor C_1, transistors $M_{3,4}$ are added.

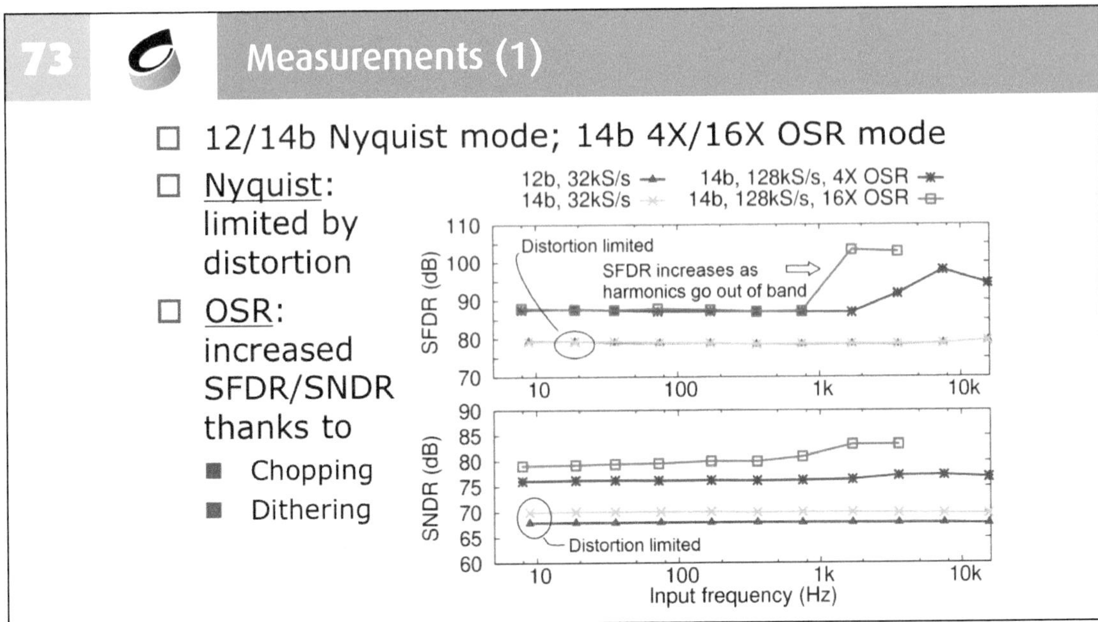

73
Measurements (1)
12/14b Nyquist mode; 14b 4X/16X OSR mode
Nyquist: limited by distortion
OSR: increased SFDR/SNDR thanks to
Chopping
Dithering
12b, 32kS/s
14b, 32kS/s
14b, 128kS/s, 4X OSR
14b, 128kS/s, 16X OSR
Distortion limited
SFDR increases as harmonics go out of band
SFDR (dB)
SNDR (dB)
Distortion limited
Input frequency (Hz)

Measurement results show that power consumption increases with the N_c parameter, while the input-referred noise of the ADC decreases, as expected. When operating at 12b Nyquist mode, this noise improvement is limited by the quantization noise. The measured SFDR and SNDR versus input frequency are shown for Nyquist-rate 12b/14b operation as well as 4×/16× oversampled 14b operation. In 14b mode at 128kS/s with 16× oversampling, the application of chopping, 16-level dithering and DDNR, both the SFDR and SNDR are improved by about 8dB and 6dB respectively.

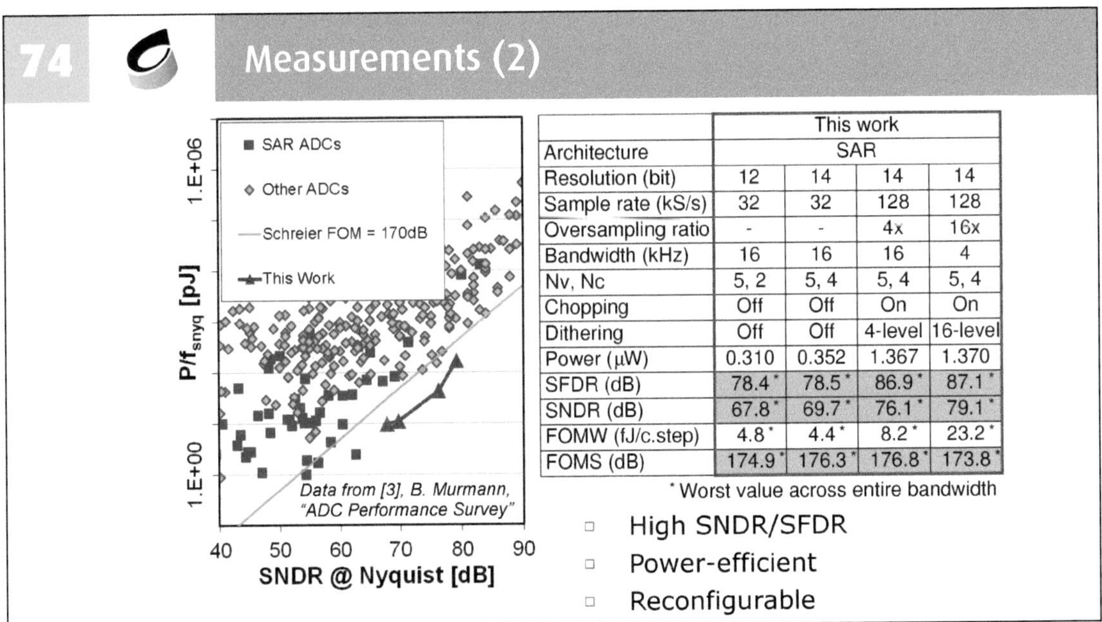

	This work			
Architecture	SAR			
Resolution (bit)	12	14	14	14
Sample rate (kS/s)	32	32	128	128
Oversampling ratio	-	-	4x	16x
Bandwidth (kHz)	16	16	16	4
Nv, Nc	5, 2	5, 4	5, 4	5, 4
Chopping	Off	Off	On	On
Dithering	Off	Off	4-level	16-level
Power (µW)	0.310	0.352	1.367	1.370
SFDR (dB)	78.4*	78.5*	86.9*	87.1*
SNDR (dB)	67.8*	69.7*	76.1*	79.1*
FOMW (fJ/c.step)	4.8*	4.4*	8.2*	23.2*
FOMS (dB)	174.9*	176.3*	176.8*	173.8*

* Worst value across entire bandwidth

The power-efficient enhancement techniques enable reaching an SNDR between 67.8dB and 79.1dB, dependent on the selected mode. This is relatively high as compared to previously reported low-power SAR ADCs. Moreover, all 4 modes of operation achieve state-of-the-art power-efficiencies in both Walden and Schreier-based FoMs as compared to ADCs with similar SNDRs.

Outline

75

1. Introduction to ADCs for IoT
2. Basic SAR ADC Architecture
3. Circuit Design Considerations
4. Advanced Examples for IoT
5. **Summary and Conclusions**

Finally, a brief summary of the topics discussed is presented. Also, some conclusions are drawn regarding SAR ADCs and their applicability to IoT devices.

Summary and Conclusions

76

- Summary
  - System-level aspects of SAR ADCs
  - Circuit-level design and considerations
  - Advanced examples for IoT
    - Low-power; reconfigurability; high-resolution
- Conclusions on SAR ADCs
  - Wide application range (resolution & speed)
  - Power-efficient choice
  - Process-scaling friendly architecture
  - Suitable for reconfigurability / on-demand operation

This chapter initially presented the essential system-level aspects of SAR ADCs (such as the binary search algorithm) and its four core blocks: the T&H, comparator, DAC and logic. Considerations on the design of each of these blocks were presented, regarding mainly noise and linearity. Finally, examples of SAR ADCs particularly designed for IoT applications are shown, focusing on three key aspects: low-power, reconfigurability and high-resolution. Focusing on low power supplies, several techniques and circuits are employed to ensure moderate-to-high resolutions, with exceptionally low distortion and moderate speed.

77  Suggested Textbooks

1. **P.G.A. Jespers, "Integrated Converters"**
2. **F. Maloberti, "Data Converters"**
3. **G. Manganaro, "Advanced Data Converters"**
4. **M.J.M. Pelgrom, "Analog-to-Digital Conversion"**
5. **R. van de Plassche, "CMOS Integrated AD and DA Converters"**
6. **B. Razavi, "Principles of Data Conversion System Design"**

SAR ADCs characterize themselves as an extremely viable A/D converter, benefitting from its process-scaling architecture making it one of, if not the most, power-efficient architecture among data converters. Its key analog blocks, the comparator and DAC, make this architecture particularly suited for several applications such as sensors and radios that demand reconfigurable features such as speed and resolution.

78  References

1. [1] H. Gao, et al., "A 71GHz RF Energy Harvesting Tag with 8% Efficiency for Wireless Temperature Sensors in 65nm CMOS," RFIC 2013, pp. 403-406.
2. [2] P. Harpe, et al., "A 7-to-10b 0-to-4MS/s Flexible SAR ADC with 6.5-to-16fJ/conversion-step," ISSCC 2012, pp. 472-473.
3. [3] B. Murmann, "ADC Performance Survey 1997-2014," [Online]. Available: http://web.stanford.edu/~murmann/adcsurvey.html.
4. [4] P. Harpe, et al., "An Oversampled 12/14b SAR ADC with Noise Reduction and Linearity Enhancements Achieving up to 79.1dB SNDR," ISSCC 2014, pp. 194-195.
5. [5] P. Harpe, et al., "A 3nW Signal-Acquisition IC Integrating an Amplifier with 2.1 NEF and a 1.5fJ/conv-step ADC," ISSCC 2015, pp. 382-383.
6. [6] T. B. Cho, et al., "A 10 b, 20 MSample/s, 35 mW pipeline A/D converter," JSSC, vol. 30, no. 3, pp. 166–172, Mar. 1995.
7. [7] A. M. Abo , et al., "A 1.5-V, 10-bit, 14.3-MS/s CMOS Pipeline Analog-to-Digital Converter," JSSC, vol. 34, no. 5, pp. 599-606, May 1999.
8. [8] M. Ding, P. Harpe, et al., "A 5.5fJ/conv-step 6.4MS/s 13b SAR ADC Utilizing a Redundancy-Facilitated Background Error-Detection-and-Correction Scheme," ISSCC 2015, pp. 460-461.
9. [9] J. Xu, P. Harpe, et al., "A Low Power Configurable Bio-Impedance Spectroscopy (BIS) ASIC with Simultaneous ECG and Respiration Recording Functionality," ESSCIRC 2015.
10. [10] B. P. Ginsburg, et al., "A 500MS/s 5b ADC in 65nm CMOS," IEEE Symp. VLSI Circuits, 2006, pp. 140-141.
11. [11] C.-C. Liu, et al., "A 10-bit 50-MS/s SAR ADC With a Monotonic Capacitor Switching Procedure," JSSC, vol. 45, no. 4, pp. 731-740, Apr. 2010.
12. [12] V. Hariprasath, et al., "Merged Capacitor Switching Based SAR ADC with Highest Switching Energy-Efficiency," IET Electronics Letters, vol. 46, no. 9, pp. 620-621, Apr. 2010.
13. [13] C.-Y. Liou, et al., "A 2.4-to-5.2fJ/conversion-step 10b 0.5-to-4MS/s SAR ADC with Charge-Average Switching DAC in 90nm CMOS," ISSCC 2013, pp. 280-281.
14. [14] H.-Y. Tai, et al., "A 0.85fJ/conversion-step 10b 200kS/s Subranging SAR ADC in 40nm CMOS," ISSCC 2014, pp. 196-197.
15. [15] A. Agnes, et al., "A 9.4-ENOB 1V 3.8μW 100kS/s SAR ADC with Time-Domain Comparator," ISSCC 2008, pp. 246-247.
16. [16] P. Harpe, et al., "A 30fJ/conversion-step 8b 0-to-10MS/s asynchronous SAR ADC in 90nm CMOS," ISSCC 2010, pp. 388-389.
17. [17] P. Harpe, et al., "A 26μW 8bit 10MS/s Asynchronous SAR ADC for Low Energy Radios," JSSC, vol. 46, no. 7, pp. 1585-1595, July 2011.
18. [18] P. Harpe, et al., "A 10b/12b 40kS/s SAR ADC with Data-Driven Noise Reduction achieving up to 10.1b ENOB at 2.2fJ/conversion-step," JSSC, vol. 48, no. 12, pp. 3011-3018, Dec. 2013.
19. [19] M. van Elzakker, et al., "A 1.9μW 4.4fJ/Conversion-step 10b 1MS/s Charge-Redistribution ADC," ISSCC 2008, pp. 244-245.
20. [20] P. Harpe, et al., "A 0.7V 7-to-10bit 0-to-2MS/s Flexible SAR ADC for Ultra Low-Power Wireless Sensor Nodes," ESSCIRC 2012, pp. 373-376.
21. [21] M. Liu, P. Harpe, et al., "A 0.8V 10b 80kS/s SAR ADC with Duty-Cycled Reference Generation," ISSCC 2015, pp. 278-279.

CHAPTER 3

Sigma-Delta Modulators for Internet of Things

Nuno Paulino

Universidade NOVA de Lisboa,
NOVA School of Science and Technology
and CTS-UNINOVA - Portugal

This chapter will cover the basics and recent innovations in the field of SAR ADCs for IoT. Even though SAR ADCs have been known for a very long time, they receive a lot of attention recently thanks to their power-efficiency and beneficial scaling with technology. This lecture starts with a basic overview of SAR ADC design, and continues with discussing several recent innovations aiming for better power efficiency, improved performance, and more versatility

2
Why use ΔΣMs
Compared to Nyquist rate ADCs
Advantages:
Resolutions from 10 to 18 bits without calibration.
Simple anti-alias-filter due to the high oversampling ratio.
Large dynamic range.
Disadvantages:
Higher clock frequency.
ΔΣMs require a digital decimation filter.
Each output code does not correspond to a single input sample.

ΔΣMs can achieve large resolutions (> 10 bits) without requiring calibration, which is an advantage over Nyquist-rate ADCs. Such high resolutions correspond to a larger dynamic range of the A/D converter. Also, as a consequence of the high oversampling ratio (osr), the design of their anti-alias-filter (AAF) can be greatly relaxed. However, high osr's also imply higher clock frequencies, while the modulator itself requires a decimation filter to eliminate the quantization noise and down-sample the signal. Also, there is no one-to-one correspondence between each input sample and the output code, a feature that can be deemed unacceptable by some applications.

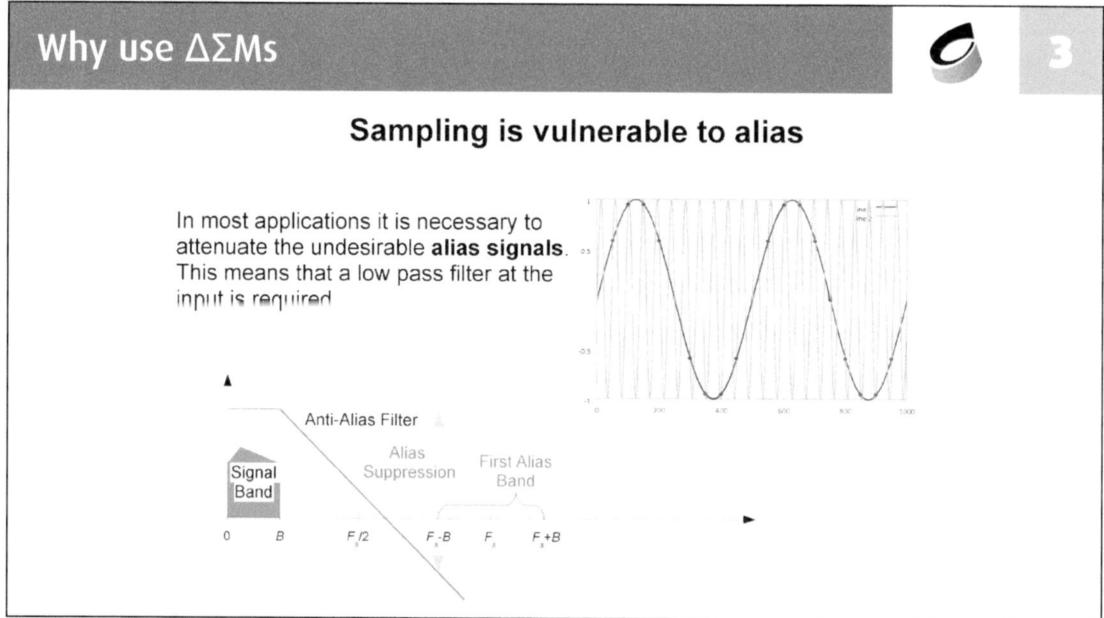

Alias occurs when two signals sampled at the same rate produce the same output samples, resulting in being impossible to distinguish one from the other. This can be an issue because IoT devices can be placed in environments with unknow interferer signals. Therefore it is necessary to use an Anti Alias Filter (AAF) to suppress the possible alias signals.

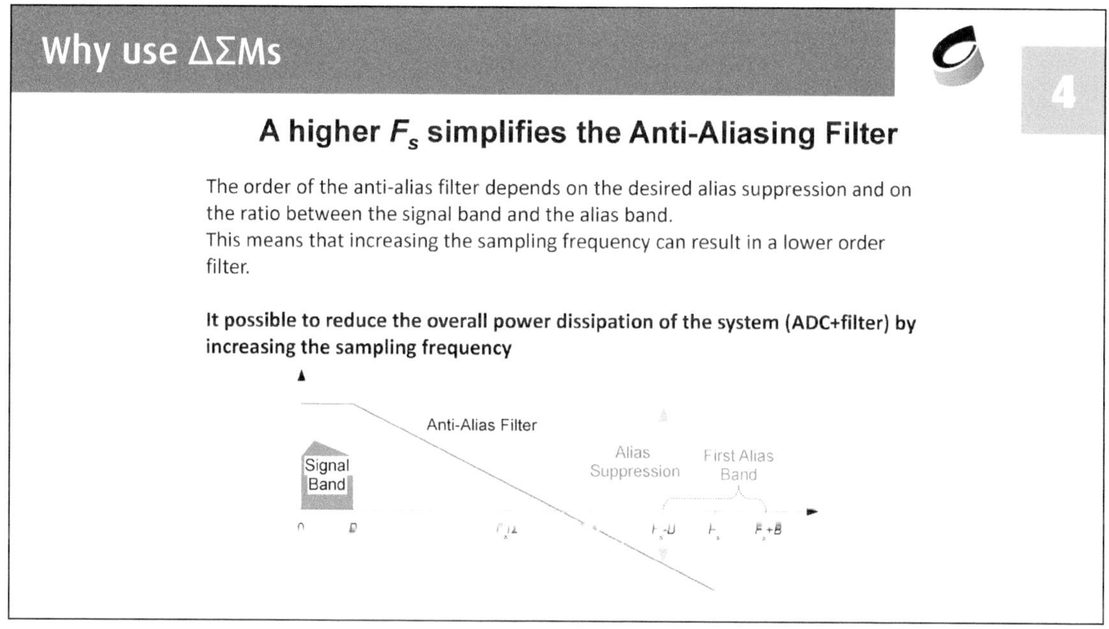

This AAF filter's order is directly related with the sampling rate Fs used (as the first alias band is centered at Fs and has a band of 2B), i.e., the lower F_s is, higher the slope of the filter, resulting in extra power consumption. So, there is an inherent advantage when using ΔΣMs, since the first alias band is placed much higher in frequency and a simple 1st order RC circuit can be used as filter. In the particular case of continuous-time (CT) implementations, the transfer function can be used to provide this anti-aliasing suppression (within certain limits).

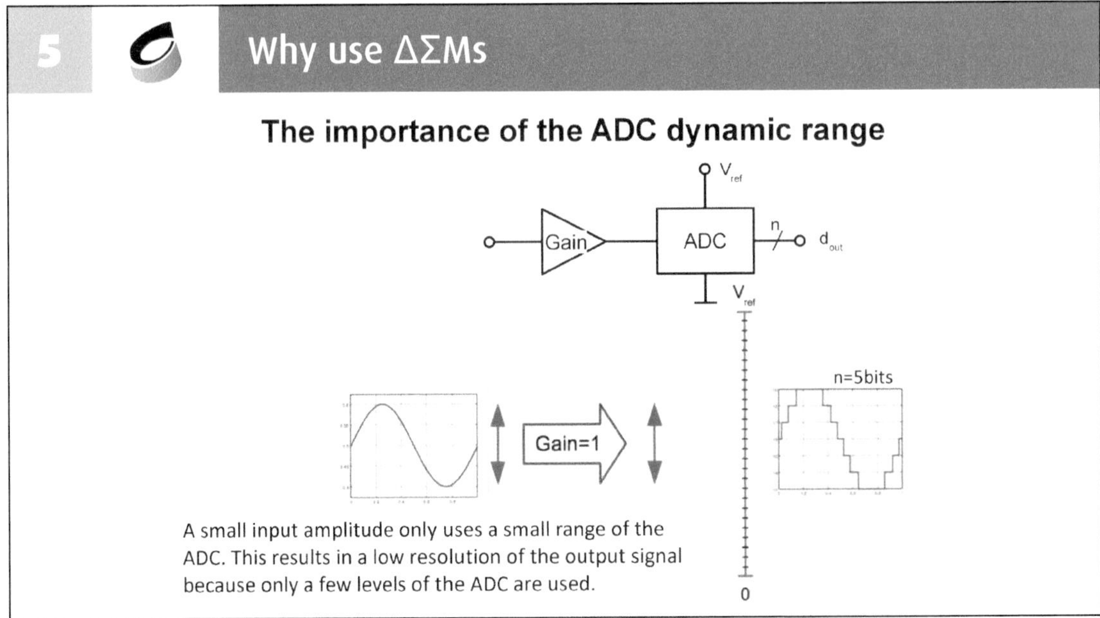

A small amplitude input signal only uses a limited range of the ADC resulting in a low resolution output signal, since few quantization steps of the ADC are used (represent here by the ruler).

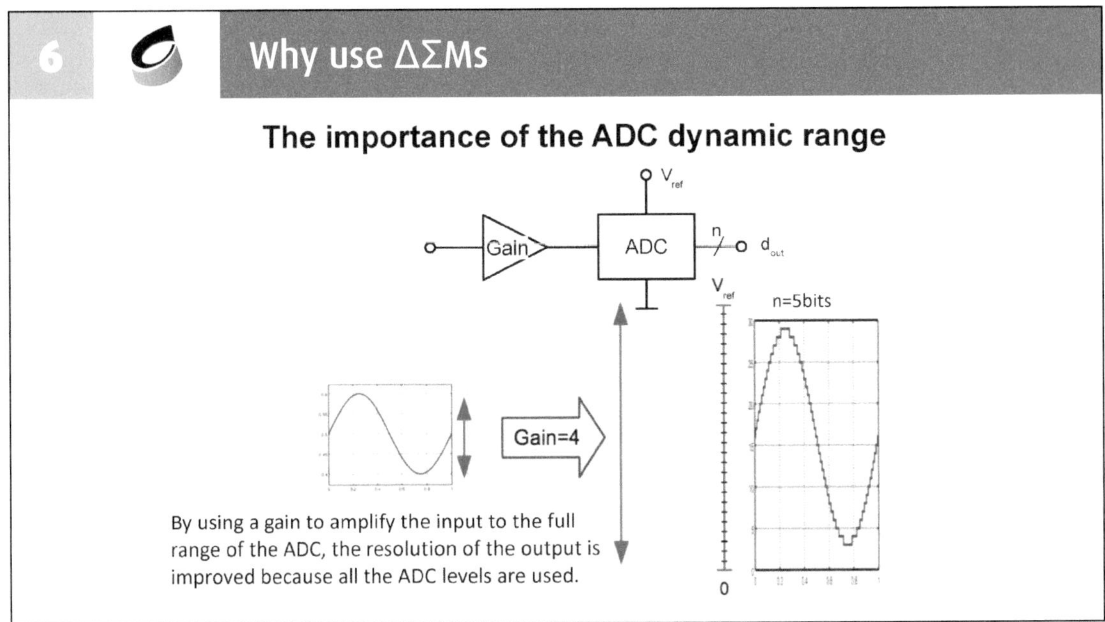

To improve resolution a gain block can be used to amplify the input signal by factor G, in which case the signal range will fit the entire set of intervals in the ADC. Therefore, a more accurate representation of the input signal can be obtained. Because all the quantization steps of the ADC are used.

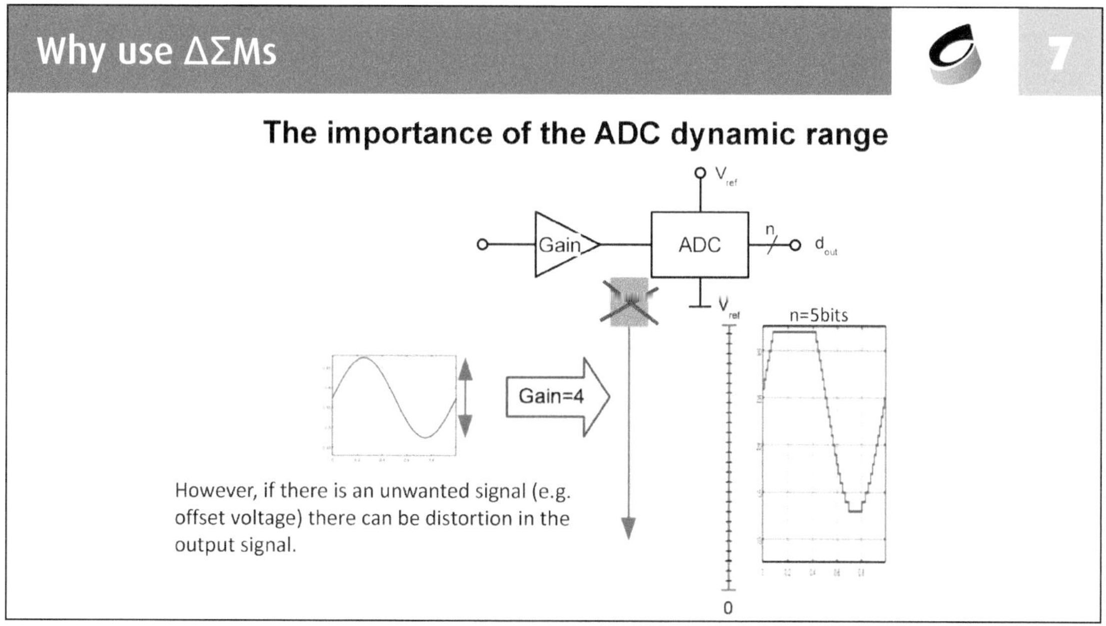

However, in several applications and in particular in IoT devices, several unwanted signals might occur, for example an offset can be added to the input signal. This leads to significant added distortion, due to clipping and other effects. So, despite the gain block being helpful in extending the input signal to cover the entire dynamic range of the ADC, it cannot be adjusted without taking into account these effects, at the cost of a significant drop in performance.

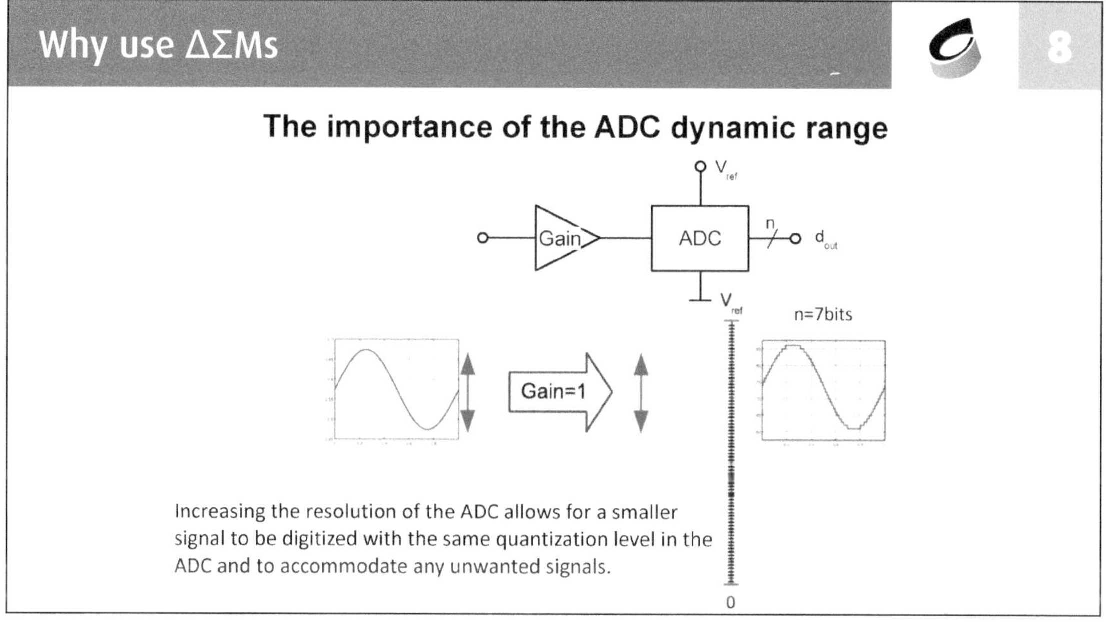

One of the most obvious solutions relies on increasing the resolution of the ADC itself, corresponding to a much finer quantization step (a smaller LSB). Thus, small signals can still be accurately represented and unwanted signals can be accommodated. In short, the trade-offs between each block within the system need to be accounted for, and if done properly, the power dissipation of the entire device can be reduced.

Outline

- Why use ΔΣMs
- **Principle of Operation of ΔΣMs**
- Active vs. Passive ΔΣMs
- Passive based ΔΣMs
  - UIS SC integrator based
  - Continuous time integrator based
- Conclusions

Next, the basics behind the operation of ΔΣMs are reviewed and an analysis is performed for 1st, 2nd and higher order modulators. The mathematical justification for the improvement of the signal-to-noise ratio (SNR) when using oversampling is also shown.

10 Principle of operation of ΔΣMs

Quantization channel using a ΔΣM

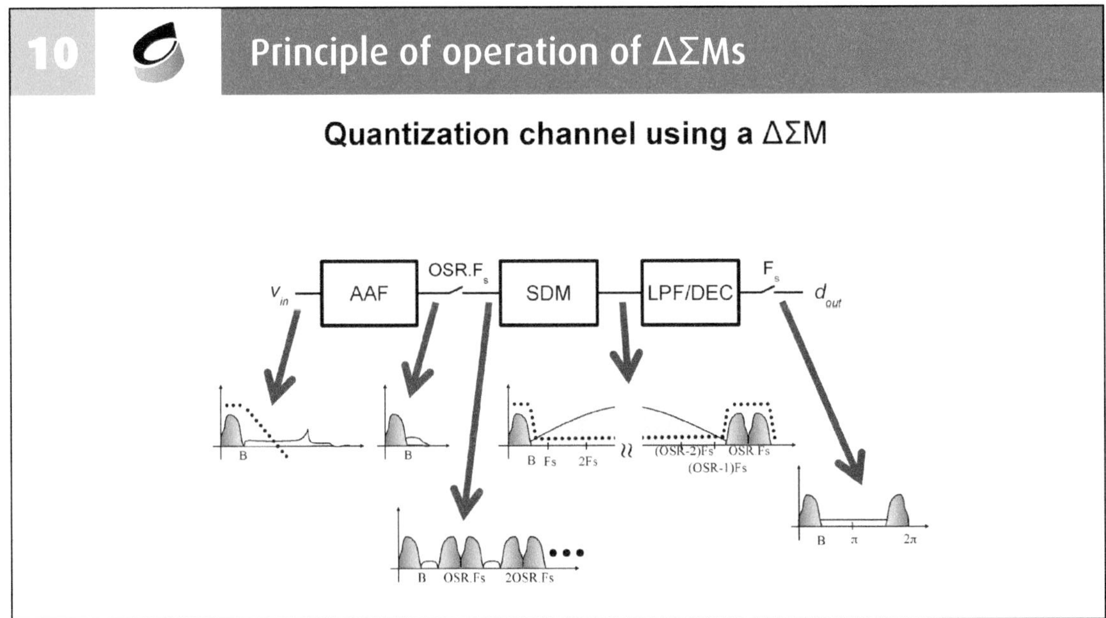

Data acquisition channels where a ΔΣMs is used, work as follows: 1) At the forefront of the channel is the AAF which, ideally, will suppress any present alias signal. 2) Next, the input signal will be sampled at a much higher rate than its Nyquist frequency, here represented by F_s, by a factor of osr. 3) The ΔΣM generates a bit stream signal where the quantization noise is located at high frequencies. 4) Afterwards, a digital low-pass decimation filter is used to filter the out-of-band quantization noise and other unwanted signals, and also lowers the sampling frequency to closer to F_s. 5) The signal can then be safely resampled at a lower frequency, since most of the unwanted noise was eliminated/sufficiently attenuated.

Principle of operation of ΔΣMs

11

Ideal 1-bit ΔΣM

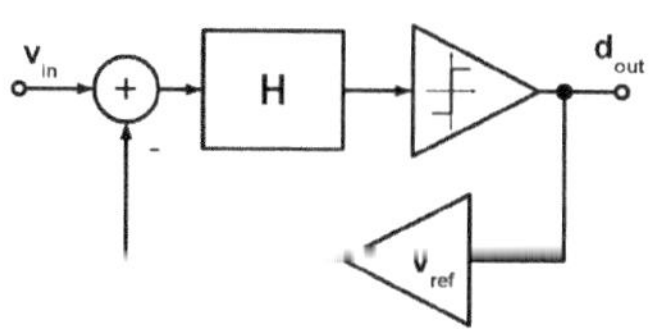

H is a transfer function with large gain in the frequency band of the input signal

$$\frac{V_{out}}{V_{in}} = \frac{H}{1+H} \approx 1$$

$$\frac{V_{out}}{V_Q} = \frac{1}{1+H} \approx 0$$

The feedback loop causes the quantization error to be very small
(assuming that the system is stable)

The simplest implementation of ΔΣM a consists of filter block (H) and a 1 bit quantizer. The core idea is to use negative feedback to attenuate quantization noise. The input signal goes through H, is subsequently quantized and the digital output is then subtracted at the input, by means of a DAC on the feedback loop. If the Mason's rule is applied to this negative feedback system, considering that H has a large gain in the signal's frequency band, it follows that the signal itself is barely affected while the quantization noise is deeply attenuated. The caveat here is that this only occurs if the system is stable.

Principle of operation of ΔΣMs

12

Ideal ΔΣM linear model

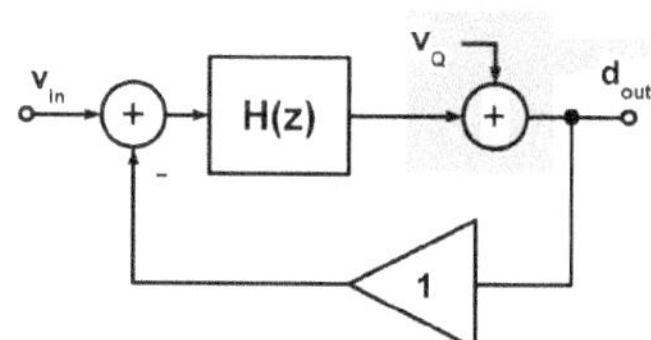

Linear model of the quantizer

Considering that: $H(z) = \dfrac{z^{-1}}{1-z^{-1}}$

(Ideal discrete time integrator)

$$\frac{V_{out}}{V_{in}} = \frac{H(z)}{1+H(z)} = z^{-1} = STF(z)$$

$$\frac{V_{out}}{V_Q} = \frac{1}{1+H(z)} = 1 - z^{-1} = NTF(z)$$

Replacing the quantizer by its linear model, where the quantization noise (VQ) is assumed to be independent from its input signal, and the the H block is assumed to be an ideal discrete time integrator. Applying once gain Mason's rule, it follows that the signal transfer function (STF) is equal to just a delay. Thus, only the phase of the signal is altered and nothing else. However, the noise transfer function (NTF) shows that at DC the quantization noise can be entirely suppressed.

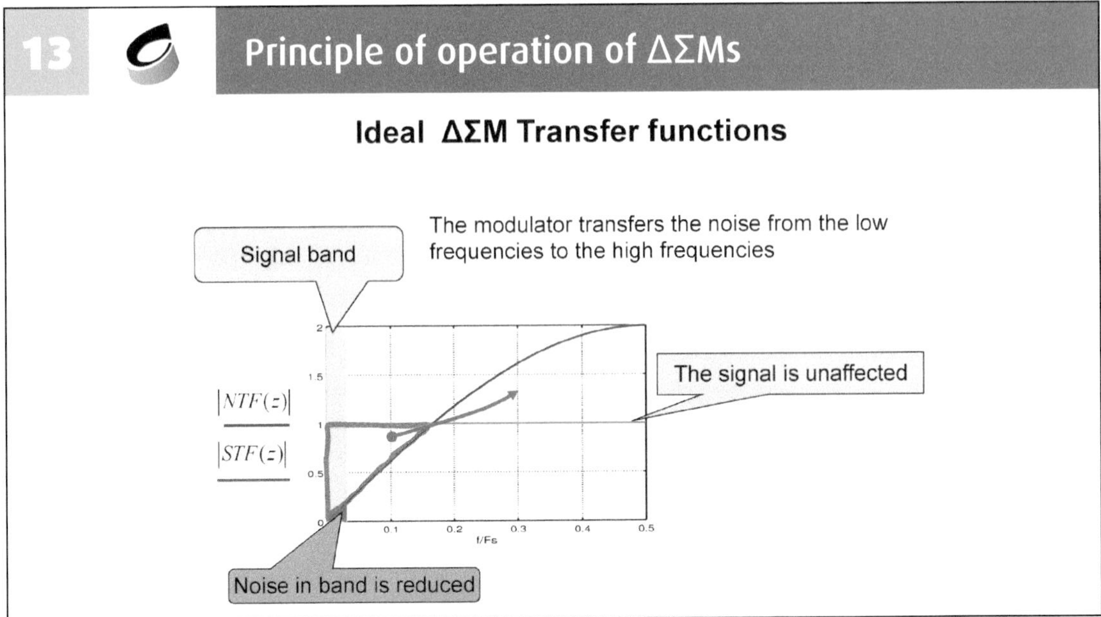

This graph represents the transfer functions of a ΔΣM. When plotting both transfer functions, it becomes apparent that when a high sampling ratio is used the signal's frequency band is very small and the quantization noise inside this band is greatly attenuated, with most of it being transferred to higher frequencies. This effect is commonly known as noise-shaping. Again, the entire signal is unaffected.

Principle of operation of ΔΣMs

SNR of the 1st order ideal ΔΣM

The total noise power is given by the integral of the PSD of the quantization noise modulated by the noise transfer function of the modulator within the signal band:

$$V^2_{QF(rms)} = \int_{-B}^{+B} S_{V_Q}(f).\left|NTF(e^{j2\pi f/F_S})\right|^2 df \approx \frac{V^2_{lsb}}{12}\frac{\pi^2}{3}\left(\frac{1}{osr}\right)^3$$

The signal to quantization noise ratio of a sinewave is given by:

$$SNR = 10.\log\left(\frac{V^2_{in(rms)}}{V^2_{QF(rms)}}\right) = 20.\log\left(\frac{V_{REF}}{2.\sqrt{2}}\times\frac{\sqrt{12}}{V_{lsb}}\times\left(\frac{3}{\pi^2}.osr^3\right)^{1/2}\right) = 20.\log\left(2^n.\left(\frac{3}{\pi^2}.osr^3\right)^{1/2}\sqrt{3/2}\right)$$

$$\boxed{SNR = 6.02\times n + 1.76 + 30.\log(osr) - 5.17\ \text{(dB)}}$$

Increasing the ***osr*** **2×** results in an improvement of **9dB** in the ***SNR***

When determining the SNR of a 1st order ideal ΔΣM, the quantization noise at its output can be obtained by calculating the area (triangle shape in the previous plot) that it occupies in the signal band. The result shows that there is a direct correlation between the osr and the noise power. When looking at the expression of the SNR itself, factor 30·log(osr) indicates that by increasing the osr two-fold results in an SNR improvement of roughly 9 dB.

Principle of operation of ΔΣMs

15

Ideal ΔΣM transfer functions

At DC

$$H(z=1)=\frac{1}{1-1}\approx\infty$$

$$STF(z=1)=\frac{\infty}{1+\infty}=1$$

$$NTF(z=1)=\frac{1}{1-\infty}=0$$

At $F_s/2$

$$H(z=-1)=\frac{-1}{1+1}\approx-\frac{1}{2}$$

$$STF(z=-1)=\frac{-\frac{1}{2}}{1-\frac{1}{2}}=-1$$

$$NTF(z=-1)=\frac{1}{1+-\frac{1}{2}}=2$$

Note that at DC the ideal integrator has **infinite** gain, which results in **infinite** quantization noise attenuation.

However, it is impossible to build a real circuit with **infinite gain**

At DC the signal is unaffected and the noise is completely suppressed, as shown before. At Fs/2, the quantization noise is amplified. However, it is physically impossible to design a real circuit with infinite gain, rendering this set of equations impractical for actual designs.

Principle of operation of ΔΣMs

16

Real SC integrator circuit

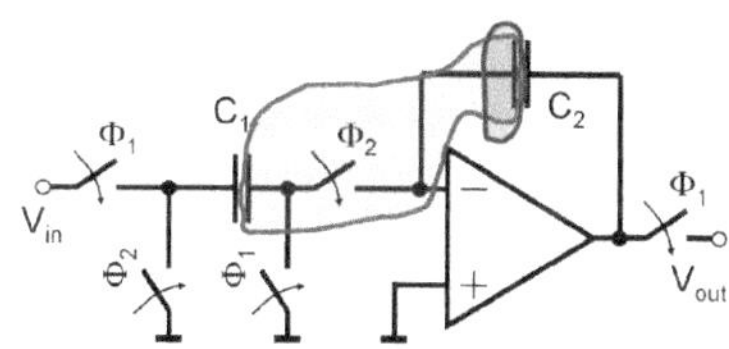

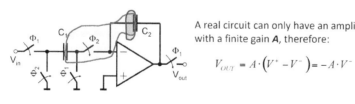

A real circuit can only have an amplifier with a finite gain **A**, therefore:

$$V_{OUT}=A\cdot\left(V^{+}-V^{-}\right)=-A\cdot V^{-}$$

time	Phase	Q_{C1}	Q_{C2}
(n-1)T	Φ1	$V_{in}[n-1]\cdot C_1$	$V_{OUT}[n-1]\cdot(A+1)/A\cdot C_1$
(n-1/2)T	Φ2	$V_{OUT}[n-\frac{1}{2}]/A\cdot C_1$	$V_{OUT}[n-\frac{1}{2}]\cdot(A+1)/A\cdot C_1$
nT	Φ1	$V_{in}[n]\cdot C_1$	$V_{OUT}[n]\cdot(A+1)/A\cdot C_1$

Charge conservation in C_1 and C_2

Charge conservation in C_2

A real switched-capacitor (SC) integrator circuit using an amplifier with finite gain is shown. It works with two phases, based on the principle of charge conservation. During Φ_1 the input signal is sampled into C_1, and in the following phase (Φ_2) the charge stored in it is transferred to C_2, since the inverting input of the amplifier is seen as a "virtual ground". Since the amplifier has a finite gain A, the output voltage will be equal to the input times A, with a 180° degrees phase shift.

Principle of operation of ΔΣMs

Real SC integrator circuit

Charge conservation in C_2

$$Q^2_{C2} = Q^1_{C2} \Leftrightarrow$$
$$\Leftrightarrow V_{OUT}[n] = V_{OUT}[n - 1/2]$$

Charge conservation in C_1 and C_2

$$Q^1_{C1} + Q^1_{C2} = Q^2_{C1} + Q^2_{C2} \Leftrightarrow$$
$$\Leftrightarrow V_{OUT}[n - 1/2] = V_{in}[n-1]\cdot\frac{C_1}{C_2}\cdot\frac{A}{A+1+C_1/C_2} + V_{OUT}[n-1]\cdot\frac{A+1}{A+1+C_1/C_2}$$

$$V_{OUT}[n] = V_{in}[n-1]\cdot\frac{C_1}{C_2}\cdot\frac{A}{A+1+C_1/C_2} + V_{OUT}[n-1]\cdot\frac{A+1}{A+1+C_1/C_2}$$

$$V_{OUT}[n] = V_{in}[n-1]\cdot\frac{C_1}{C_2}\cdot\alpha + V_{OUT}[n-1]\cdot\beta$$

$$A >> 1 \Rightarrow \begin{cases}\alpha \approx 1\\ \beta \approx 1\end{cases}$$

$$H(z) = \frac{C_1}{C_2}\frac{\alpha\cdot z^{-1}}{1-\beta\cdot z^{-1}}$$

A real integrator circuit has the pole located inside of the unit circle.

Applying the Z-transform to the final equation (obtained from solving the charge conversation equations), results that this circuit can accurately depict an integrator. For amplifiers with higher gains, the equation further approaches the ideal transfer function of an integrator. The β factor is critical, as it is responsible for moving the circuit pole from DC to low frequencies.

Principle of operation of ΔΣMs

Real ΔΣM transfer functions

At DC

$$H(z=1) = \frac{C_1}{C_2}\frac{\alpha}{1-\beta} \approx \frac{1}{1-\beta} \approx A$$

$$STF(z=1) = \frac{A}{1+A} \approx 1$$

$$NTF(z=1) = \frac{1}{1-A} \approx \frac{-1}{A}$$

At $F_s/2$

$$H(z=-1) = \frac{C_1}{C_2}\frac{-\alpha}{1+\beta} \approx \frac{1}{2}$$

$$STF(z=-1) = \frac{-1/2}{1-1/2} = -1$$

$$NTF(z=-1) = \frac{1}{1+-1/2} = 2$$

Note that at DC, a real integrator only attenuates the quantization noise by ***A***.

When replacing the ideal integrator TF with the real integrator TF, it follows that at DC, the signal is still pretty much unaltered, but the quantization noise is now only attenuated by a factor of A, instead of by an infinite amount. At Fs/2 the equations remain quite similar. This non-ideality must then be accounted for when designing the integrator block.

Principle of operation of ΔΣMs

19

Real ΔΣM transfer functions for A=100

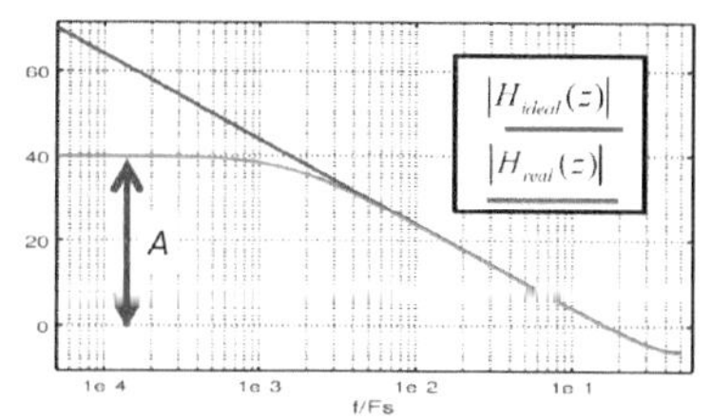

Note that the transfer functions are the same for f>fp

$$\frac{f_p}{F_S} \approx \frac{1-\beta}{2\pi \cdot \beta} \approx \frac{1}{2\pi \cdot A}$$

The quantization noise is only attenuated by $1/A \approx -40$dB at lower frequencies when using the real integrator

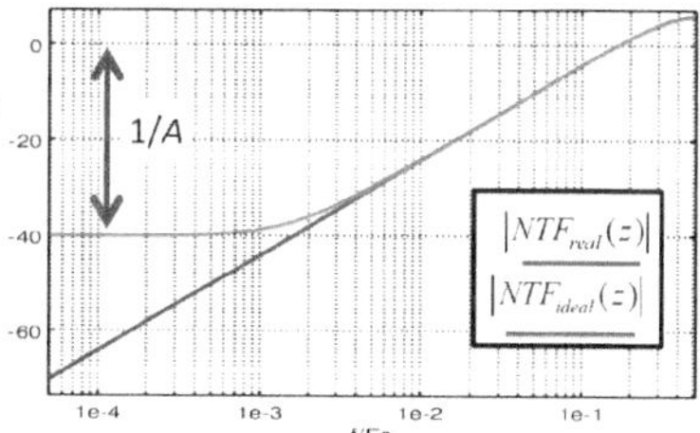

Inspecting the plots of both the ideal and real integrator TFs for a gain of 100, as well as the corresponding NTFs, it becomes clear that after a certain frequency they are essentially the same. However, at lower frequencies the quantization noise is only attenuated by about 40 dB when a real integrator is considered.

Principle of operation of ΔΣMs

20

Comparing the quantization noise

When using an ideal integrator the quantization noise power is given by:

$$V_{QF(rms)}^2 = \int_{-B}^{+B} S_{V_Q}(f).\left|NTF(e^{j2\pi f/F_S})\right|^2 df \approx \int_{-B}^{+B} S_{V_Q}(f).\left(4\pi^2\left(\frac{f}{F_S}\right)^2\right) df \approx \frac{V_{lsb}^2}{12}\frac{\pi^2}{3}\left(\frac{1}{osr}\right)^3$$

When using a real integrator the quantization noise power is given by:.

$$V_{QF(rms)}^2 = \int_{-B}^{+B} S_{V_Q}(f).\left|NTF(e^{j2\pi f/F_S})\right|^2 df \approx \int_{-B}^{+B} S_{V_Q}(f).\left(\frac{1}{A}\right)^2 df \approx \frac{V_{lsb}^2}{12}\frac{\pi^2}{A^2}\frac{1}{osr}$$

Comparing the previous expressions it is possible to conclude that if **A>osr** the SNR should be the same in both cases.

Comparing the expression of the in-band quatization noise for the ideal a real integrator case it is possible to conclude that the result will be the same if the amplifier gain is larger than osr.

21 

Principle of operation of ΔΣMs

2nd order ΔΣM

Adding a second integrator to the feedback loop results in a 2nd order ΔΣM

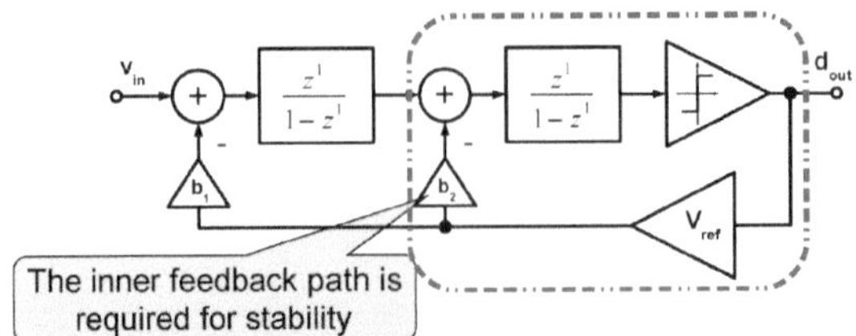

Vin
dout
b1
b2
Vref
The inner feedback path is required for stability

The second integrator doubles the quantization noise attenuation in the signal band, resulting in an improved SNR compared to the first order modulator.

$$SNR = 6.02 \times n + 1.76 + 50.\log(osr) - 12.9 \ \ (\text{dB})$$

Increasing the ***osr* 2×** results in an improvement of **15dB** in the ***SNR***

Another obvious way of increasing the noise suppression is to add another stage to the modulator. This second integrator requires another feedback path in the loop for stability reasons, as without it, we would be left with an oscillator. An additional integrator reduces the quantization noise by two-fold and by redoing the SNR calculations shows a 15dB improvement of the SNR for each doubling of the OSR.

22 

Principle of operation of ΔΣMs

The stability problem of higher order modulators

The noise transfer function of an ideal higher order modulator is given by:

$$NTF(z,n)=(1-z^{-1})^n$$

$$NTF(-1,1)=(1+1^{-1})^1=2 \quad V_{NQtotal}=\int_{-F/2}^{F/2}\frac{V_{LSB}^2}{12\cdot F_s}\cdot|NTF(e^{j\cdot 2\cdot\pi\cdot f/F})|^2 df=\frac{V_{LSB}^2}{12\cdot F_s}\times 2\cdot F_s=\frac{V_{LSB}^2}{6}$$

$$NTF(-1,2)=(1+1^{-1})^2=4 \quad V_{NQtotal}=\int_{-F/2}^{F/2}\frac{V_{LSB}^2}{12\cdot F_s}\cdot|NTF(e^{j\cdot 2\cdot\pi\cdot f/F})|^2 df=\frac{V_{LSB}^2}{12\cdot F_s}\times 6\cdot F_s=\frac{V_{LSB}^2}{2}$$

$$NTF(-1,3)=(1+1^{-1})^3=8 \quad V_{NQtotal}=\int_{-F/2}^{F/2}\frac{V_{LSB}^2}{12\cdot F_s}\cdot|NTF(e^{j\cdot 2\cdot\pi\cdot f/F})|^2 df=\frac{V_{LSB}^2}{12\cdot F_s}\times 20\cdot F_s=\frac{V_{LSB}^2\cdot 20}{12}$$

For n>2 the total noise power can exceed the amount of correction available in the feedback path

Although adding more stages seems tempting, increasing the order (> 2) of the system leads to inevitable stability issues. For 3rd order or higher, the total quantization noise power begins to exceed the amount of correction available in the path (± 1 LSB). So, even if all the system closed loop poles are placed on left half complex plane, the fact that the quantization noise is being amplified at Fs/2, which translates into an amount of overall quantization power that may lead to instability in the modulator.

Principle of operation of ΔΣMs

23

Implementations of higher order modulators

- There are two approaches utilized to implement higher order modulators:
  - Reduction of the NTF gain at high frequencies by adding low frequency complex poles in the transfer function. This approach reduces the maximum SNR value achievable by the modulator and requires a complex design of the modulator.
  - Cascading first and second order modulators: **MASH** – **M**ulti st**A**ge noise **SH**aping. This approach requires almost ideal integrators in the modulator which can be challenging for an analog circuit.

There are two options for addressing the stability issue. First, the NTF gain at high frequencies can be reduced by placing the transfer function poles at low frequencies. This can be done by means of a Butterworth or Chebyshev approach. However, this makes the design more complex and the maximum achievable SNR is reduced, as the gain contribution added by each integrator stage becomes lesser each time. Second, the MASH approach (which ensures stability) combines the output of several cascaded modulators to cancel the quantization noise of the previous one. However, close-to-ideal integrators are required, leading to a significant design challenge in deep submicron technologies.

Outline

24

- Why use ΔΣMs
- Principle of Operation of ΔΣMs
- **Active vs. Passive ΔΣMs**
- Passive based ΔΣMs
  - UIS SC integrator based
  - Continuous time integrator based
- Conclusions

Next, an overview of both passive and active ΔΣM is presented. The removal of gain blocks in between stages and how this affects the overall performance is analyzed. Special attention is given to the comparator, as in passive implementations it becomes the main source of gain in the overall circuit. The addition of low-gain inter-stage blocks is also discussed.

25 

Active vs. Passive ΔΣMs

Active vs Passive integrators

$$H_A(z)=\frac{z^{-1}}{1-\beta_A\cdot z^{-1}};\quad \beta_A=1-\alpha_A;\quad \alpha_A\approx\frac{1}{A}$$

$$H_P(z)=\frac{\alpha_P\cdot z^{-1}}{1-\beta_P\cdot z^{-1}};\quad \beta_P=1-\alpha_P;\quad \alpha_P<<1$$

The only difference between the transfer functions of an active and a passive integrator is α in the numerator (assuming $\alpha_P=\alpha_A$).

The frequency response of the active and passive integrators is the same, except for the gain difference of $1/\alpha$. Note that the difference in the gain at DC and at $F_s/2$ is the same ($1/\alpha$)

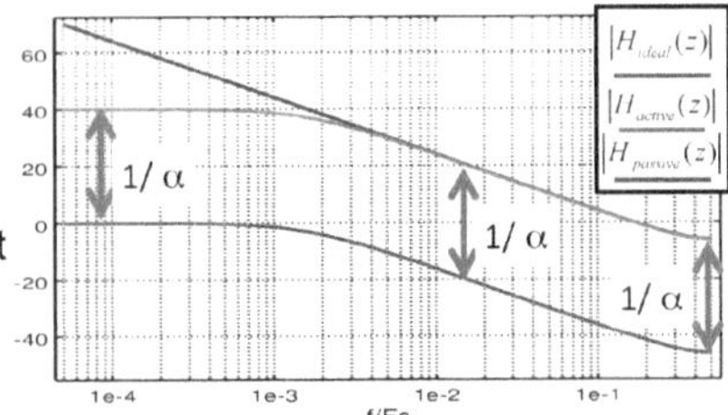

The main difference between active and passive implementations is the DC gain value of the integrator. By inspecting the graph, it is clear that the frequency response of a passive integrator is exactly the same as the active one, only shifted down to 0 dB (as now there is no amplifier), from DC to Fs/2. But can ΔΣM still operate as expected, with such a gain loss?

26 

Active vs. Passive ΔΣMs

Generic 1st order ΔΣM behavior

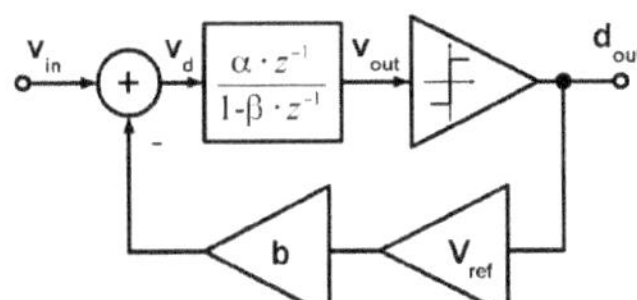

$$vout[n]=\alpha\cdot vin[n-1]+\beta\cdot vout[n-1]$$

$$d_{out}=\begin{cases}1 & if \quad vout[n]>0\\ -1 & if \quad vout[n]<0\end{cases}$$

Considering that the input is equal to δ and that the comparator is capable of producing a valid output digital level from its input signal:

vin	vd	Vout[n]	Vout[n-1]	Dout[n-1]
δ	δ−Vref	(δ−Vref)α	0+	+1
δ	δ+Vref	(δ+Vref+(δ−Vref)β)α	(δ−Vref)α	-1
δ	δ−Vref	(δ−Vref +(δ+Vref+(δ−Vref)β) β)α	(δ+Vref+(δ−Vref)β)α	+1
δ	δ+Vref	(δ+Vref +(δ−Vref +(δ+Vref+(δ−Vref)β) β) β)α	(δ−Vref +(δ+Vref+(δ−Vref)β) β)α	-1
...	...	...	...	...
δ	δ−Vref	$(\delta(1+\beta+\beta^2+...-\beta^{(n-1)})-Vref(1-\beta+\beta^2+...-\beta^{(n-1)}))\alpha$	$(\delta(1+\beta+\beta^2+...+\beta^{(n-2)})-Vref(1-\beta+\beta^2+...+\beta^{(n-2)}))\alpha$	+1

Analyzing the behavior of a generic 1st order ΔΣM, for a small input signal δ, results in an output that alternates between ±1.

Active vs. Passive ΔΣMs

27

Generic 1st order ΔΣM behavior

Output of the integrator

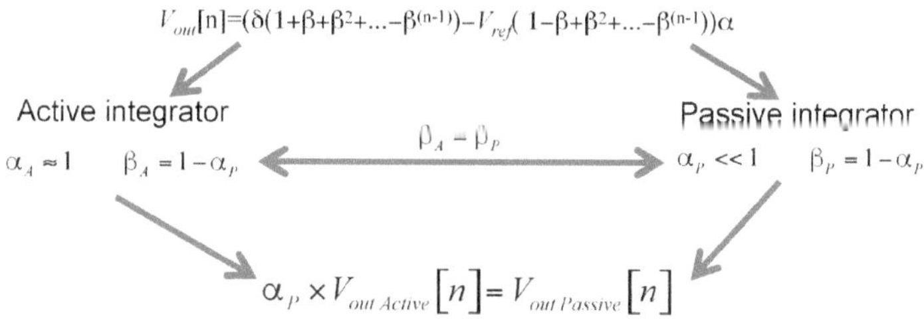

$V_{out}[n]=(\delta(1+\beta+\beta^2+...-\beta^{(n-1)})-V_{ref}(1-\beta+\beta^2+...-\beta^{(n-1)})\alpha$
Active integrator
$\alpha_A \approx 1 \quad \beta_A = 1-\alpha_P$
$\beta_A = \beta_P$
Passive integrator
$\alpha_P << 1 \quad \beta_P = 1-\alpha_P$
$\alpha_P \times V_{out\ Active}[n] = V_{out\ Passive}[n]$

This result shows that the output of the ΔΣM is the same, independently of the type of integrator, as long as both integrators have the same β value. The only difference being the output amplitude of each type of integrator.

Assuming that the comparator always produces a valid digital output, the generic expression for the output of the integrator is accurate for both active and passive designs, i.e., regardless of the values for α and β. Then, if both β_A and β_P are equal, the amplitude of a passive 1st order ΔΣM is just α_P times smaller than the amplitude of an active 1st order ΔΣM.

Active vs. Passive ΔΣMs

28

Generic 1st order ΔΣM linear model

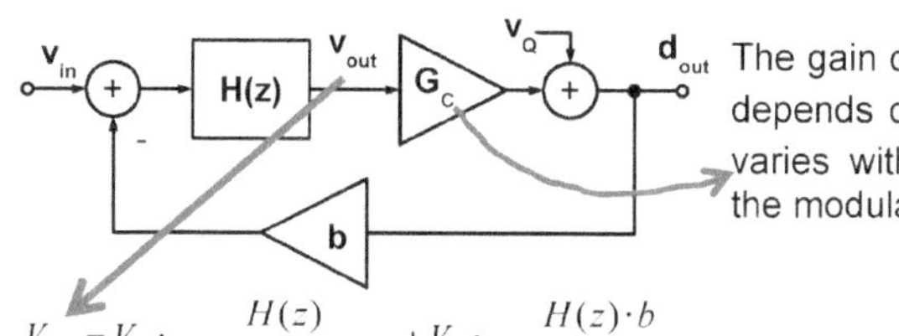

The gain of the comparator (G_C) depends on V_{out} and d_{out} and it varies with the signal levels in the modulator.

$$V_{out} = V_{in} \cdot \frac{H(z)}{1+G_C \cdot H(z) \cdot b} + V_Q \cdot \frac{H(z) \cdot b}{1+G_C \cdot H(z) \cdot b}$$

The previous analysis shows that comparing passive and active modulators with the same inputs and with $\beta_P=\beta_A$, results :

$$\begin{cases} d_{out\ Passive} = d_{out\ Active} \\ V_{Q\ Passive} = V_{Q\ Active} \\ V_{out\ Passive} = \alpha_P \cdot V_{out\ Active} \end{cases} \longrightarrow G_{CP} = \frac{G_{CA}}{\alpha_P} >> 1$$

When linearizing the model, the comparator becomes represented by the Gc block (which represents the comparator's processing gain) and an additional independent input VQ (which represents the quantization noise). This Gc gain is, much like the 1-bit quantizer, non-linear. In other words, for small amplitude signals this gain will be quite large, while for larger amplitudes the gain will be close to unity. Since the only difference between active and passive implementations is the signal amplitude, this leads to the conclusion that in passive implementations this Gc gain is larger when compared to active designs.

29 

Active vs. Passive ΔΣMs

Comparator gain of a 1st order ΔΣM

The exact value of $\boldsymbol{G_C}$ is very difficult to calculate, however it is possible to find the exact ratio between the active and passive versions:

$$\begin{cases} \alpha_P \cdot V_{in} \cdot \dfrac{H_A(z)}{1+G_{CA}\cdot H_A(z)\cdot b} = V_{in}\cdot \dfrac{H_P(z)}{1+G_{CP}\cdot H_P(z)\cdot b} \\ \alpha_P \cdot V_Q \cdot \dfrac{H_A(z)\cdot b}{1+G_{CA}\cdot H_A(z)\cdot b} = V_Q\cdot \dfrac{H_P(z)\cdot b}{1+G_{CP}\cdot H_P(z)\cdot b} \end{cases}$$

$$H_P(z) = \alpha_P \cdot H_A(z)$$

$$G_{CA}\cdot H_A(z) = G_{CP}\cdot H_P(z)$$

$$G_{CP} = \frac{G_{CA}}{\alpha_P} >> 1$$

This shows that in a passive ΔΣM the gain of the loop is concentrated in the comparator instead of the integrator. This allows to reduce the power dissipation of the modulator.

Although it is not trivial to realize the obtain the value of G_C, the ratio between each version can be precisely calculated. Inspecting the result, it follows that the G_C gain, in passive implementations, is given by the ratio between the G_C gain of the active version and the α_P factor. Recalling that α_P is quite small (<< 1), G_{CP} is then quite large. As a result, the comparator can be used as the main source of gain in passive implementations, removing the need for amplifiers in the design which are more difficult to design and dissipate additional power.

30 

Active vs. Passive ΔΣMs

Estimation of the comparator gain

The value of $\boldsymbol{G_C}$ is the ratio between the *rms* values of the output and input of the comparator and it depends on the signal level. It can be estimated for the case of small input amplitudes considering:

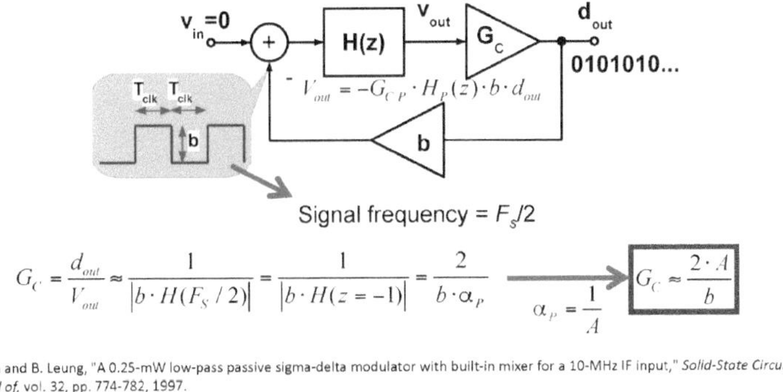

$$G_C = \frac{d_{out}}{V_{out}} \approx \frac{1}{|b\cdot H(F_S/2)|} = \frac{1}{|b\cdot H(z=-1)|} = \frac{2}{b\cdot\alpha_P} \qquad \alpha_P = \frac{1}{A} \longrightarrow G_C \approx \frac{2\cdot A}{b}$$

F. Chen and B. Leung, "A 0.25-mW low-pass passive sigma-delta modulator with built-in mixer for a 10-MHz IF input," *Solid-State Circuits, IEEE Journal of*, vol. 32, pp. 774-782, 1997.

To obtain an estimative of the comparator gain, a 0V amplitude input is assumed. In this case, the output is known to alternate between logic '0' and '1', which is essentially a square wave with a frequency of Fs/2. It follows that Gc is inversely proportional to the feedback gain times α_P. Essentially, in a passive implementation there is a gain from the filtering stages to the comparator stage.

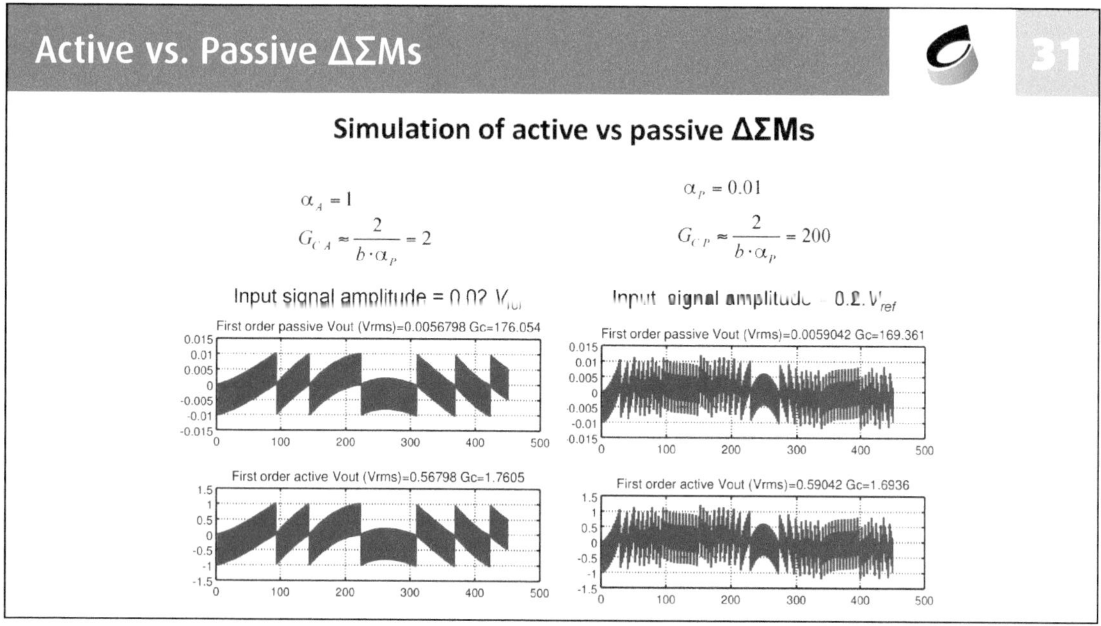

Simulation results prove that the output voltage of both active and passive 1st order ΔΣM tracks fairly well, with G_{cP} being 100 times larger than G_{cA}. It should also be noticed that for a larger input signal, the estimation of the G_c becomes less accurate, as expected.

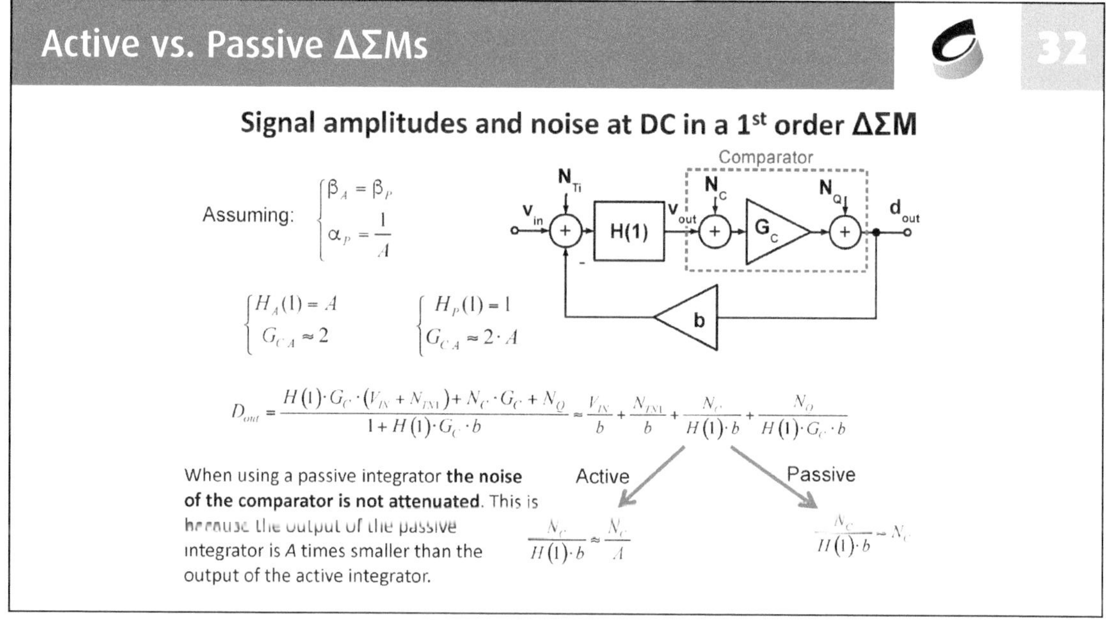

Now the consequences from a system point of view of a passive implementation are analyzed. Assuming the gain at DC, the output expression raises the importance behind the comparator noise. This noise is greatly reduced due to the gain of the amplifier, meaning that in passive implementations (no amplifier gain) this factor will carry much more weight. Thereby, the design of the low-noise comparator is key when considering a passive ΔΣM. This stringent specification can still be an acceptable trade-off, as the extra power required for the comparator might be smaller than that required by a high gain amplifier.

33 

Active vs. Passive ΔΣMs

Conclusions about passive integrator based 1st order ΔΣM

- The output signals of active integrator and passive integrator based ΔΣMs are the same as long as $\beta_A=\alpha_P$.
- In a passive ΔΣM the loop gain is concentrated in the comparator instead of the integrator.
- This avoids the need for a high gain amplifier, which can be replaced by the comparator.
- The only required active gain circuit is the comparator.
- The main drawback is that the output amplitude of the passive integrator is small, which results in the comparator noise becoming more important.
- Since $A \geq osr$ in order to not have a degradation of the SNR then $\alpha_P \leq 1/osr$

In short, a passive ΔΣM will have the same output signal as the active approach as long as the β_s are the same. Passive implementations transfer the loop gain from the integrator stages to the comparator, being the sole gain block in the entire architecture. As a result, its noise contribution becomes significantly larger and therefore low-noise comparators are required. As long as α_P is smaller or equal to 1/osr, the same SNR performance is ensured.

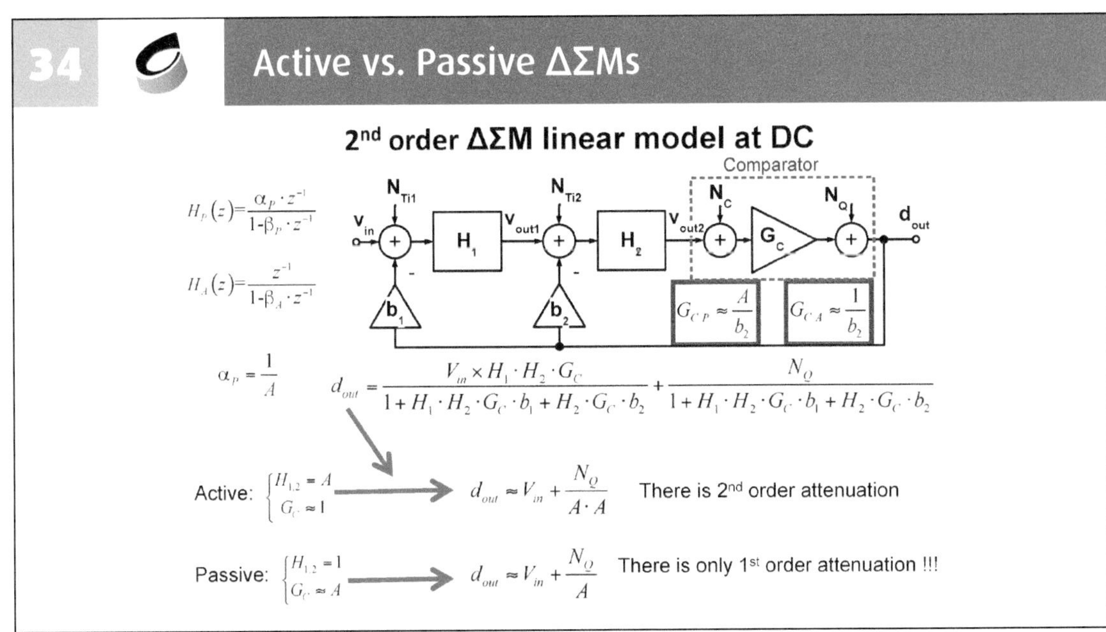

Moving on to a 2nd order ΔΣM, analyzing the output expression for both the passive and active implementations at DC, it follows that the passive version only attenuates the quantization noise by a factor of A, instead of A^2 as in the active version. Thus, the passive 2nd order ΔΣM acts a 1st order ΔΣM (+20 dB/dec slope, instead of +40 dB/dec). This occurs because all the processing gain G_c is due to the inner feedback loop, as the outer feedback loop is only subjected to a gain of 1, as the first integrator stage has no gain.

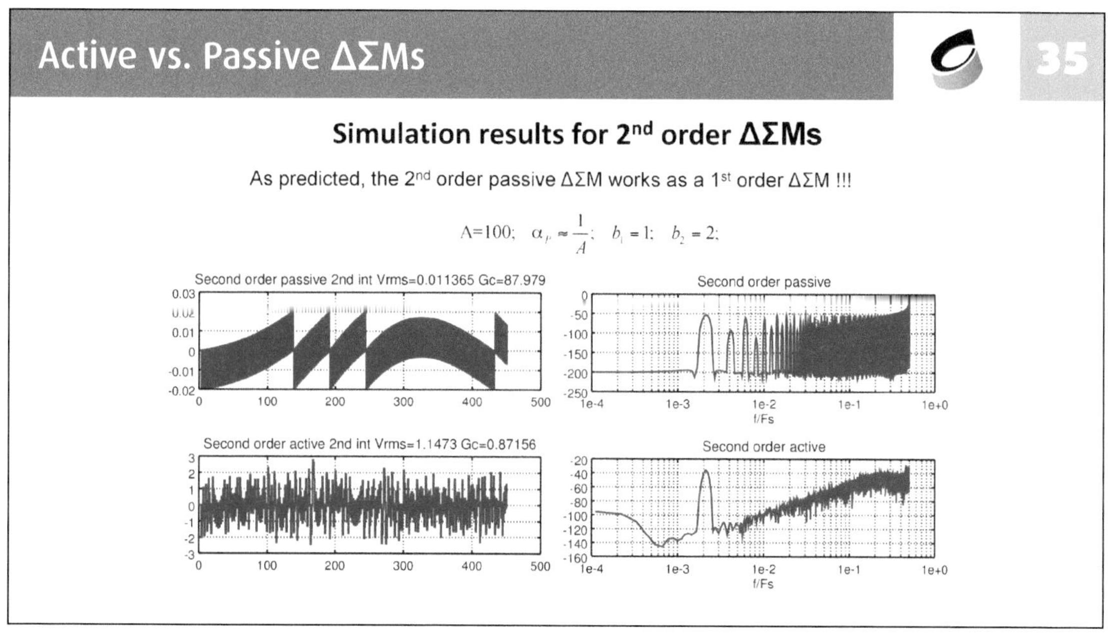

Active vs. Passive ΔΣMs
35
Simulation results for 2nd order ΔΣMs
As predicted, the 2nd order passive ΔΣM works as a 1st order ΔΣM !!!
A=100; $\alpha_P \approx \frac{1}{A}$; $b_1 = 1$; $b_2 = 2$;
Second order passive 2nd int Vrms=0.011365 Gc=87.979
Second order passive
f/Fs
Second order active 2nd int Vrms=1.1473 Gc=0.87156
Second order active
f/Fs

Running simulations for both systems with a gain A of 100 and feedback coefficients b_1 and b_2 equal to 1 and 2, respectively, it's clear from the output voltage of the second integrator that its behavior resembles that of the first integrator. This can be confirmed by inspecting the output spectrum, where the 2nd order active modulator has a +40 dB/dec slope as expected, but the 2nd order passive modulator (masked behind all the noise and harmonics) only has a slope characteristic of a 1st order ΔΣM.

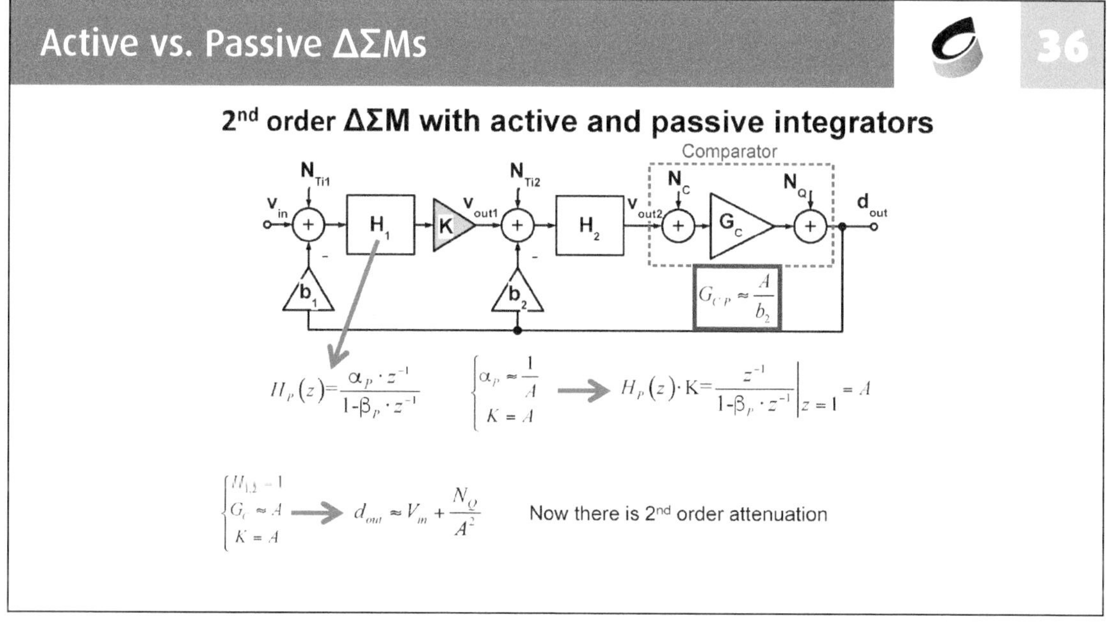

If a gain block is added to the outer loop, then the output expression of the passive version becomes similar to that of the active implementation. With the quantization noise now being attenuated by A_2, a 2nd order behavior is ensured.

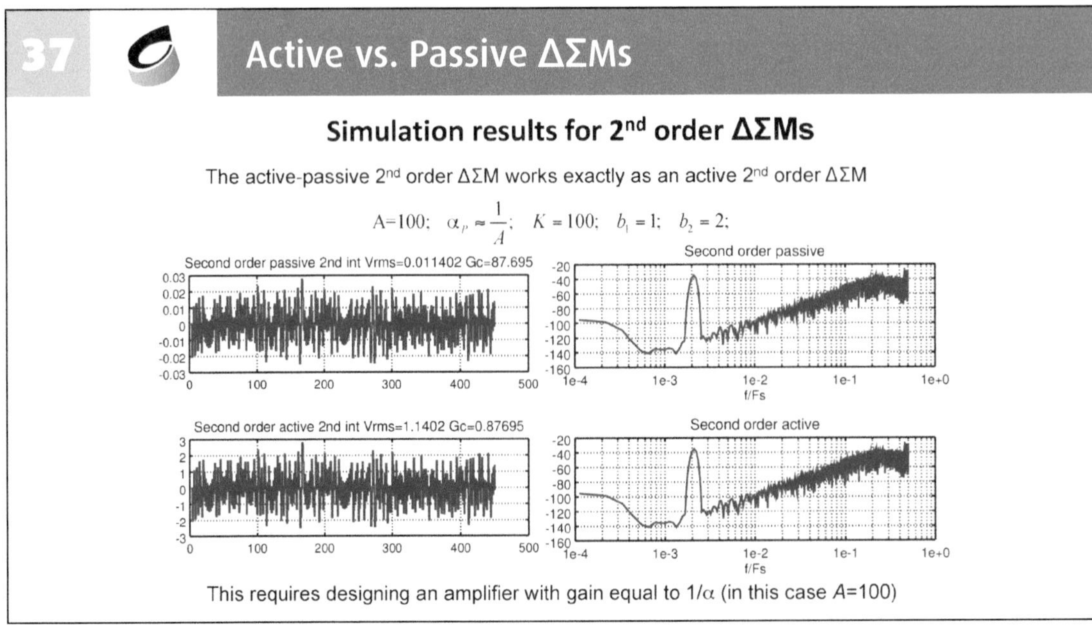

This assumption can be proven by, once again, simulating both systems, where in the passive implementation a gain K of 100 is introduced in the outer loop. However, this would require designing an amplifier with a gain K of 40 dB.

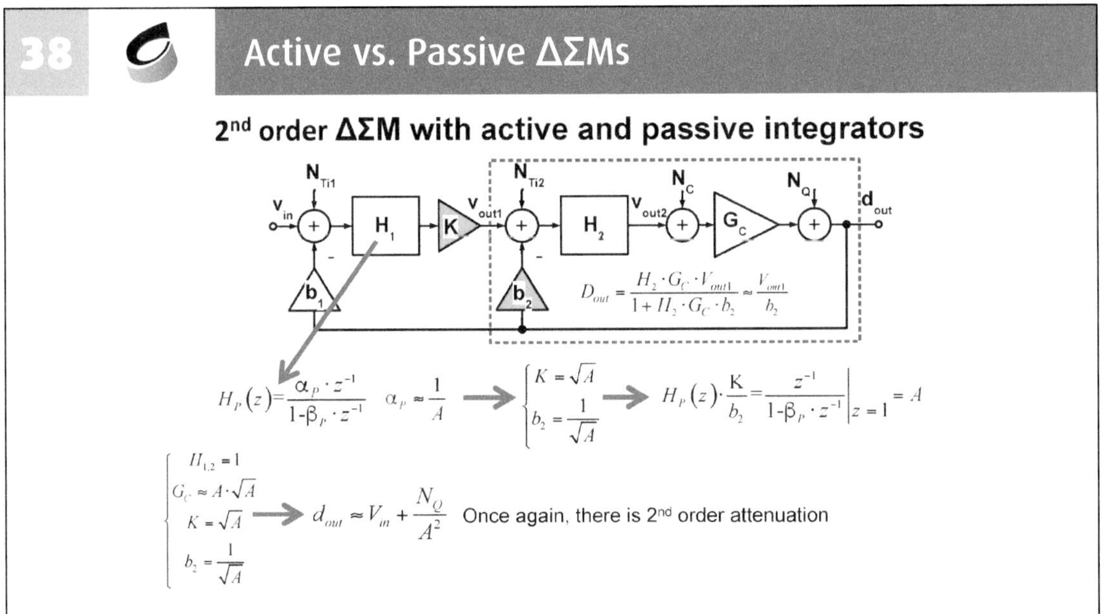

However, when analyzing the output expression of the modulator as a function of the output signal of the first integrator, it becomes apparent that by lowering the feedback coefficient b2, the loop gain is increased. Therefore, it is not mandatory to design an amplifier with a gain K of 100, just as long as the feedback coefficients are properly adjusted. For instance, if the gain K is set to $\sqrt{(A)}$, and the b2 coefficient is set to $1/\sqrt{(A)}$, the transfer function of the passive integrator still holds true, as well as the output expression of a generic 2nd order ΔΣM.

Active vs. Passive ΔΣMs

39

Simulation results for 2nd order ΔΣMs

The low gain active-passive 2nd order ΔΣM works exactly as an active 2nd order ΔΣM

$$A=100; \quad \alpha_p \approx \frac{1}{A}; \quad K=10; \quad b_1 = 1; \quad b_2 = 0.2;$$

Second order passive 2nd int Vrms=0.0011334 Gc=882.187

Second order passive

Second order active 2nd int Vrms=1.1334 Gc=0.88219

Second order active

f/Fs

This approach only requires designing an amplifier with gain equal to 10

By once again running the simulation for both systems, but with a gain K of 10 and a b_2 coefficient of 0.2 (instead of 2), it follows that a 2nd order ΔΣM behavior is maintained. However, an amplifier with a gain of 20 dB suffices, which is far less power-hungry and can be as simple as a differential pair.

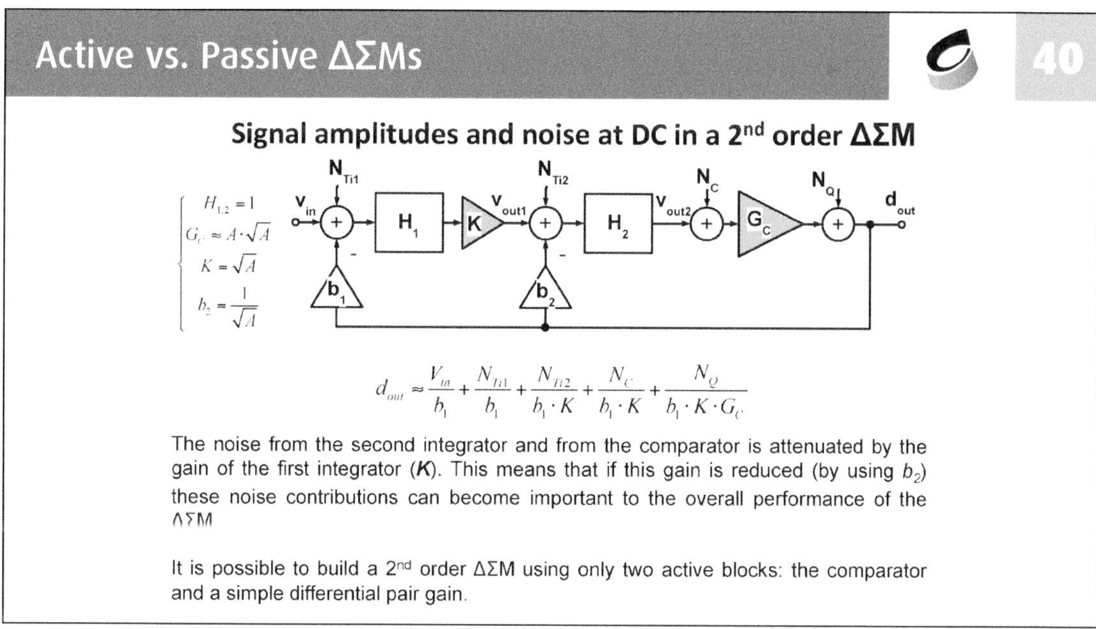

Active vs. Passive ΔΣMs
40
Signal amplitudes and noise at DC in a 2nd order ΔΣM
$H_{1,2}=1$
$G_C \approx A\cdot\sqrt{A}$
$K=\sqrt{A}$
$b_2=\frac{1}{\sqrt{A}}$
N_{Ti1}
N_{Ti2}
N_C
N_Q
V_{in}
V_{out1}
V_{out2}
d_{out}
H_1
H_2
K
G_C
b_1
b_2
$$d_{out} \approx \frac{V_{in}}{b_1} + \frac{N_{Ti1}}{b_1} + \frac{N_{Ti2}}{b_1 \cdot K} + \frac{N_C}{b_1 \cdot K} + \frac{N_Q}{b_1 \cdot K \cdot G_C}$$
The noise from the second integrator and from the comparator is attenuated by the gain of the first integrator (*K*). This means that if this gain is reduced (by using b_2) these noise contributions can become important to the overall performance of the ΔΣM
It is possible to build a 2nd order ΔΣM using only two active blocks: the comparator and a simple differential pair gain.

As a result, a 2nd order passive implementation of a ΔΣM only requires two active blocks: a low-gain amplifier, that burns much less power when compared with a high gain amplifier, and a low-noise comparator. The output expression of the modulator, when considering this low-gain block, shows that the attenuation of the noise contributions of both the second integrator and the comparator depends on the gain of the first integrator. To cope with the thermal noise contribution of the integrator stages, larger capacitors can be used. The only penalty of doing so being the larger area occupied, i.e, the power dissipation does not increase.

41 

Outline

- Why use ΔΣMs
- Principle of Operation of ΔΣMs
- Active vs. Passive ΔΣMs
- **Passive based ΔΣMs**
  - **UIS SC integrator based**
  - Continuous time integrator based
- Conclusions

42

Concept of Ultra Incomplete Settling (UIS)

V_C in an *RC* circuit after a step input with amplitude V_{in}

$$v_c(t) = V_{in} \cdot \left(1 - e^{-t/R_{total} \cdot C_1}\right) + V_{C0} \cdot e^{-t/R_{total} \cdot C_1}$$

R_{on} R C_1

Complete Settling

Incomplete Settling

Ultra Incomplete Settling (UIS)

$T_{clk} << R_{total} \cdot C_1$

Voltage (v)

Time (s)

The concept behind the UIS consists on purposely adding an extra resistor in between a switched-capacitor branch. As opposed to most common designs that operate assuming complete settling, here the region where the charging of the capacitor is in its early stages is considered. In it, the voltage inside the capacitor v_c is related not only to the input voltage V_{in}, but also to the previous voltage in the capacitor, V_0. Thus, at that region it can be considered that there is memory effect.

Concept of Ultra Incomplete Settling (UIS)

43

•Assuming $T_{clk} << R_{total} \cdot C$ the following approximation holds:

$$x << 1 \quad => \quad e^{x} \simeq 1 + x$$

$$v_c(T_{clk}) = V_{in} \cdot \left(1 - e^{-T_{clk}/R_{total}\cdot C}\right) + V_{C0} \cdot e^{-T_{clk}/R_{total}\cdot C} \approx V_{in} \cdot \frac{T_{clk}}{R_{total}\cdot C} + V_{C0} \cdot \left(1 - \frac{T_{clk}}{R_{total}\cdot C}\right)$$

•Considering discrete time instants:

$$v_c\left[n \cdot T_{clk}\right] = V_{in}\left[(n-1/2)\cdot T_{clk}\right] \cdot \frac{T_{clk}}{R_{total}\cdot C} + v_c\left[(n-1)\cdot T_{clk}\right] \cdot \left(1 - \frac{T_{clk}}{R_{total}\cdot C}\right)$$

If the exponent is assumed to be much smaller than 1 (which can be done by increasing the value of the extra resistor), then the discrete-time equivalent expression resembles that of an integrator, in which the voltage at instant n is given by the input voltage at instant n-1/2 and by the capacitor voltage in the previous clock cycle.

Discrete time integrator using UIS

44

Z transfer function of the integrator

$$H(z) = \frac{V_C(z)}{V_{in}(z)} = \frac{\frac{T_{clk}}{R_{total}\cdot C}}{z - \left(1 - \frac{T_{clk}}{R_{total}\cdot C}\right)} = \frac{\alpha}{z - \beta}$$

$$\begin{cases} \alpha = \frac{T_{clk}}{R_{total}\cdot C} \\ \beta = 1 - \alpha \end{cases}$$

$$H_{max} = |H(z=1)| = \left|\frac{\alpha}{1-\beta}\right| = 1$$

$$H_{min} = |H(z=-1)| = \left|\frac{\alpha}{-1-\beta}\right| \approx \frac{\alpha}{2}$$

- A SC circuit using UIS behaves as a discrete time integrator with losses.
- The output amplitude is very small.
- The circuit is passive (does not dissipate static power)

Applying the Z-transform to the time domain equation, results in the transfer function of the circuit. Inspecting this result, it follows that an SC branch operating in UIS behaves as a lossy integrator, where the output amplitude is small.

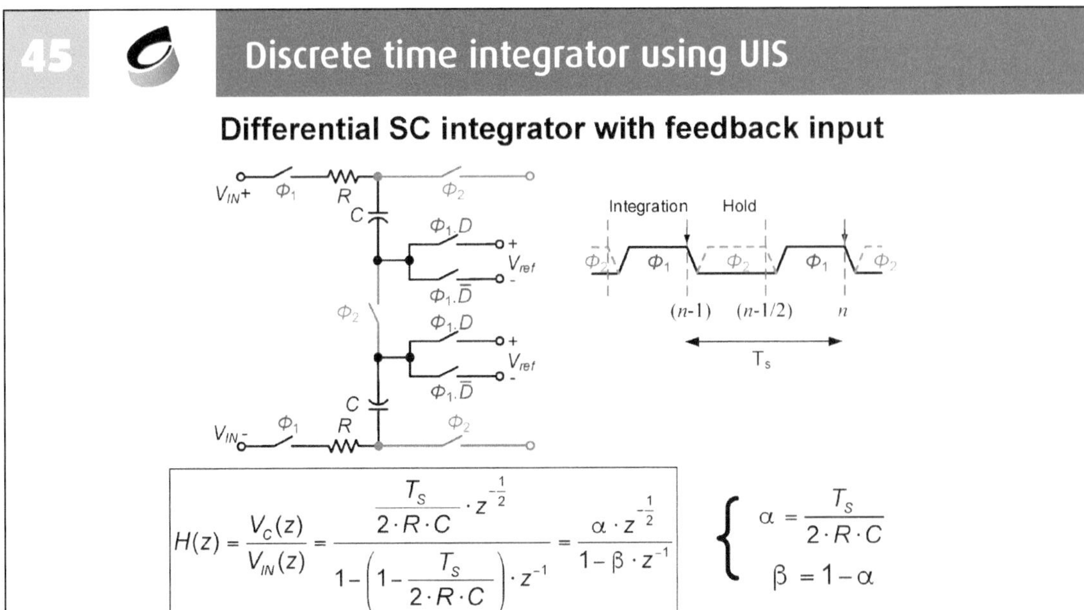

As an example, a fully differential SC integrator is shown. This SC branch consists of two resistors, two capacitors and a set of switches. Some of these switches are used to implement the feedback path of the ΔΣM. In the first phase the capacitors sample the input voltage minus the reference voltage V_{Ref}, while in the second phase both voltages are added in series for the next stage.

46 Discrete time integrator using UIS

Thermal Noise Analysis

- The thermal noise originates in the switch resistance which is connected to the capacitor only during T_{clk}
- Due to the incomplete settling the circuit is not in equilibrium $V_{Nthermal} \neq \sqrt{\frac{K\cdot T}{C}}$
- This means that the thermal noise power must be recalculated
- The impulse response of this RC circuit is given by:

$$h(t)=\begin{cases}\frac{1}{R_{ON}\cdot C}e^{-\frac{t}{R_{ON}\cdot C}} & 0\leq t\leq T_{clk}\\ 0 & t>T_{clk}\end{cases}\approx\begin{cases}\frac{1}{R_{ON}\cdot C} & 0\leq t\leq T_{clk}\\ 0 & t>T_{clk}\end{cases}$$

B. Nowacki, N. Paulino, and J. Goes, "Analysis and the design of a first - order delta sigma modulator using very incomplete settling," presented at the Mixed Design of Integrated Circuits and Systems (MIXDES), 2011 Proceedings of the 18th International Conference, 2011.

Since the circuit never reaches charge equilibrium, the thermal noise is not given by the well-known expression of kT/C. So, the impulse response of this circuit is used to recalculate the thermal noise power during this UIS region. Since *t* can be considered very small, the impulse response can be approximated by $1/R_{on}\cdot C$.

Discrete time integrator using UIS

47

Thermal Noise Analysis

- The transfer function of the RC circuit is given by:

$$H(f)=\int_{-\infty}^{\infty} h(t)\cdot e^{-2\pi\cdot f\cdot t}dt=\frac{\sin(\pi\cdot f\cdot T_{clk})}{\pi^2\cdot f\cdot C\cdot R_{ON}}$$

- The thermal noise power is obtained by:

$$P_{NT}=\int_0^{\infty} 4\cdot k\cdot T\cdot R_{ON}\cdot|H(f)|^2 df=\frac{2\cdot k\cdot T\cdot T_{clk}}{C^2\cdot R_{ON}}$$

- Finally the input referred thermal noise power is given by:

$$P_{NTinput}=\frac{P_{NT}}{\alpha^2}=\frac{2\cdot k\cdot T\cdot R_{ON}}{T_{clk}}=\frac{2\cdot k\cdot T}{C\cdot\alpha}$$

- Since there is a digital low-pass output filter, the noise power is:

$$P_{NTinput\ filtered}=\frac{P_{NTinput}}{osr}=\frac{2\cdot k\cdot T}{C\cdot\alpha\cdot osr}$$

The transfer function is similar to the well known *sin x / x* function, and so the total noise power is given by the integral of the power spectral density of the noise from the resistor times the squared module of the RC circuit transfer function. The obtained expression indicates that this noise is α times smaller than the kT/C noise. Therefore, in an UIS approach, the input referred noise is somewhat larger than in a regular SC circuit. Again, this can be dealt with by increasing the capacitor sizes.

UIS SC passive integrator ΣΔM

48

Block diagram of the UIS based ΔΣM

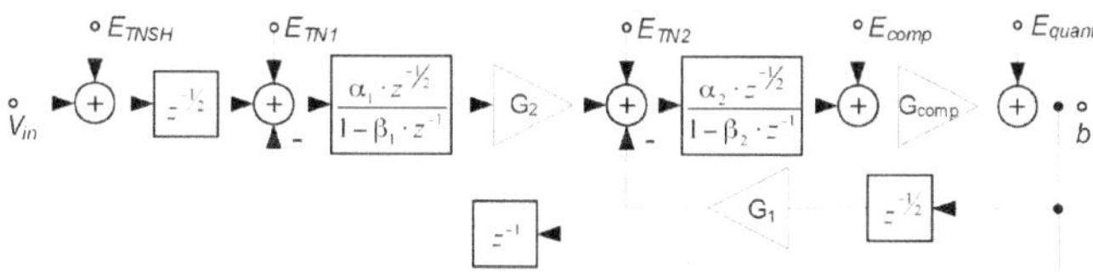

$G_{comp}=1/(G_1*\alpha_2)$ These parameters are optimized in order to maximize the SNDR of the modulator

$G_1=V_{ref2}/V_{ref1}$

$G_2=gm*R_2$

Parameter	F_{CLK}	C_{SH}	R_1	R_2	C_1	C_2	α_1	α_2	G_1	G_2
Value	100MHz	7pF	140kΩ	140kΩ	3.5pF	10pF	0.01	0.0036	0.2	6

The block diagram of a 2nd order passive ΔΣM using UIS is presented, where only two active blocks exist. The outer loop low-gain G2, which is a differential pair and is also used to saturate both stages, is still present. Also notice the size of the capacitors which are larger than usual ΔΣM designs, due to the reasons presented before.

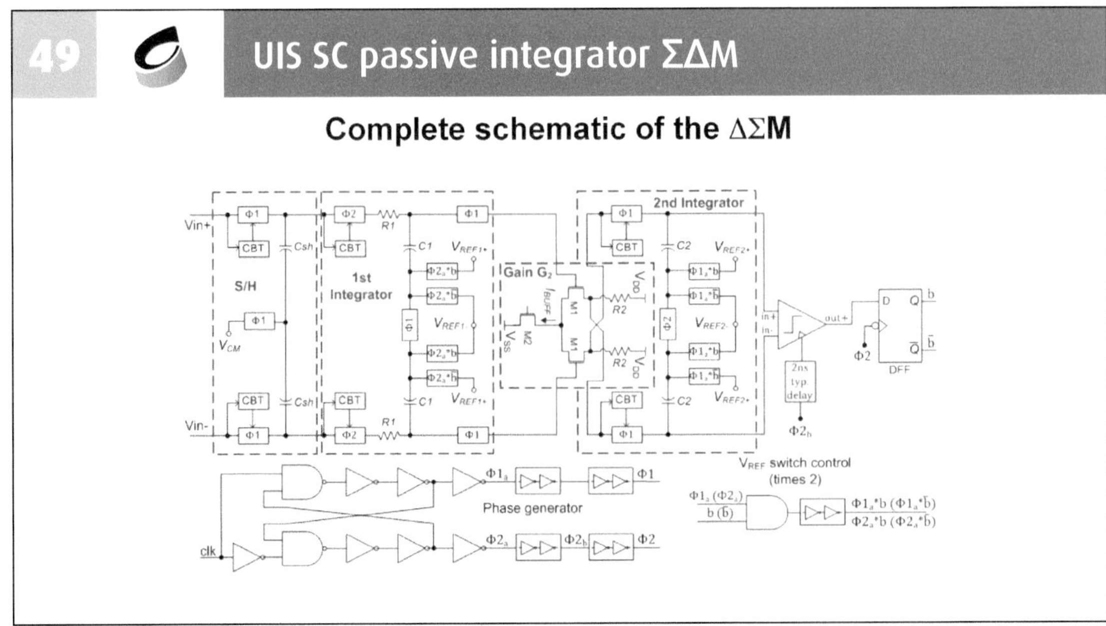

The complete schematic is shown next. One of the major features is its simplicity. Clock boosting is applied to the switches in series with the input signal in order to minimize distortion. Since the switches in the SC branch of the first integrator are in series with resistors R_1, they end up being quite small.

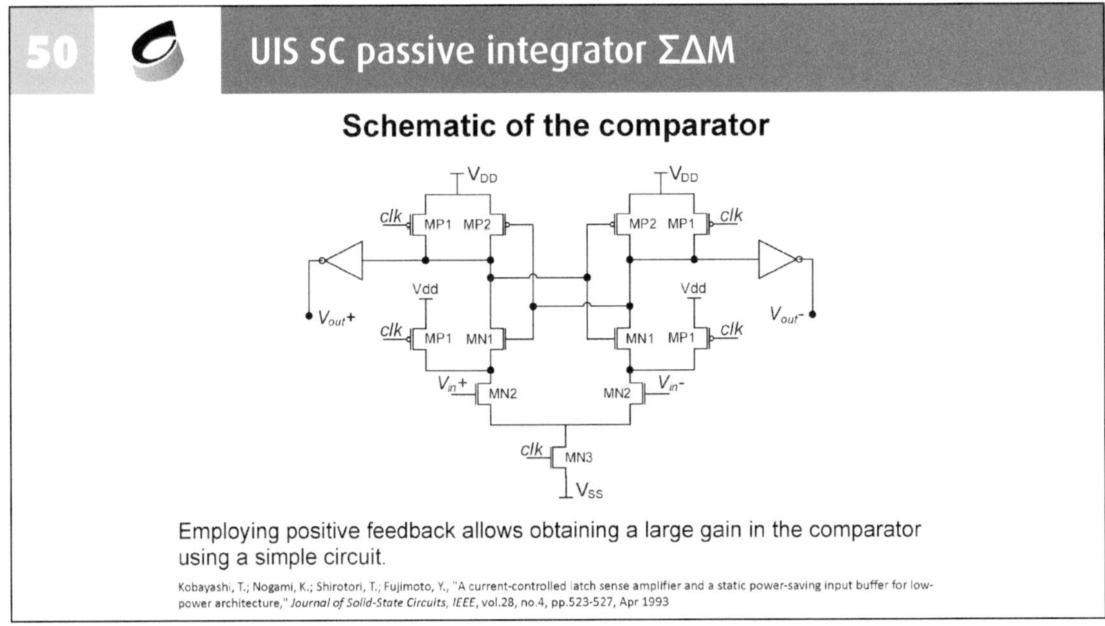

50
UIS SC passive integrator ΣΔM
Schematic of the comparator
VDD
VDD
clk
MP1
MP2
Vdd
MN1
MN2
MN3
clk
Vout+
Vout-
Vin+
Vin-
VSS
Employing positive feedback allows obtaining a large gain in the comparator using a simple circuit.
Kobayashi, T.; Nogami, K.; Shirotori, T.; Fujimoto, Y., "A current-controlled latch sense amplifier and a static power-saving input buffer for low-power architecture," Journal of Solid-State Circuits, IEEE, vol.28, no.4, pp.523-527, Apr 1993

The comparator is just a regular latch, with no need for a preamplifier, positive feedback is used to obtain a large gain with such a simple circuit.

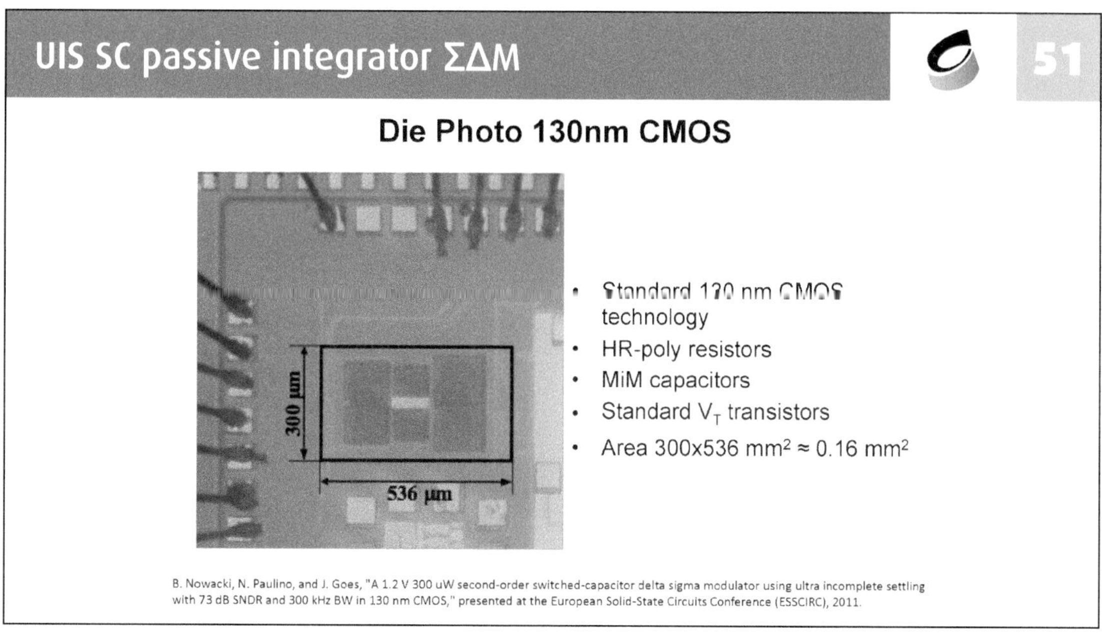

The layout of the circuit clearly illustrates the advantage in terms of area when using a passive approach, as no moderate-to-high gain amplifier is needed. Despite the use of larger capacitors, the chip only occupies an area of 0.16 mm².

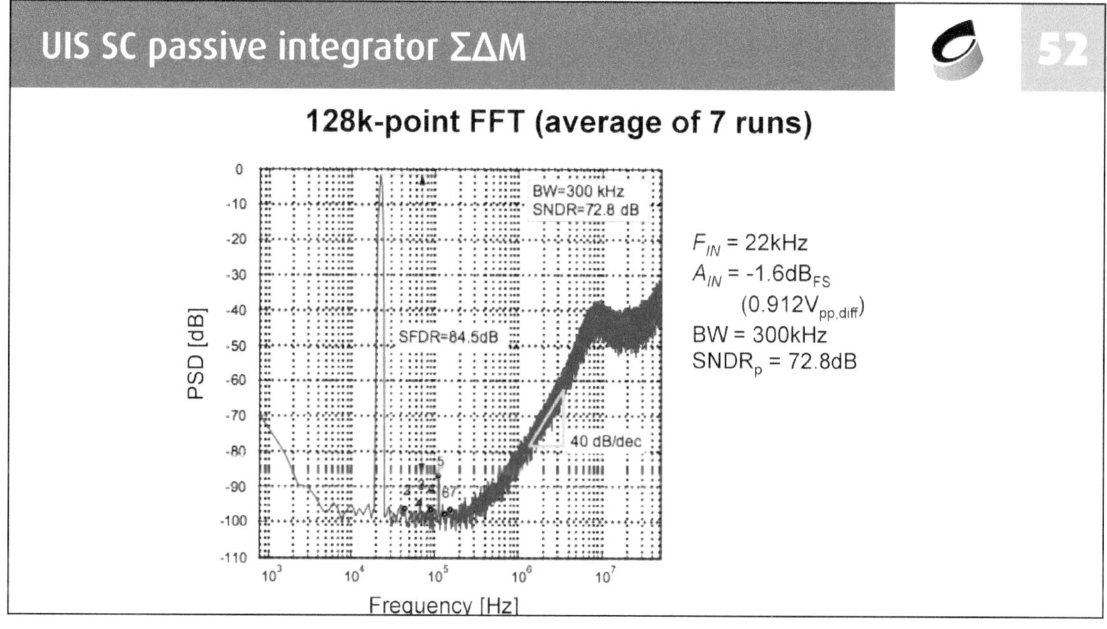

Measured results show that for a 300 kHz bandwidth, an SNDR of 72.8 dB and a SFDR of 84.5 dB are obtained, when an input signal close to 1 V is applied. The +40 dB/dec slope proves that this passive approach can indeed behave similar to its active counterpart.

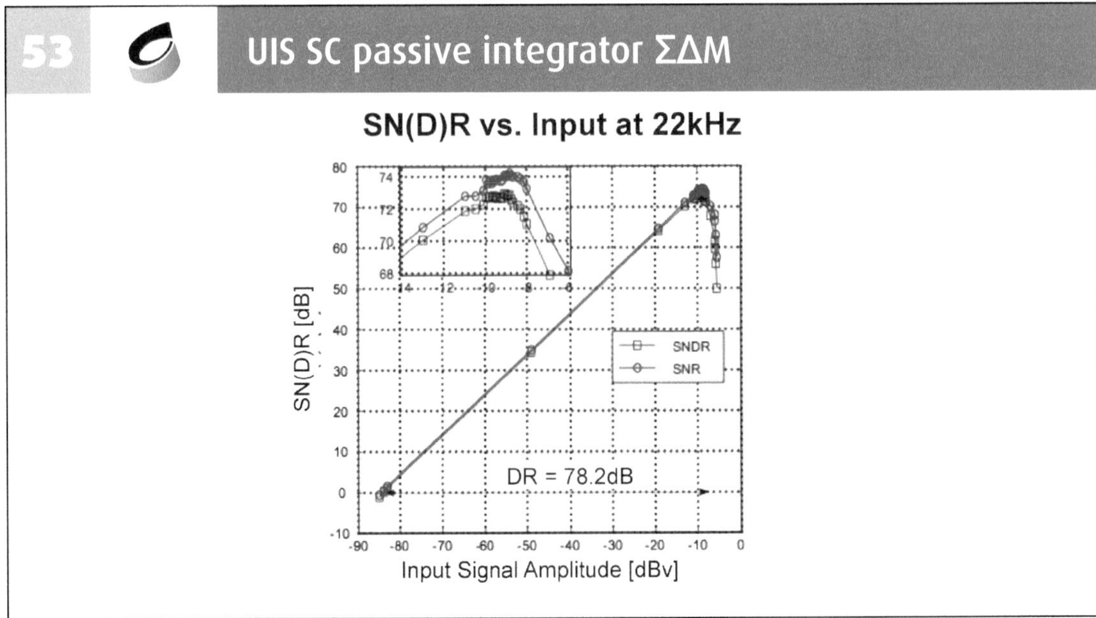

Both SNR and SNDR vs Input amplitude curves yield a dynamic range of 78.2 dB.

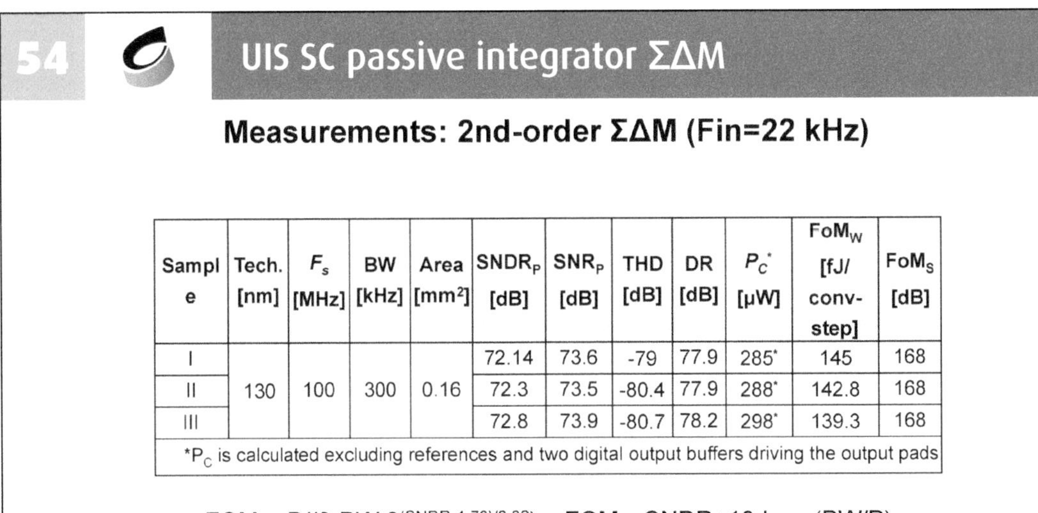

Sample	Tech. [nm]	F_s [MHz]	BW [kHz]	Area [mm²]	$SNDR_P$ [dB]	SNR_P [dB]	THD [dB]	DR [dB]	P_C^* [µW]	FoM_W [fJ/conv-step]	FoM_S [dB]
I					72.14	73.6	-79	77.9	285*	145	168
II	130	100	300	0.16	72.3	73.5	-80.4	77.9	288*	142.8	168
III					72.8	73.9	-80.7	78.2	298*	139.3	168

*P_C is calculated excluding references and two digital output buffers driving the output pads

$FOM_W = P/(2 \cdot BW \cdot 2^{(SNDR-1.76)/6.02})$ $FOM_S = SNDR + 10 \cdot \log_{10}(BW/P)$

Since the circuit is based on RC time constants, three distinct samples were measured, in order to prove the consistency of the circuit, ensuring its robustness for this kind of resolutions.

Outline

55

- Why use ΔΣMs
- Principle of Operation of ΔΣMs
- Active vs. Passive ΔΣMs
- **Passive based ΔΣMs**
  - UIS SC integrator based
  - **Continuous time RC integrator based**
- Conclusions

A consequence of using a discrete-time approach is that for higher clock frequencies a significant amount of power will be wasted when turning the switches ON and OFF. Instead, a continuous-time integrator can be used which, in passive implementations, is nothing more than a simple RC circuit.

Continuous time RC integrator

56

Differential RC integrator with SC feedback input

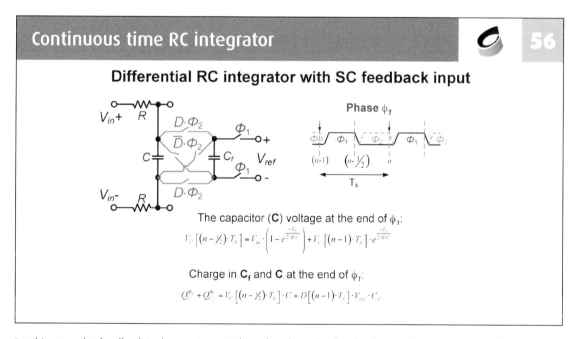

The capacitor (**C**) voltage at the end of ϕ_1:

$$V_C\left[\left(n-\tfrac{1}{2}\right)\cdot T_S\right] = V_{in}\cdot\left(1-e^{\frac{-T_S}{2\cdot R\cdot C}}\right)+V_C\left[\left(n-1\right)\cdot T_S\right]\cdot e^{\frac{-T_S}{2\cdot R\cdot C}}$$

Charge in $\mathbf{C_f}$ and **C** at the end of ϕ_1:

$$Q_C^{\Phi_1}+Q_{C_f}^{\Phi_1} = V_C\left[\left(n-\tfrac{1}{2}\right)\cdot T_S\right]\cdot C + D\left[\left(n-1\right)\cdot T_S\right]\cdot V_{ref}\cdot C_f$$

In this case, the feedback is done using a SC branch, where the feedback capacitor C_f is quite small. As a result, the feedback voltage does not change much and therefore these switches are smaller and easier to design. In order to obtain the transfer function of this RC circuit, the voltage in capacitor C and the charge in both capacitors C and C_f are calculated, at the end of phase φ1.

57  Continuous time RC integrator

Differential RC integrator with SC feedback input

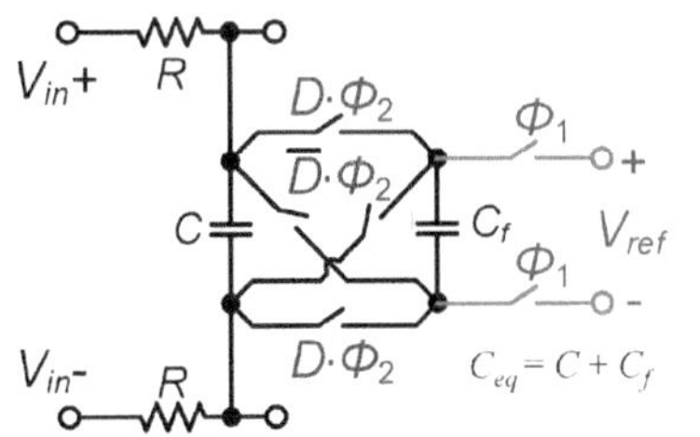

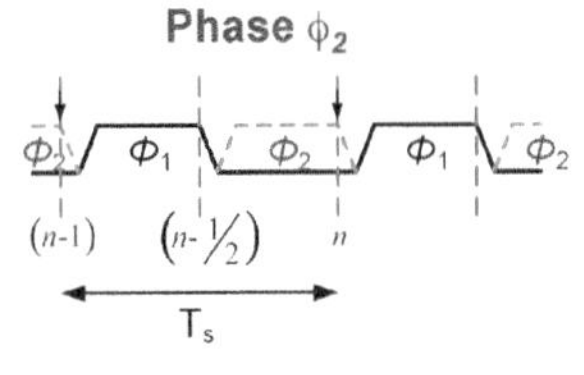

Assuming that there is instant charge transfer between $\mathbf{C_f}$ and **C**:

$$Q_C^{\Phi_2} + Q_{C_f}^{\Phi_2} = Q_C^{\Phi_1} + Q_{C_f}^{\Phi_1} \Rightarrow V'_C\left[\left(n-\tfrac{1}{2}\right)\cdot T_S\right]\cdot(C+C_f) = V_C\left[\left(n-\tfrac{1}{2}\right)\cdot T_S\right]\cdot C + D\left[(n-1)\cdot T_S\right]\cdot V_{ref}\cdot C_f$$

$$\Downarrow$$

$$V'_C\left[\left(n-\tfrac{1}{2}\right)\cdot T_S\right] = V_C\left[\left(n-\tfrac{1}{2}\right)\cdot T_S\right]\cdot\frac{C}{C+C_f} + D\left[(n-1)\cdot T_S\right]\cdot V_{ref}\cdot\frac{C_f}{C+C_f}$$

In order to facilitate the analysis during phase $_2$, it is assumed that the charge redistribution between the capacitors occurs instantly. Using the charge-conservation method again, the voltage across the capacitor C (right after the feedback switches close) is obtained. Inspecting the equation, it is possible to conclude that the voltage across the capacitor is a function of the voltage that was already there, plus a factor proportional to the feedback signal.

58  Continuous time RC integrator

Differential RC integrator with SC feedback input

The capacitor (**C+ $\mathbf{C_f}$**) voltage at the end of ϕ_2:

$$V_C\left[n\cdot T_S\right] = V_{in}\cdot\left(1-e^{\frac{-T_S}{2\cdot R\cdot C_{eq}}}\right) + \left[V_C\left[\left(n-\tfrac{1}{2}\right)\cdot T_s\right]\cdot\frac{C}{C_{eq}} + D\left[(n-1)\cdot T_s\right]\cdot V_{ref}\cdot\frac{C_f}{C_{eq}}\right]\cdot e^{\frac{-T_S}{2\cdot R\cdot C_{eq}}}$$

$$\Downarrow$$

$$V_C(n\cdot T_s) = V_{in}\cdot\left[1-e^{\frac{-T_S}{2\cdot R\cdot C_{eq}}} + \frac{C}{C_{eq}}\cdot\left(1-e^{\frac{-T_S}{2\cdot R\cdot C}}\right)\cdot e^{\frac{-T_S}{2\cdot R\cdot C_{eq}}}\right] + V_C\left[(n-1)\cdot T_S\right]\cdot\frac{C}{C_{eq}}\cdot e^{\frac{-T_S}{2\cdot R\cdot C}}\cdot e^{\frac{-T_S}{2\cdot R\cdot C_{eq}}} + D\left[(n-1)\cdot T_S\right]\cdot V_{ref}\cdot\frac{C_f}{C_{eq}}\cdot e^{\frac{-T_S}{2\cdot R\cdot C_{eq}}}$$

Assuming that $T_S << R\cdot C$ then: $x<<1 \Rightarrow e^x \simeq 1+x$

$$V_C(n\cdot T_s) \approx V_{in}\cdot\frac{T_S}{2R\cdot C_{eq}} + V_C\left[(n-1)\cdot T_S\right]\cdot\frac{C}{C_{eq}}\cdot\left(1-\frac{T_S}{2\cdot R\cdot C_{eq}}-\frac{T_S}{2\cdot R\cdot C}\right) + D\left[(n-1)\cdot T_S\right]\cdot V_{ref}\cdot\frac{C_f}{C_{eq}}\cdot\left(1-\frac{T_S}{2\cdot R\cdot C_{eq}}\right)$$

With the total capacitor voltage at the end of phase φ_2 and assuming that the sampling period is much smaller than the RC constant, the expression that relates the voltage across capacitor C in instant n with the input signal, the previous capacitor voltage and the feedback loop is obtained.

Continuous time RC integrator

59

RC integrator with SC feedback input TF

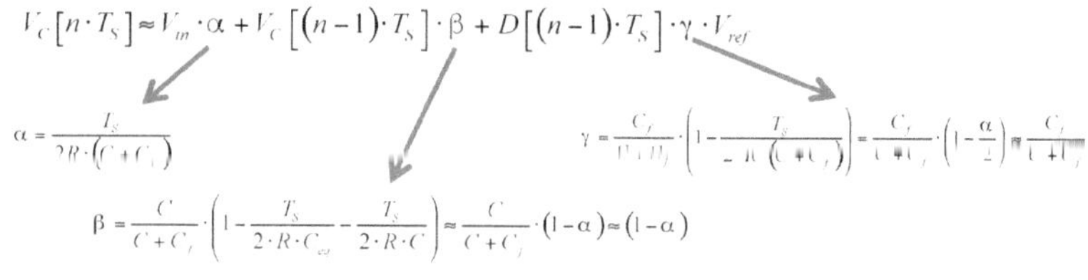

$$V_C[n\cdot T_S]\approx V_{in}\cdot\alpha+V_C[(n-1)\cdot T_S]\cdot\beta+D[(n-1)\cdot T_S]\cdot\gamma\cdot V_{ref}$$

$$\alpha=\frac{T_S}{2R\cdot(C+C_f)}$$

$$\gamma=\frac{C_f}{C+C_f}\cdot\left(1-\frac{T_S}{2\cdot R\cdot(C+C_f)}\right)=\frac{C_f}{C+C_f}\cdot\left(1-\frac{\alpha}{2}\right)\approx\frac{C_f}{C+C_f}$$

$$\beta=\frac{C}{C+C_f}\cdot\left(1-\frac{T_S}{2\cdot R\cdot C_{eq}}-\frac{T_S}{2\cdot R\cdot C}\right)\approx\frac{C}{C+C_f}\cdot(1-\alpha)\approx(1-\alpha)$$

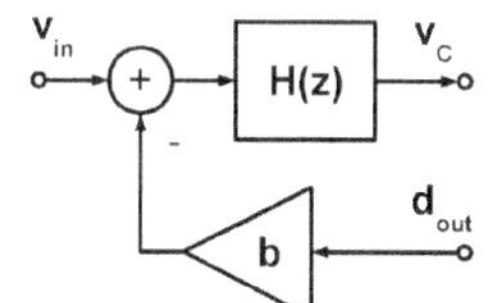

$$H(z)=\frac{V_C(z)}{V_{in}(z)}=\frac{\alpha}{1-\beta\cdot z^{-1}}$$

$$b=\frac{\gamma}{\alpha}=\frac{R\cdot C_f}{T_S}$$

From this expression, factors α, β, and γ are obtained. The equivalent block diagram is shown, and the feedback coefficient b can be given by the ratio between γ and α. This, in turn is directly related to the resistor R, the feedback capacitor C_f and the sampling period T_S. So, when designing the integrator, α can be adjusted by changing the values of R and C, while β depends on the value of C_f.

Continuous time RC integrator

60

Thermal Noise Analysis

- The thermal noise in the integrator is due to the SC feedback switches' On resistance and due to the RC resistor.
- The first noise contribution can be calculated using traditional methods:

$$P_{NT1}=\left(\frac{k\cdot T}{C_f}+\frac{k\cdot T}{C_{eq1}}\right)\cdot\left(\frac{C_f}{C+C_f}\right)^2\approx\frac{2\cdot k\cdot T\cdot C_f}{C^2}\qquad C_{eq1}=(C\cdot C_f)/(C+C_f)\qquad \text{Output noise}$$

$$\boxed{P_{NT1\_input}=\frac{P_{NT1}}{\alpha^2}\approx\frac{2\cdot k\cdot T\cdot C_f}{C^2\cdot\alpha^2}}\quad\boxed{P_{NT1\_input\_filt}=\frac{P_{NT1\_input}}{osr}\approx\frac{2\cdot k\cdot T\cdot C_f}{C^2\cdot\alpha^2\cdot osr}}\qquad\text{Input referred noise}$$

- The noise contribution of the RC resistor is calculated using the method previously used for the SC UIS integrator, resulting in:

$$P_{NT2}=\frac{4\cdot k\cdot T\cdot T_S}{R\cdot(C+C_f)^2}\qquad\text{Output noise}$$

$$\boxed{P_{NT2\_input}=\frac{P_{NT2}}{\alpha^2}=\frac{4\cdot k\cdot T}{(C+C_f)\cdot\alpha}}\quad\boxed{P_{NT2\_input\_filt}=\frac{P_{NT2\_input}}{osr}=\frac{4\cdot k\cdot T}{(C+C_f)\cdot\alpha\cdot osr}}\qquad\text{Input referred noise}$$

When performing a thermal noise analysis on the circuit, this has to be split into two main contributions: one from the on-resistance of the SC branch in the feedback path and the other from the resistance in the RC integrator. The contribution from the feedback is pretty straightforward from a SC analysis as it operates with complete settling, while the RC circuit operates in UIS. Following the same method used for the discrete-time structure, both input referred noise equations can be obtained.

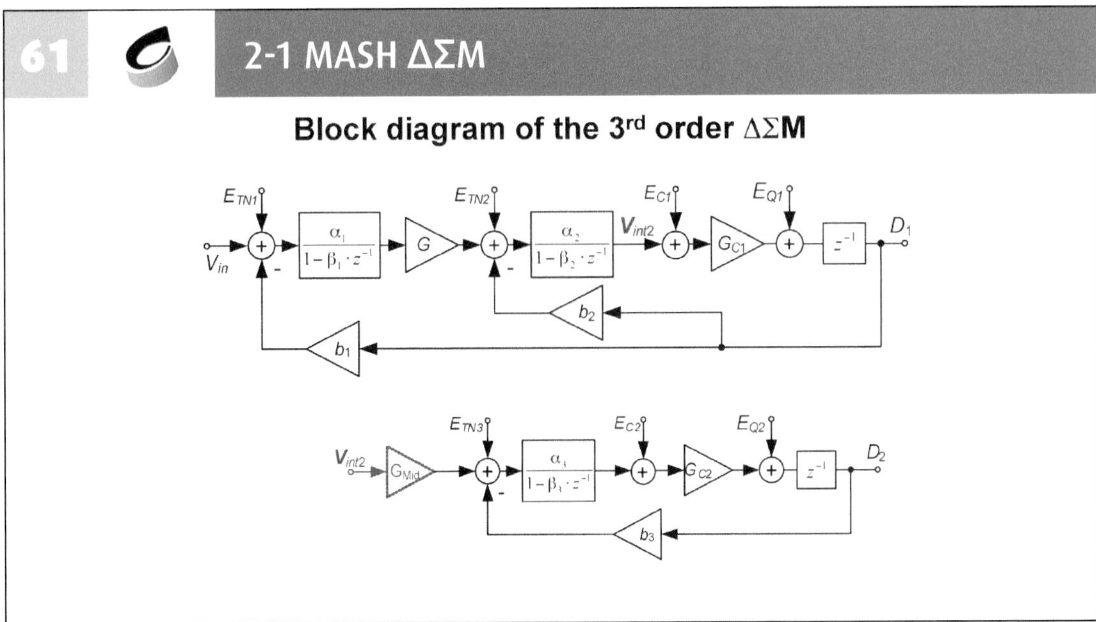

For this continuous time approach, a passive MASH topology was designed, where a 2nd order ΔΣM was cascaded with a 1st order ΔΣM, leading to a 3rd order ΔΣM. The input of the 1st order ΔΣM is taken from the output of the second integrator in the 2nd order ΔΣM. In order to suppress the noise contributions of both the previous integrator stages and the comparator, a gain block G_{mid} is added. This has the same specifications as other low-gain amplifiers in the passive approach (i.e., a differential pair), with the advantage of acting as a buffer between stages.

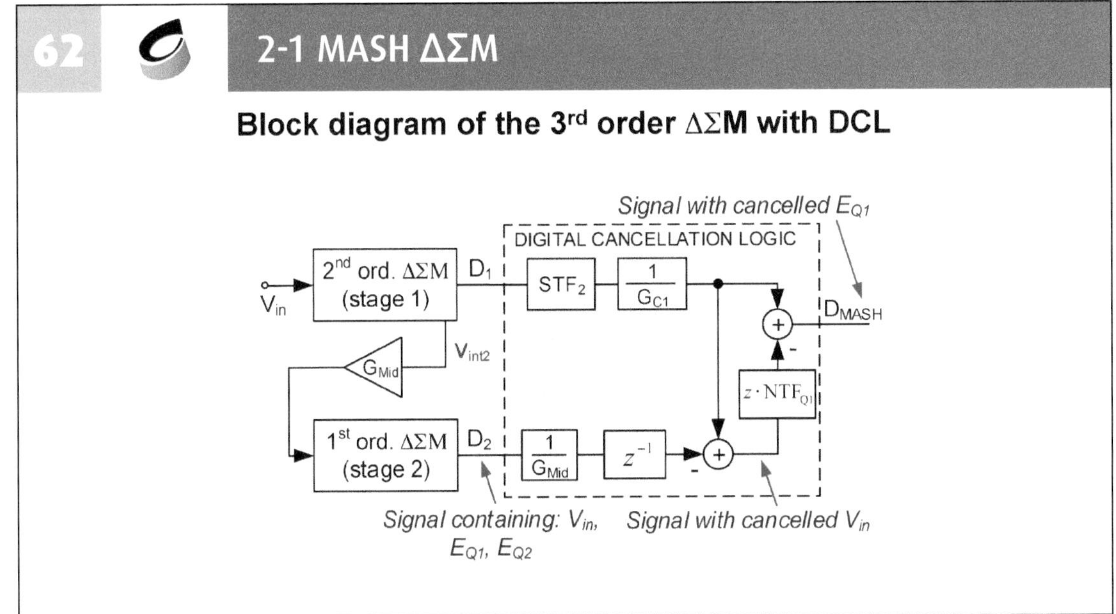

The block diagram of the 3rd order ΔΣM is now expanded to include the Digital Cancellation Logic (DCL). Since the design relies on a non-ideal ΔΣ, both the STF and NTF blocks inside the DCL use the α and β of the analog circuit.

2-1 MASH ΔΣM

Digital Cancellation Logic

- NTF_{Q1}, STF_2 → denominators with poles located outside BW only required to use their DC gain factors
- Necessary to cancel the signal in the BW → the IIR response can be ignored

$$NTF_{Q1} = \frac{Numerator\{NTF_{Q1}(z)\}}{Denominator\{NTF_{Q1}(z=1)\}}$$

$$STF_2 = \frac{Numerator\{STF_2(z)\}}{Denominator\{STF_2(z=1)\}}$$

Simplified DCL: FIR filter

(270 logic gates and 4 DFFs)

✓ no multipliers

In order to simplify the DCL, the number of multiplications must be reduced. The primary objective is to cancel the quantization noise in-band. To do so, only the values of the NTF and STF in-band are required, and with those in hand the digital cancelation logic design is greatly relaxed to just a small number of logic gates and a few D-type flip-flops.

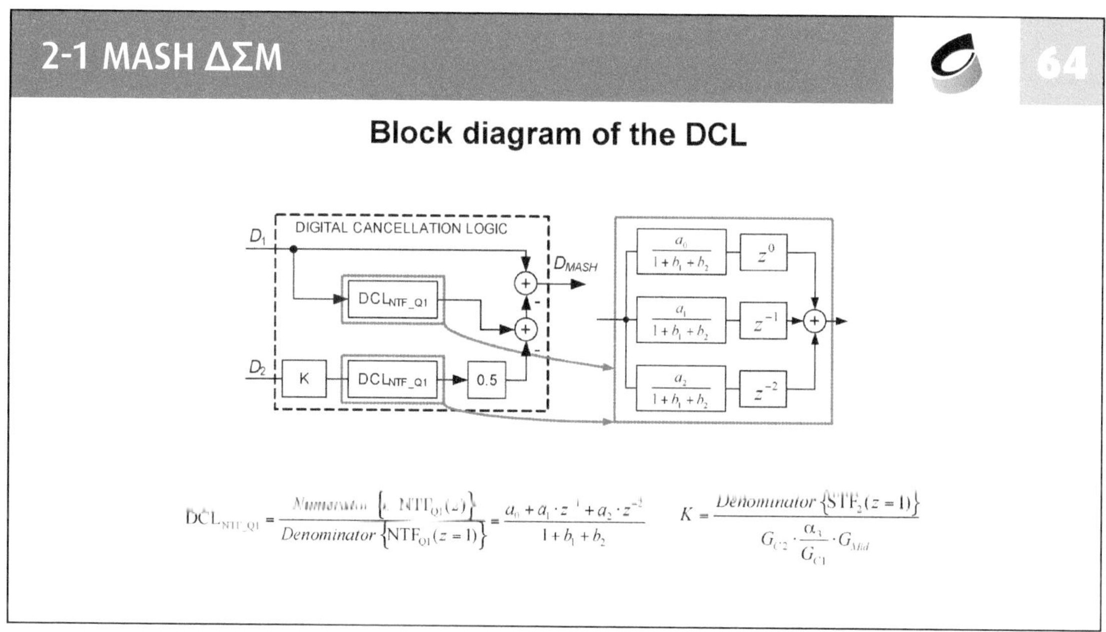

This block diagram exemplifies how the design of DCL is simplified.

65 

2-1 MASH ΔΣM

Design of the parameters of the ΔΣM

- All the blocks are equally important in determining the performance of the circuit (low gain value in the front of the signal processing chain)
- Component variations have to be considered during the design phase
- This means that there are many, equally important, variables during the design procedure
- It is difficult to obtain the best possible combination of the design variables without using an optimization algorithm

J. L. A. de Melo, B. Nowacki, N. Paulino, and J. Goes, "Design methodology for sigma-delta modulators based on a genetic algorithm using hybrid cost functions," presented at the Circuits and Systems (ISCAS), 2012 IEEE International Symposium on, 2012

C_1 [pF]	C_2 [pF]	C_3 [pF]	R_1 [kΩ]	R_2 [kΩ]	R_3 [kΩ]	C_{f1} [fF]	C_{f2} [fF]	C_{f3} [fF]	G_1	G_{Mid}	$V_{ref1,2}$ [V]	V_{ref3} [V]
13.7	2.1	2.3	2.1	10	20	220	13	7	9.5	9.5	0.9	0.44

C_{CM1} [pF]	C_{CM2} [pF]	C_{CM3} [pF]		α_1	α_2	α_3	G_{c1}	G_{c2}	b_1	b_2	b_3	Z_{IN} [kΩ]
2.5	0.47	0.5		0.015	0.02	0.01	46dB	59dB	0.84	0.23	0.12	2.1

The major problem with this circuit is the design phase, since there are no high-gain amplifiers and the noise contribution of each contribution is more or less on the same level. Moreover, the sizing of the components should be made in such a way that the circuit is less sensitive to variations. With all these parameters taken into consideration, it is fairly easier to design such a circuit through the use of an optimization tool based on a genetic algorithm (GA), instead of doing it by hand.

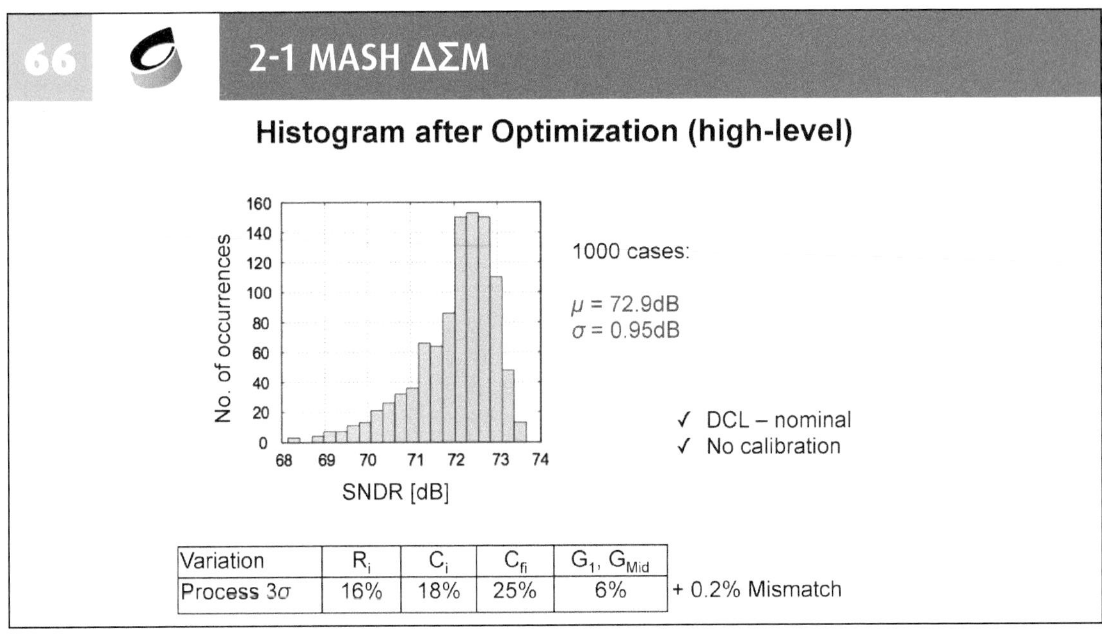

After optimization, a 1000-case Monte Carlo (MC) analysis assuming 3σ =16% for the resistors R, 18% for the capacitors C, 25% for the feedback capacitors C_f and 6% for G_{Mid}, resulted in a mean SNDR value of 72.9dB with a standard deviation of 1.3%. For the large majority of cases the performance surpasses an SNDR of 70 dB. Therefore, no calibration circuit is required. Here it is assumed that the digital coefficients were calculated for their nominal value.

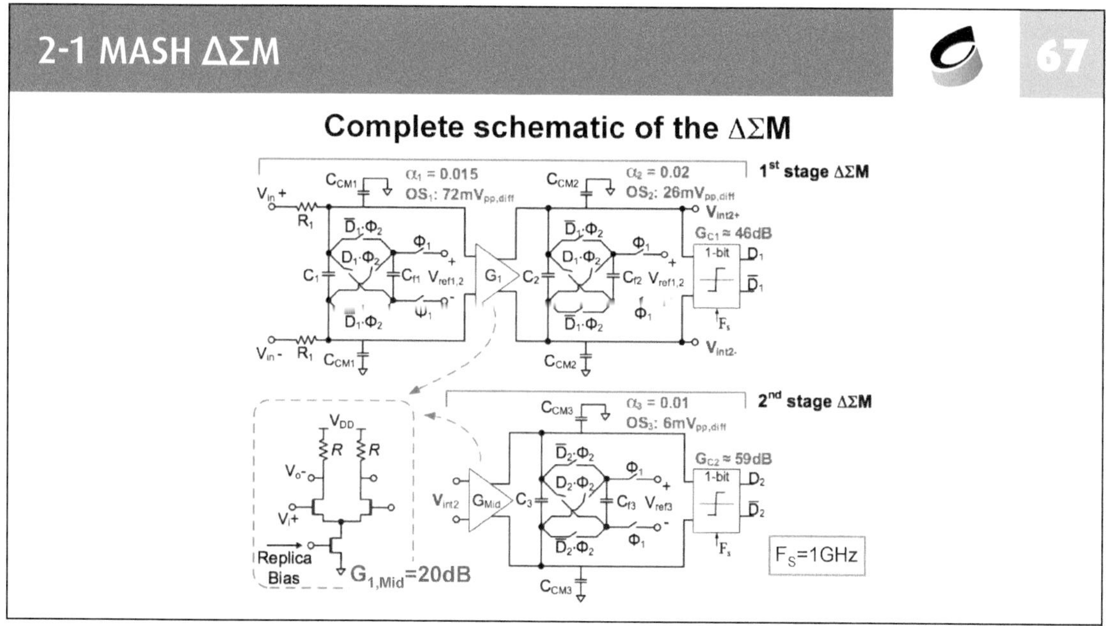

Each stage of the MASH modulator consists of RC integrators, gain block and 1-bit quantizer (clocked comparator + DFF).

Blocks G_1 and G_{Mid} are differential pairs loaded by resistors. These R's are also part of the RC time constants of the 2nd and 3rd integrators.

Additional capacitors C_{CMi} are used to reduce the common-mode voltage swing at the integrator's outputs. As mentioned before, the SNR is limited by the integrators thermal noises, which are defined by values of C's.

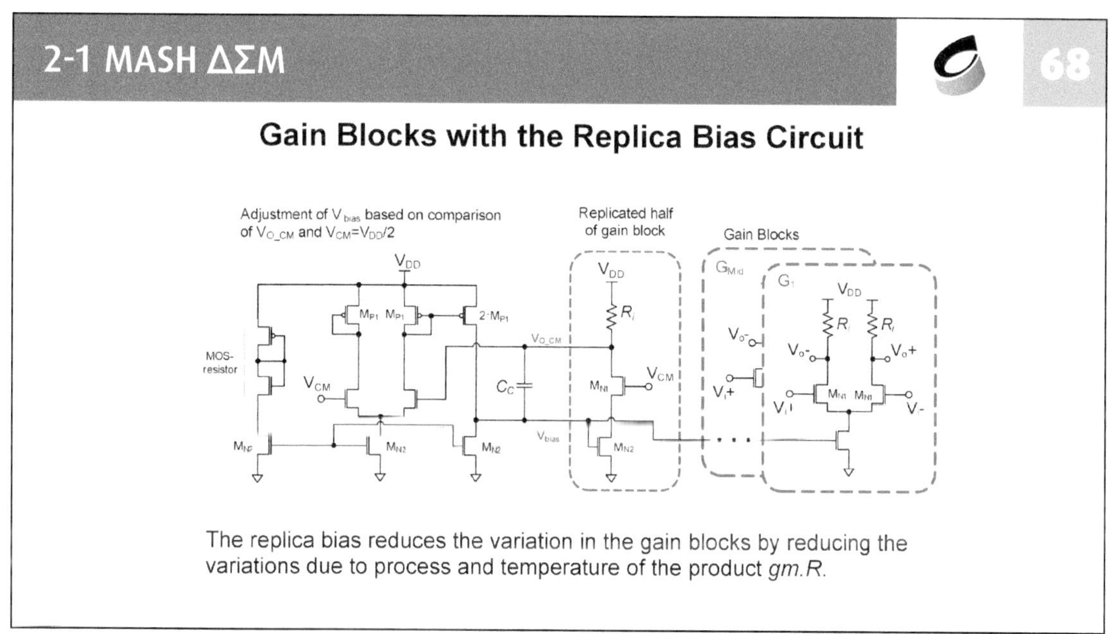

The replica bias circuit ensures that voltage V_{bias} that biases the differential pair's current source is kept constant, improving the robustness of the gain block to process and temperature variations. As a result, a variation as low as 6% for the gain can be estimated across process and temperature.

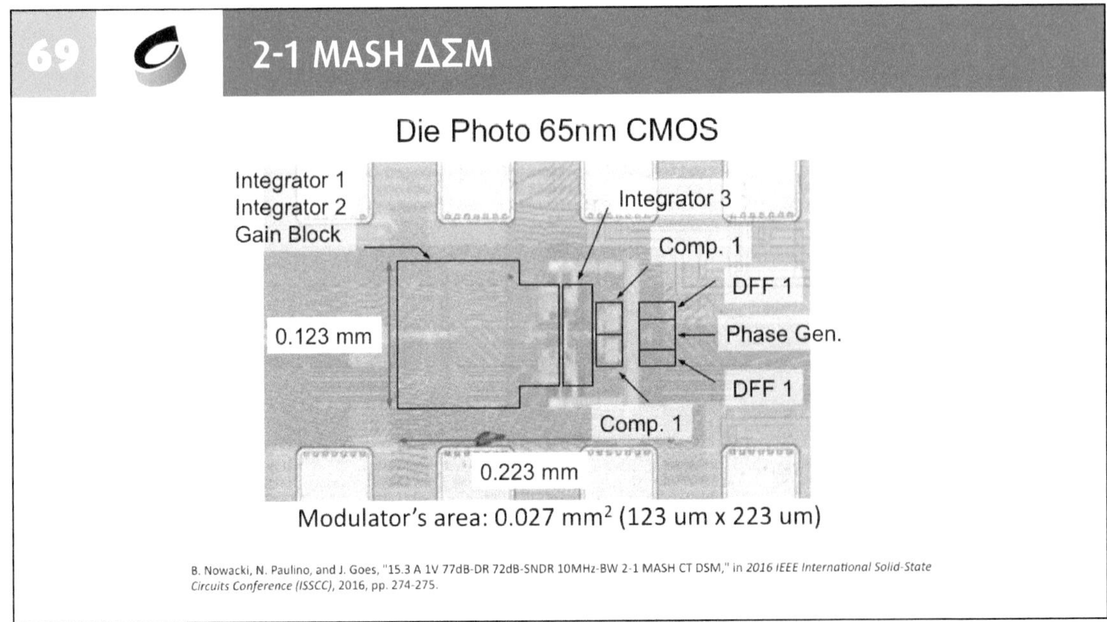

The die photo of the circuit is shown. The circuit was fabricated in 65 nm CMOS with a 1V supply. Again, despite the use of large capacitors to lower the thermal noise contribution, the total area is still quite small, falling below 0.03 mm^2.

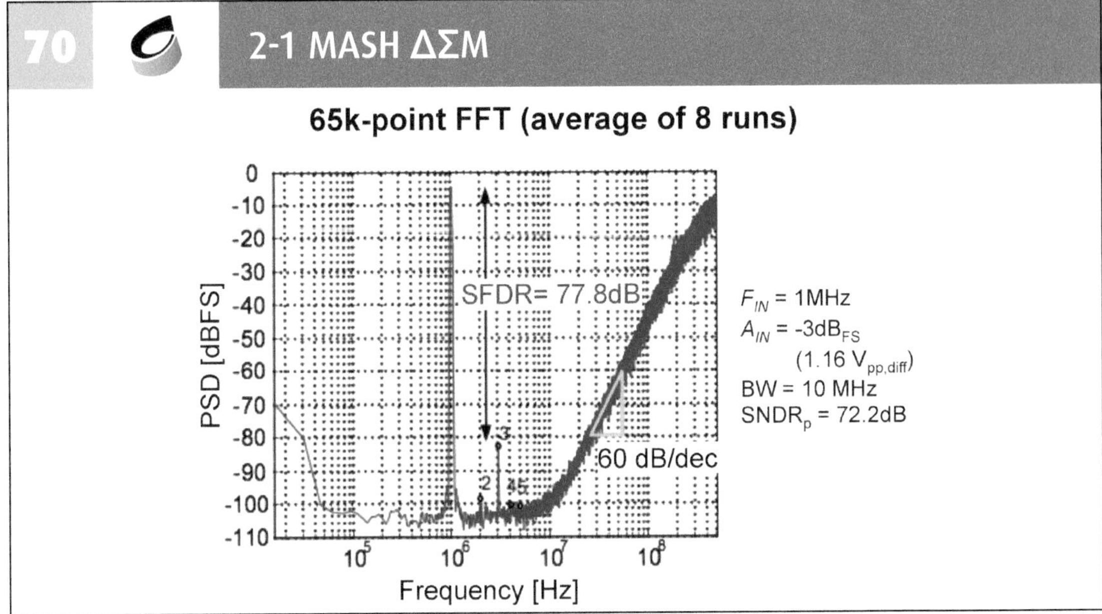

Measured results show that for a 10 MHz bandwidth, an SNDR of 72.2 dB and a SFDR of 77.8 dB are obtained, when an input signal above 1 V@1MHz is applied. Here, the +60 dB/dec slope is clearly visible and ensures a 3rd order ΔΣM behavior. The performance is mainly dominated by the third harmonic, originated mainly in the first differential pair.

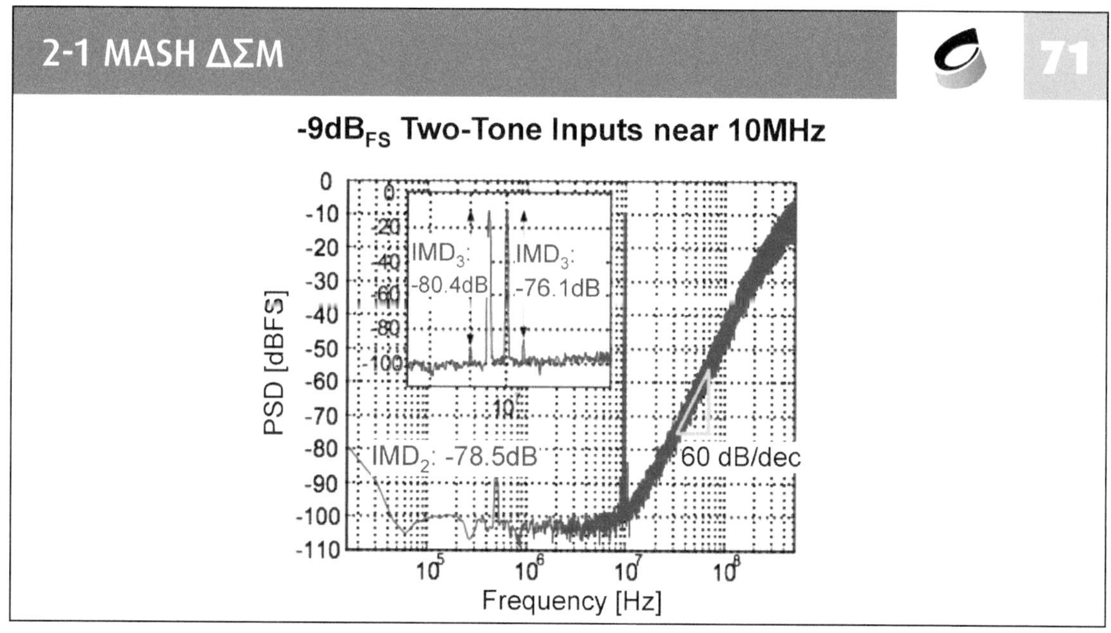

The two tone test near the band edge shows IMD3 of -76,1 and IMD2 of 78,5dB.

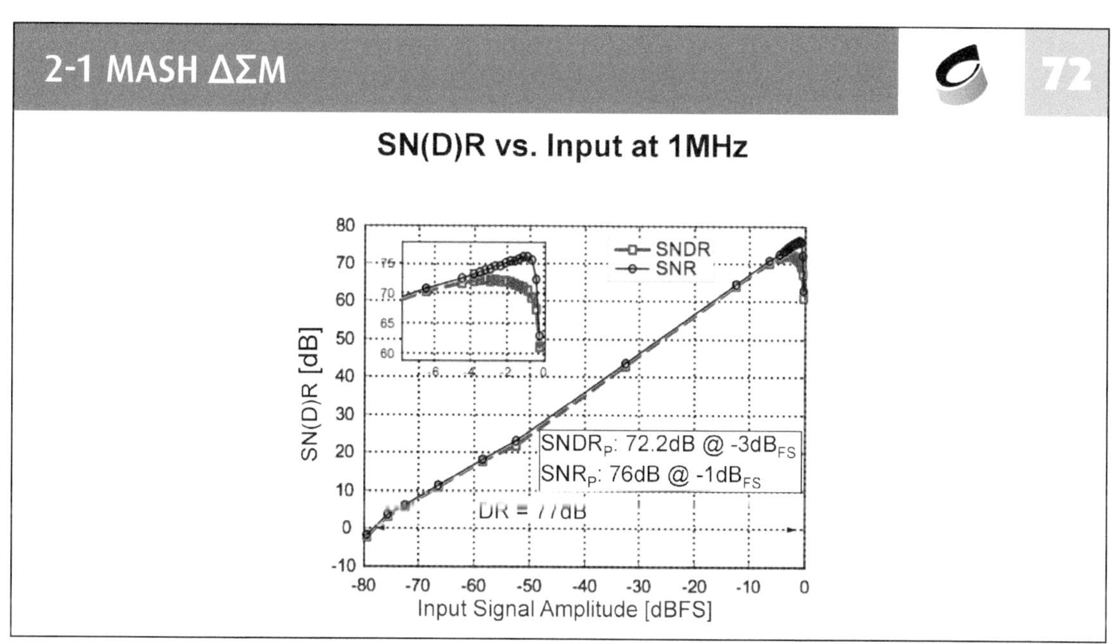

The SN(D)R vs. Input signal test resulted in DR of 77dB.

Measurements for +-5% V_{DD} and V_{ref} variations resulted in worst case SNDR of 70.5dB. The alias suppression (measured for various F_{in} from 980MHz to 1.02GHz) is ~51dB.

73 

2-1 MASH ΔΣM

	This work	Do-Yeon ISSCC '15	Zeller JSSC '14	Shu ISSCC '13	Loeda VLSI '15	Wei VLSI '15
Process [nm]	**65**	28	65	28	40	28
Area [mm^2]	**0.027**	0.34	0.039	0.08	0.0194	0.066
Supply [V]	**1**	1.2/1.5	1.1	1.2 / 1.5	-	0.9/1.8
P [mW]	**1.57***	78	1.82	3.9	1.94	3.16
Fs [GHz]	**1**	1.8	0.65	0.64	0.6	0.432
BW [MHz]	**10**	50	10	18	10	5
DR [dB]	**77**	85	71.2	78.1	68.7	83.9
SNDR [dB]	**72.2**	74.6	68.6	73.6	67.4	80.5
FOM_W [fJ/step]	**23.6**	177.7	41.4	27.7	50.5	36.4
FOM_S [dB]	**170.2**	162.7	166	166	161.5	172.5

* Measured 1.3mW (two MASH stages) + 0.27mW (DCL - simulated)

$FOM_W=P/(2\cdot BW\cdot 2^{(SNDR-1.76)/6.02})$ $FOM_S=SNDR+10\cdot\log_{10}(BW/P)$

Combination of MASH topology and passive RC integrators with low gain blocks results in

- Reduction of the power dissipation
- Small the chip size
- High energy efficiency
- DR of 77dB and SNDR of 72,2dB

Due to the design methodology no calibration was required.

74 

Outline

- Why use ΔΣMs
- Principle of Operation of ΔΣMs
- Active vs. Passive ΔΣMs
- Passive based ΔΣMs
  - UIS SC integrator based
  - Continuous time RC integrator based
- Conclusions

Conclusions

- It is possible to use passive integrators in the loop filter of SDM.
- When using passive integrators the signal comparator gain is larger than 1 and can be used to replace high gain amplifiers in the loop filter.
- Since the comparator uses positive feedback it is much simpler and more energy efficient than a normal amplifier.
- It is possible to build a 2nd order ΔΣM using only two active blocks: a comparator and a simple differential pair gain block.
- Higher order ΔΣM can be built using MASH structures.
- The low gain amplifier in the loop results in a similar noise contribution from all the integrators.
- By using a genetic optimization algorithm it is possible to obtain designs that have a good performance and are energy efficient.

References

- B. Nowacki, N. Paulino, and J. Goes, "Analysis and the design of a first - order delta sigma modulator using very incomplete settling," presented at the Mixed Design of Integrated Circuits and Systems (MIXDES), 2011 Proceedings of the 18th International Conference, 2011.
- Kobayashi, T.; Nogami, K.; Shirotori, T.; Fujimoto, Y., "A current-controlled latch sense amplifier and a static power-saving input buffer for low-power architecture," *Journal of Solid-State Circuits*, IEEE, vol.28, no.4, pp.523-527, Apr 1993.
- B. Nowacki, N. Paulino, and J. Goes, "A 1.2 V 300 uW second-order switched-capacitor delta sigma modulator using ultra incomplete settling with 73 dB SNDR and 300 kHz BW in 130 nm CMOS," presented at the European Solid-State Circuits Conference (ESSCIRC), 2011.
- J. L. A. de Melo, B. Nowacki, N. Paulino, and J. Goes, "Design methodology for sigma-delta modulators based on a genetic algorithm using hybrid cost functions," presented at the Circuits and Systems (ISCAS), 2012 IEEE International Symposium on, 2012.
- B. Nowacki, N. Paulino, and J. Goes, "15.3 A 1V 77dB-DR 72dB-SNDR 10MHz-BW 2-1 MASH CT DSM," in 2016 *IEEE International Solid-State Circuits Conference* (ISSCC), 2016, pp. 274-275.
- Y. Do-Yeon, et al., "An 85dB-DR 74.6dB-SNDR 50MHz-BW CT MASH Delta Sigma Modulator in 28nm CMOS," *ISSCC Dig. Tech. Papers*, pp. 272-273, 2015.
- S. Zeller, et al., "A 0.039 mm2 Inverter-Based 1.82 mW 68.6dB-SNDR 10 MHz-BW CT-Sigma Delta-ADC in 65 nm CMOS Using Power- and Area-Efficient Design Techniques," *IEEE J. Solid-State Circuits*, vol. 49, pp. 1548-1560, 2014
- S. Yun-Shiang, T. Jui-Yuan, C. Ping, L. Tien-Yu, and C. Pao-Cheng, "A 28fJ/conv-step CT \Σ modulator with 78dB DR and 18MHz BW in 28nm CMOS using a highly digital multibit quantizer," in *Solid-State Circuits Conference Digest of Technical Papers (ISSCC)*, 2013.
- S. Loeda, J. Harrison, F. Pourchet, and A. Adams, "A 10/20/30/40 MHz feed-forward FIR DAC continuous-time Delta Sigma ADC with robust blocker performance for radio receivers," in VLSI Circuits (VLSI Circuits), 2015 Symposium on , 2015, pp. C262-C263.
- W. Guowen, P. Shettigar, S. Feng, Y. Xinyu, and T. Kwan, "A 13-ENOB, 5 MHz BW, 3.16 mW multi-bit continuous-time Delta Sigma ADC in 28 nm CMOS with excessloop- delay compensation embedded in SAR quantizer," in *VLSI Circuits (VLSI Circuits)*, 2015 Symposium on, 2015, pp. C292-C293.

CHAPTER 4

Industrial Internet of Things

Noel O'Riordan

S3 Group

Dublin, ireland

The contents will cover two case studies of real world applications of IoT: one based on wireline IoT to control hazardous industrial equipment, and the other based on satellite radio for machine communication. The decisions concerning the architecture choices, block performance, and issues related to the IC process choice will be detailed. Selected topics on interfaces, power, IC realization and measurement will be presented.

Two examples of where IOT can be applied in industrial environment are discussed, namely:

- Satellite Systems
- Heavy duty industrial type application

Industrial IOT

Common features to IOT definition sensing, processing and communicating data, and energy management

- Many physical parameters can be measured by sensors;
- Diverse technologies can be employed to realise sensors;
- Sensors can be reduced to voltage or current sources;
- Processing data locally avoids excessive communication;
- IOT interfaces comprises:
  - -PAN, LAN, WAN;
  - -WIFI;
  - -LORA;
- Ultra low power;
- Energy Harvesting.

Industrial IOT – Satellite System

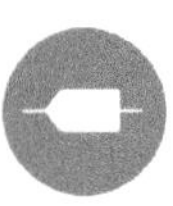

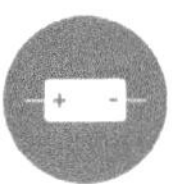

- Sense & Measure - position/motor/temperature
- Process & Interface – satellite communication
- Analyse – data/tracking/maintenance/predictive

IIOT – Satellite Topics

Interface

Architecture

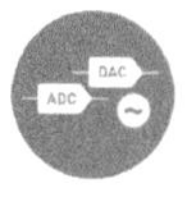

Circuitry

Economics

- System Requirements
- RFIC Architecture
- Cost impact
- RF Circuit details and measurements

These aspects of the IOT, applied to the satellite system, are further discussed.

IIOT – Satellite System (1)

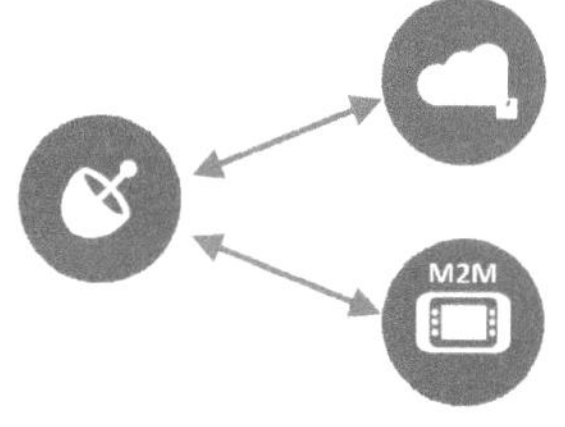

- Tracking Cargo Container
  - Logistics
  - Environmental Control
- Telematics
  - Engine Maintenance
  - Asset tracking

Example of a satellite system where IOT is used in an industrial context.

Cargo Containers:

Some of these containers have products worth 1 Million dollars.

The ships must to travel by areas where there is no communication coverage, so satellite communication is needed for:

- Tracking the data
- Control the environment
- Take action based on parameters that were measured

Telematics:

Heavy Machinery worth 5 million dollars. Some work almost 24h per day.

One important aspect is engine maintenance. The engine is constantly monitored avoiding to stop the work for a scheduled maintenance. The maintenance is done as needed saving money.

5 

IIOT – Satellite System (2)

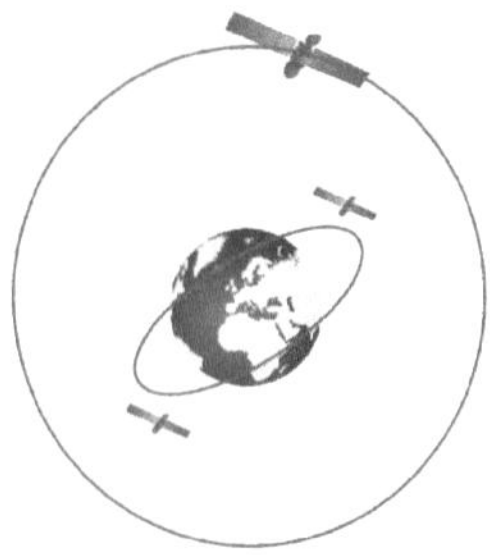

- Satellite:
- Geosynchronous
  - ~30,000 km
- Low Earth Orbit
  - ~800 km & ~30,000 km/h

Geosynchronous satellites are geostationary and can cover a large area of the earth. The time delay to communicate with a satellite this far is about 250 ms. It is a large delay to control machinery in a IOT industrial environment.

Low earth orbit satellites have an elliptic orbit around the earth and they move across the sky very fast, so they need to be tracked.

Time delay changes depending on the position and must be taken into account.

Doppler effect is also a significant factor in low earth orbit systems.

Network access is a challengein satellite systems, as users can simultaneously request access from any geographical part of the world at any time.

6 

IIOT – Satellite System (3)

Link Budget:

Ptx – Prx = Path Loss – Gtx – Grx

Path Loss = 20log(λ/4πD)

G = 4πA/λ²

- Link Budget
  - Limited by satellite's physical size and power
  - Frequency and physical form factor dependent
- Large Path Loss -180 dB at 1.55GHz
- Bandwidth and power limited channel
  - Modulation scheme
- Sharing spectrum with other users – control of emissions

Satellite link:

- Some satellites demodulate and re-modulate the data, others just repeat the signal by boosting the power.
- These satellites operate in a licenced frequency band. For very low power and small physical size, as may be required in IOT, there are frequencies more suited such as 2.4 GHz (original ISM license free band, due to microwave oven issues);
- ISM band emissions are not controlled, and therefore are more prone to blockers;
- More efficient modulation schemes can allow IOT power to be reduced as the transmitter dominates power;
- Delay latency of 250 ms is impracticable for 5G 1 ms target for IOT.

IIOT – System Architecture TXR

7

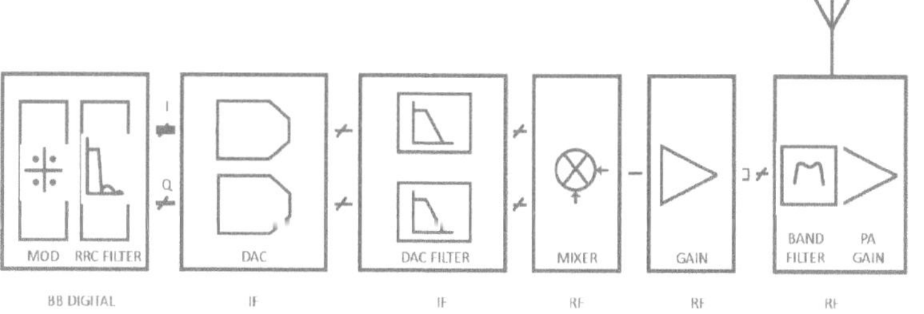

- TXR
  - Output spectrum
    - Power/Linearity OIP3/Phase Noise
  - Zero IF

Transmitter:

The transmitter dominates the power consumption, so it is worth looking into it to reduce the overall system power.

Digital baseband modulator

- Modulates the amplitude and phase of the RF signal creating a digital constellation

An orthogonal radio architecture with I/Q(In phase/Quadrature) is used.

IIOT – System Architecture RXR

8

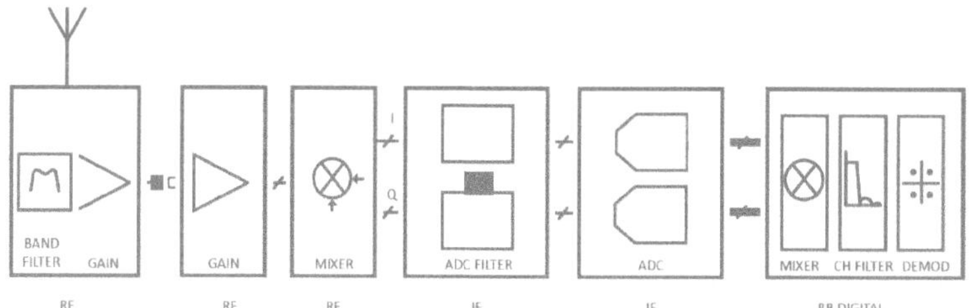

- **Down conversion**
  - Low IF /Zero IF : Blocker
  - Flexible Digital BB
- **Filtering**: In-Band / Out-Band
  - Analog : Anti Aliasing and Blocking
  - Digital : Channel filtering
- **Gain** : gain input up to ADC FS (Full Scale)

Receiver:

Software Defined Radio is not practical:

- Sampling at RF demands high power
- Accommodating Blocker power implies large dynamic range.

The use of down conversion and filtering is more practical and some degree of flexibility is possible in the digital baseband..

Low IF/Zero-IF –Image rejection ratio (IRR) and blocker type determines which down conversion technique is more adequate (Low IF issues with dynamic offset).

IIOT – RXR Spectrum

A simplified diagram of down conversion represented in the frequency spectrum is shown.

A variable local oscillator allows a down-conversion to a fixed IF, which relaxes the on chip filter requirements (Q – quality factor).

Image and phase noise, due to a noisy LO, reciprocal mixing degrade the SNR.

The overall SNR required at output for demodulation is related to the SNR at input by NF and other disturbance effects (e. g. non-linearities).

- LO variable
- IF fixed
  - On-chip filtering
- Low side/High side
- Image Rejection
- SNR at output for demodulation BER

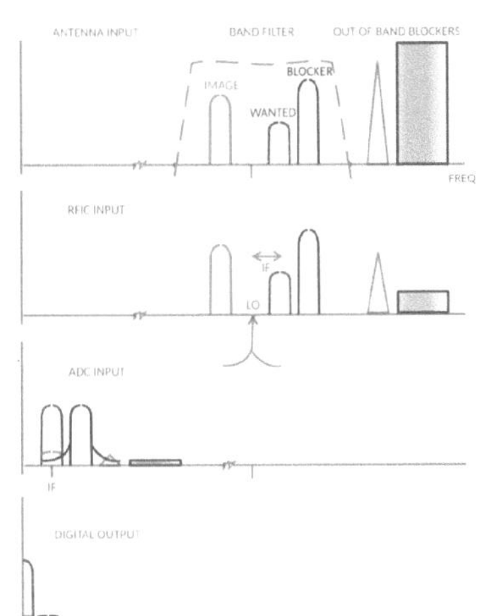

10

IIOT – RXR Design Equations

IRR is achieved by phasing of LO and/or by filtering (e.g., BPF).

IRR depends on gain and phase mismatch (mainly due to LO).

The formula for the gain include the filtering effects.

Noise Floor: Nin + NF = SENS - SNRo
Phase Noise: Pblk + PN + 10log(BW) = SENS – SNRo - M
Image IRR : Pblk – IRR = SENS – SNRo - M
3rd order IP3: 3Pblk -2IP3 = SENS – SNRo – M
2nd order IP2: 2Pblk – IP2 = SENS – SNRo – M
Gain: ADCFS – ADCDR -10log(Fs/2/BW)- Gain= SENS – SNRo – M

Impairments to SNRo

- Noise Figure , NF = SNRin/SNRo
  - RFIC device thermal & flicker noise, Nin = -174 dBm/Hz+ 10Log(BW)
  - SENSitivity , min receive signal power to be demodulated
- PN, reciprocal mixing. Margin, M, e.g 16dB increases SENS by 0.1dB
- IRR, image falls on wanted channel
- Intermodulation Distortion IP3 & IP2

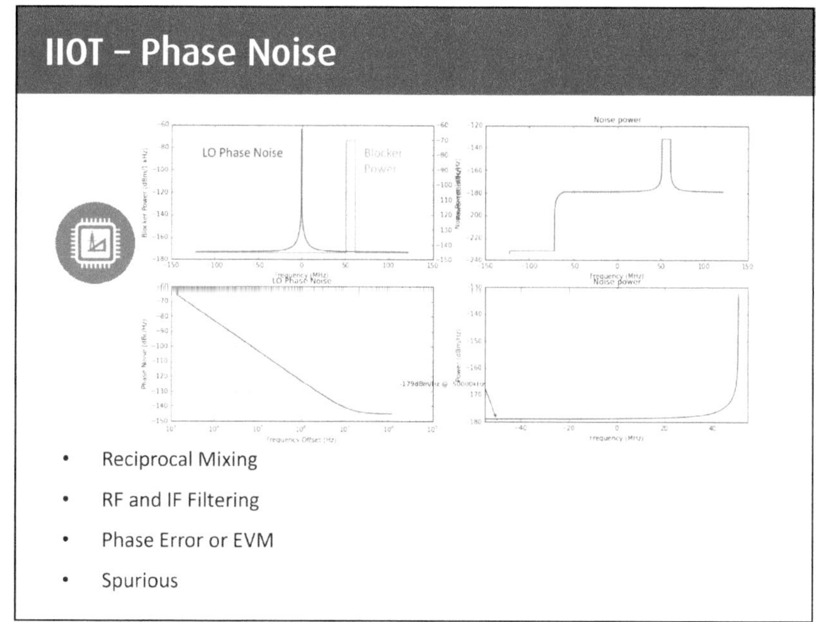

11

The IF filter does not relax the reciprocal mixing as it is after the mixer, but if the filtering is too wide then the ADC dynamic range (in band IIP3) may limit the performance rather than the close in phase noise.

The phase error limits the maximum SNR that can be achieved, no matter the signal power, especially for high order digital modulation like QAM64.

The phase noise is quantified in dBC/Hz.

Blockers have more impact regarding the phase noise in satellite,cellular, and unregulated bands than in Wi-Fi. Reciprocal mixing of the LO phase noise with a blocker generates noise on the wanted channel.

Spurs should be less than:

phase noise + 10 log (BW), where BW should be the channel bandwidth for a receiver, but can be a bandwidth for transmitter standard compliance.

IIOT – RXR LNA

RF LNA

Common source LNA with tuned load

- Board Parasitics modelled

CascodeGain DAC

Vbias

INP

INN

S11

Simulated

measured

-15 dB

1 GHz

2 GHz

12

The key aspect is that a package model with parasitics is important to predict the accurate behavior of the circuit and such features are part of IC CAD tools.

13 

IIOT – RXR Mixer

A passive mixer has a reduced flicker noise.

The passive mixer also has a switch loss of 2/Pi (if rds < Rin), so NF is limited by this and by the gain of the following active device.

A 25% duty cycle signal is typically used at LO to avoid image current.

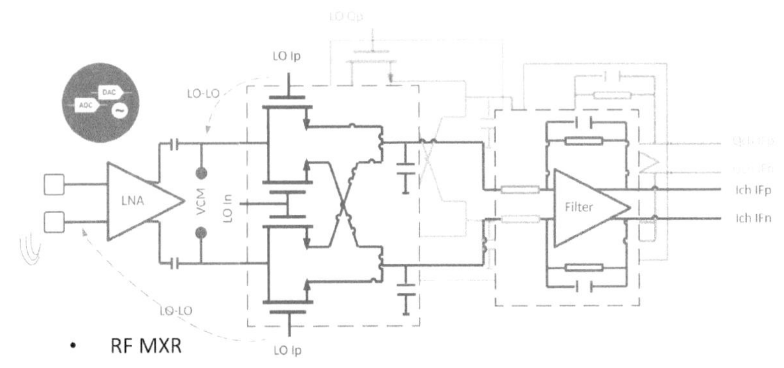

- RF MXR
  - Passive mixer
    - IP3, NF
  - Simulation issues: IP3, LO-LO

14 

IIOT – Passive Mixer IIP3 sim.

IIP3 of a Passive Mixer is difficult to simulate with BSIM3 model due to model inaccuracies, and a true Surface Potential model is needed.

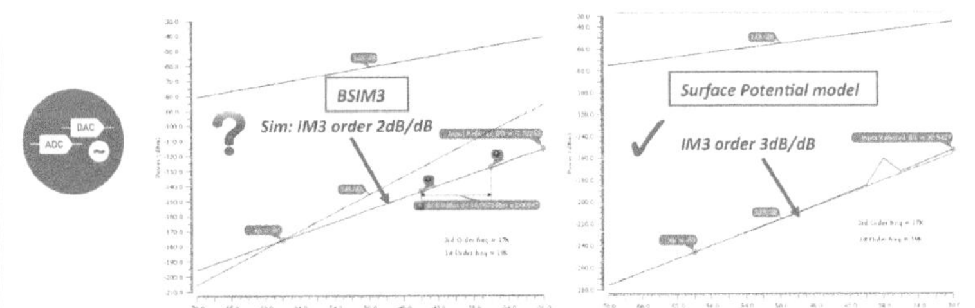

- RF MXR
  - IIP3 important specification for mixer
  - BSIM3/4 does not model passive mixer IP3
    - Discontinuity at 0V Vds results in 2nd order behaviour for IP3

IIOT – RXR Filter

15

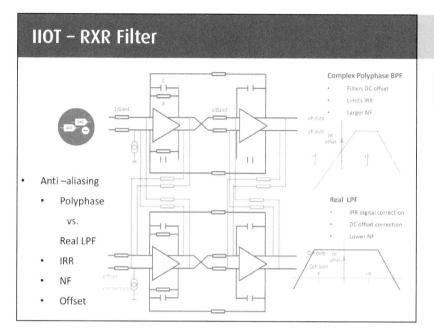

- Anti –aliasing
  - Polyphase vs. Real LPF
  - IRR
  - NF
  - Offset

Complex Polyphase BPF
- Filters DC offset
- Limits IRR
- Larger NF

Real LPF
- IRR digital correction
- DC offset correction
- Lower NF

Polyphase BPF:
- DC offset is reduced without compensation;
- Image rejection is limited and may not be improved for very low IF;
- Has a larger NF due to cross resistors.

Real LPF:
- requires DC static offset correction (if dynamic offset this approach is not effective and a good circuit IP2 is needed);
- IRR can be improved with digital correction.

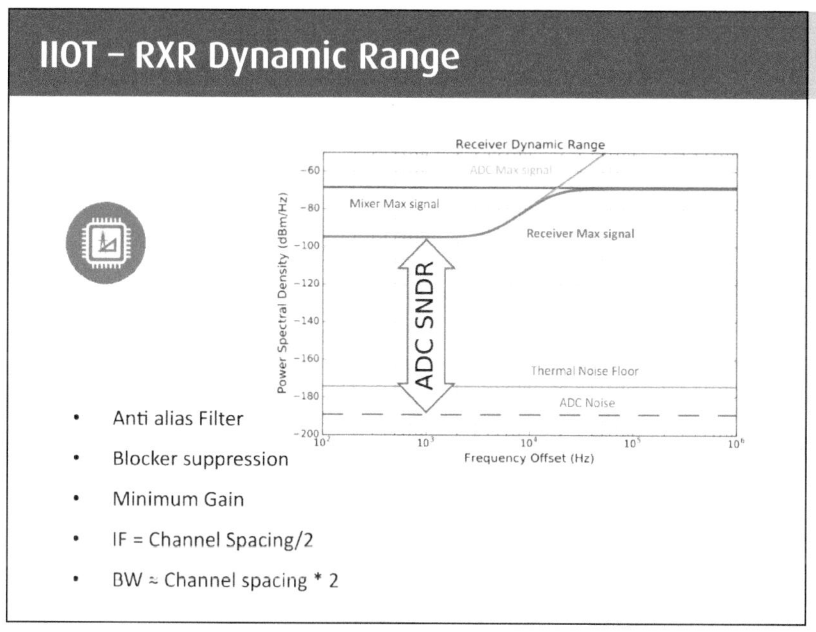

Anti-aliasing filter is required to suppress blockers from Fs/2 to IF. The dynamic range can be extended by minimizing the gain variation. The channel filtering is done digitally.

The gain is set to minimum to place the receiver noise 5 to 15dB above the ADC noise floor.

- Max input signal in band is: 10dBm - 10Log10(BW) - Gain - 95 dBm. ADC noise floor is +10dBm – (6.02*10.2+1.76) – 10Log10(Fs/2).
- ADC DR setting:
- Lower limit is due to the thermal noise plus NF;
- Upper limit is due to ADC input range and blockers (e.g., 10 dBm - 60 dB (gain) - 10log(BW=7200) = -88.6 dBm)

17 

IIOT – TXR IF

Transmitter:

DAC dynamic range set the noise floor;

The aliasing from DAC must be properly filtered to avoid the generation of spur; The built-in LPF can be used for that purpose

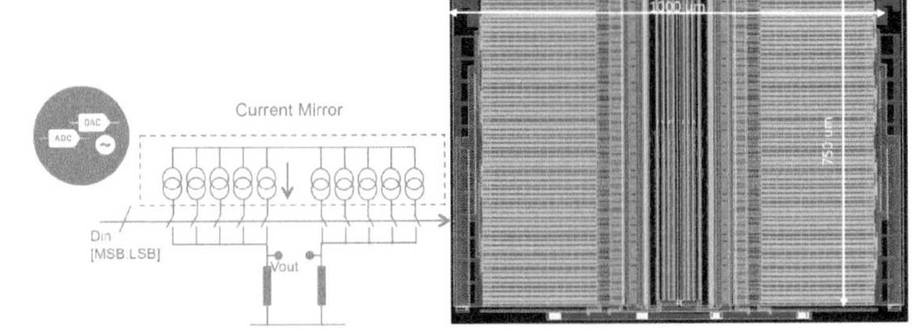

- DAC 12b / 20 MSps
  - DR noise floor
  - Alias attenuation/spur

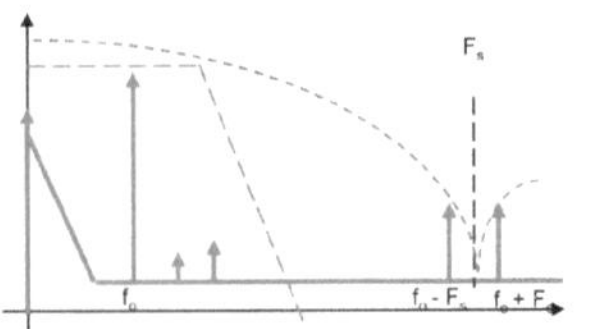

18 

IIOT – TXR Output Spectrum

Example of the output spectrum of a transmitter (BPSK).

The image shows three different sources of noise (regrowth, phase noise and thermal noise).

- TX Noise
- Phase noise
- Thermal noise
- Regrowth – non linearity

IIOT –TX EVM

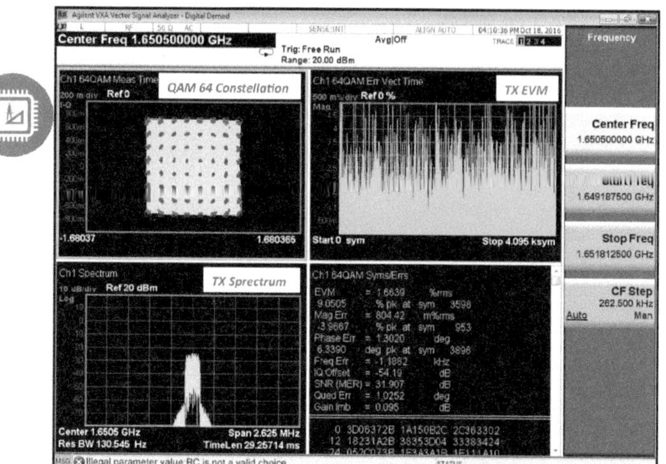

Modulated data by the transmitter.

Constellation and spectrum and error vector magnitude (EVM) measurements.

IIOT – Synthesizer

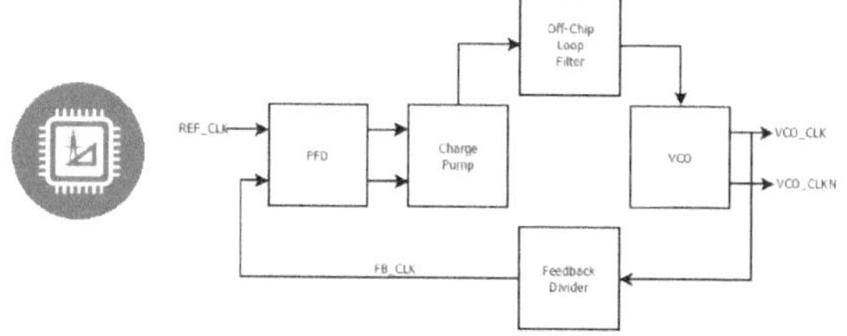

- Fractional-N Synthesizer CLKout= N.f*REF_CLK
- Narrow IF bandwidth
- LO resolution better than channel spacing
- Frequency correction, Doppler effect

Dithering the feedback divider achieves a fine frequency resolution.

Noise shaping the dither with a modulator, facilitates its suppression later by a low pass filter.

Linear PFD/CP are used to avoid shaped noise folding back in band.

IIOT – Synthesizer Phase Noise

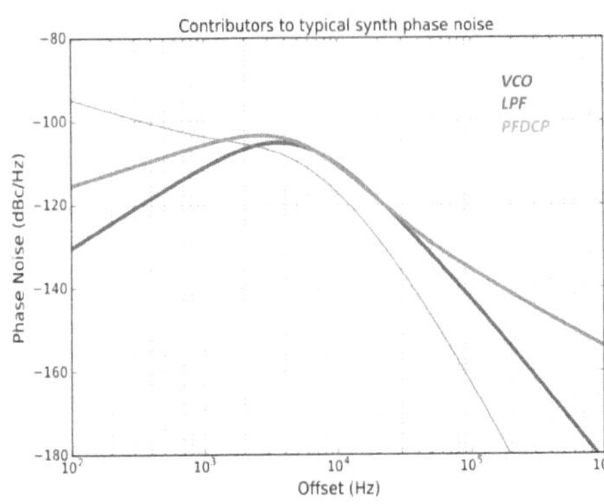

- VCO noise above loop bandwidth
- PDF/CP noise below loop bandwidth
- Optimise for phase error or blocking

Minimum phase error occurs for a bandwidth where the PFD/CP noise equals the VCO noise.

A better blocking can be achieved using a lower loop filter bandwidth to reduce the filter and PFD/CP noise just above the loop bandwidth.

Loop filter bandwidth must be low enough to suppress sigma delta modulator noise.

IIOT – Synthesizer Spurs

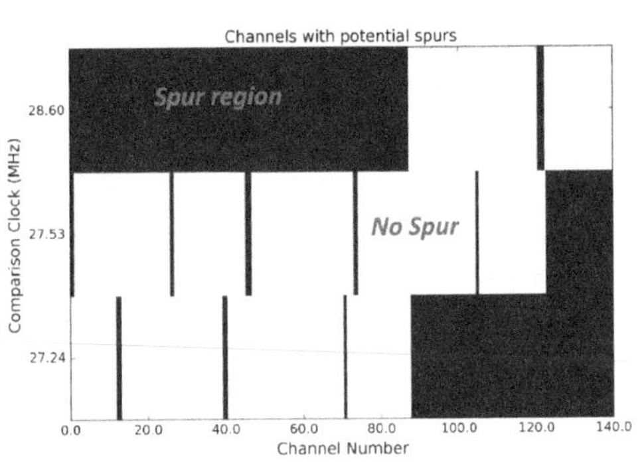

- Avoid harmonics of comparison clock Fcmp near harmonics of the LO
    - SDM ≠ 0.001, 0.5001, 0.25001 etc.
- Change reference/comparison clock frequency
    - Integer-N PLL

The LO is variable and if it falls near harmonics of other clocks in the system, there is the possibility of a spur resulting via a mixing process. These spurs can be hard to simulate, and a frequency plan can be employed to avoid them

FM modulation of the VCO when a harmonic of Fcmp falls near VCO frequency, often at twice the LO frequency.

Phase modulation of the PFD/CP by the VCO gives a spur frequency within the loop bandwidth.

Harmonics of Fcmp falling near the odd harmonics of the LO must be avoided.

If the PFD/CP is non linear the sigma-delta noise can fold back in band.

IIOT – Synthesizer VCO

23

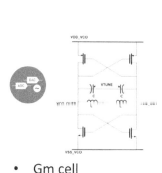

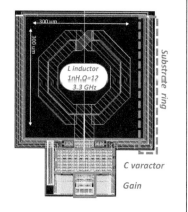

- Gm cell
- Noise
- I/Q

Vtune is a sensitive node. Accuracy of L depends of the PDK (PDK is not accurate for 5 GHz).

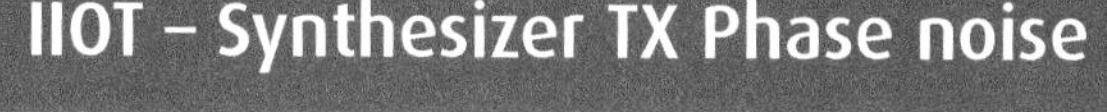

IIOT – Synthesizer TX Phase noise

24

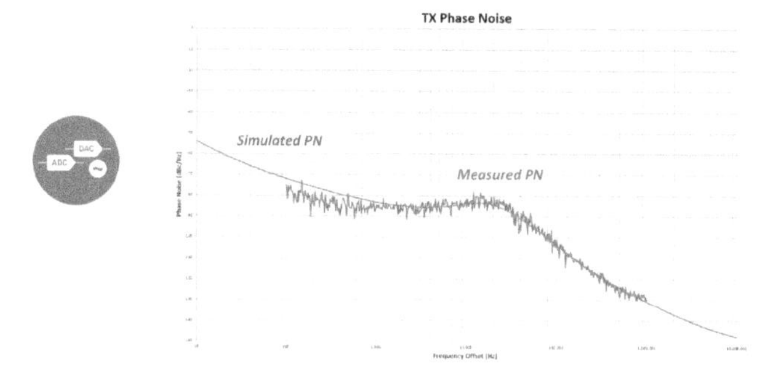

- Noise
  - Simulation method
  - Noise model accuracy
- I/Q

Flicker and thermal noise measured with open loop VCO. Settings used for simulation:

Fref = 26MHz;

N = 2x 63.46 (VCO is running at 3.3 GHz, LO=1.65 GHz).

Simulation was performed using an AC testbench representing PLL components in a closed loop. Similar testbench but in open loop mode is also used for stability purposes.

Vdc and Idc components are used to represent individual blocks.

Noise is defined in csv format in the properties Vdc and Idc components.

25

Synthesizer phase noise is measured in digital domain after the ADC via reciprocal mixing by inputting a large blocker and varying its offset from channel.

IIOT – Synthesizer RX Phase noise

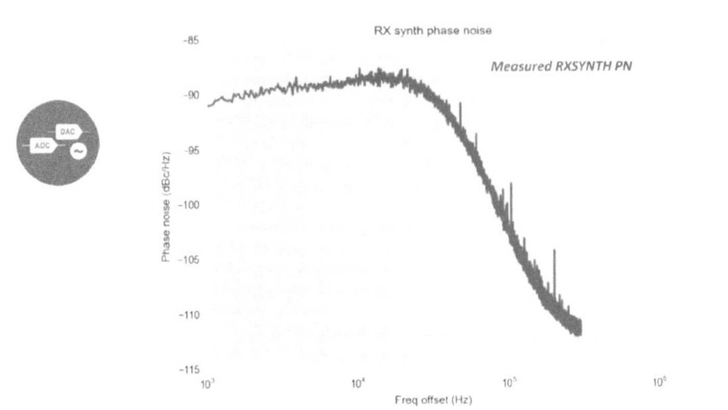

- RX Synth Phase Noise measured via reciprocal mixing of large input blocker and sweeping blocker offset

26

Sensitive nodes: Vtune, VCO_OUT and ground.

It is possible to model the substrate network with SNA (Substrate Network Analysis).

The spurs are dependent of the substrate network around VCO.

A low ohmic connection from the substrate guardring is seen to be important for spur reduction.

Without substrate network spur is around -100 dBc.

IIOT – Synthesizer Spur Measure

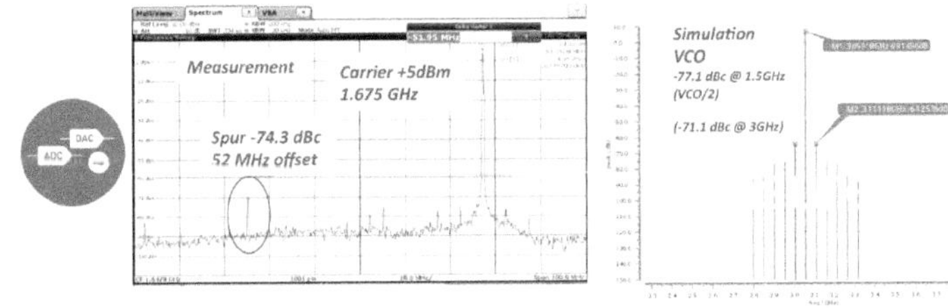

- Spurs: substrate FM of VCO
  - Substrate NW simulation
  - Cadence SNA & IC Process Stack
  - Prediction of 52MHz spur
  - Low Substrate Guardring metal resistance :

Substrate Guard ring metal resistance [Ohm]

Spur [dBc]

IIOT – Substrate Shielding

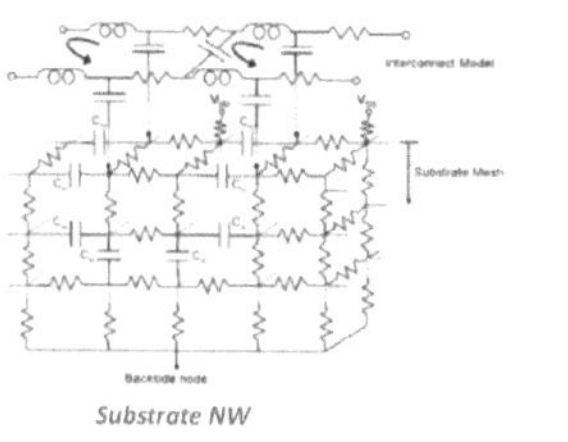

- Spurs
- DNWELL shields lower frequencies
- But not Higher harmonics of clocks
- Low Substrate Guardring metal resistance can reduce spur

Substrate shielding is dependent on resistance, it should have low resistance including also the bondwire resistance. Higher harmonics of low frequency clocks are not shielded.

IIOT – Shielding

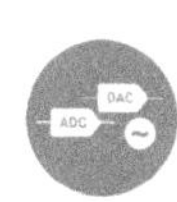

- 4 Synths
- On at same time FDD
- RX signal nVolts
- TX spurs

Substrate shielding is performed using empirical data, guidelines and simulation. RX signal is -130dBm at RFIC, approximately 100 nV.

IIOT – Shielding and Coupling

Simulation of mutual inductance (M) of two coupled inductors based on 3D field.

The closed formula is accurate and can be used instead of numerical simulation.

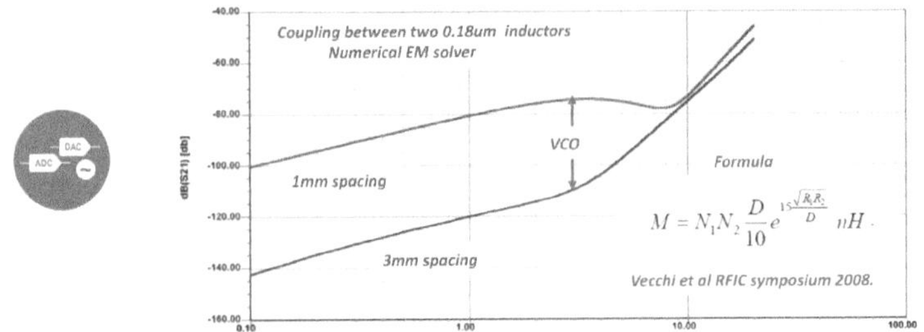

- Coupling between inductors on RFIC
- Reduce interference by
  - Spacing
  - Offset in frequency of operation (5%)
- Numerical Solution and analytic formula agree

30

IIOT – Circuitry Energy

Representation of power domains related with synthesizer and clocks. The use of OTP to store calibration data can save power at start up.

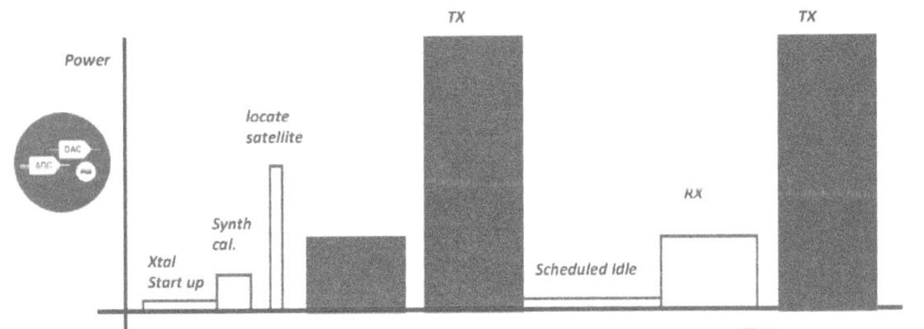

- 30 Power Domains
  - Isolation of RX from TX, clock domains
  - Isolation analog IF from RF, and digital
  - Power Monitor for TX spectrum
- Idle power saving
- Not ULP , TXR consumes power to transmit

IIOT – B.O.M.

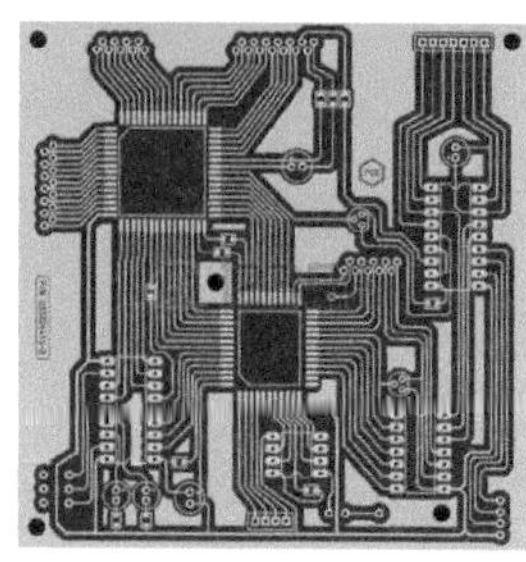

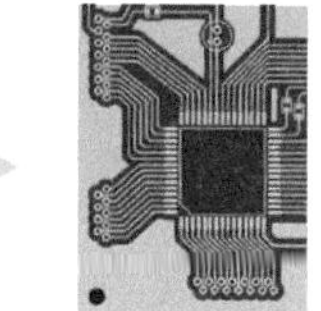

- Bill Of Materials (BOM)
- Cost reduction
  - From $100 to $10 order of magnitude
- Reliability
  - PCB reduced from 1000 components to 100 order of magnitude

Reducing costs by reducing B.O.M by integrating many of the components on chip.

IIOT – IC Process Cost

32

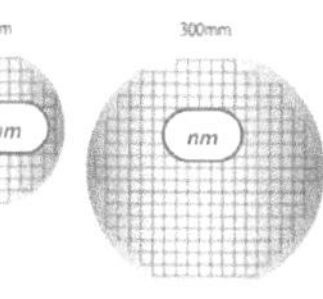

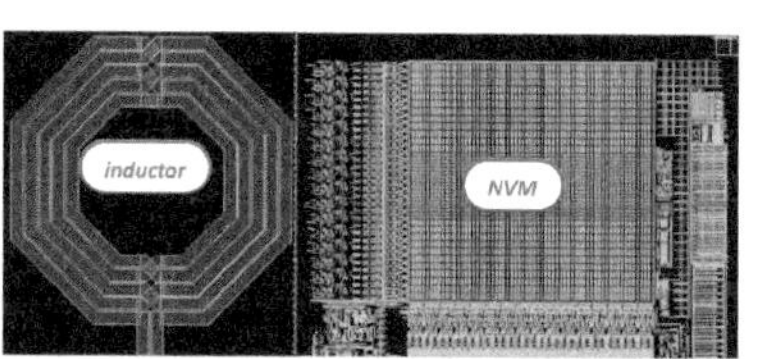

- IC Process choice :
  - IOT 'More than Moore'
  - Costs 200mm vs 300mm processes:
    - Masks ~5 times cheaper, silicon die ~25% cheaper
- Features
  - Analog , RF inductor/varactor, Low leakage, Shielding Options, models
- Calibration
  - Non Volatile Memory

IC process cost also depends the features provided by the foundry (RF options, low leakage which is important for IOT , etc...).

33 

Second Industrial IOT example: Valve controller.

Industrial IOT – Valve Controller

34 

Industrial IOT – Valve Controller

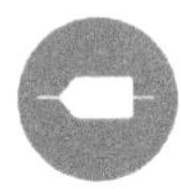

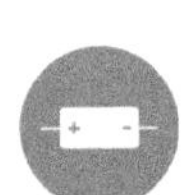

- Sense & Measure - pressure/motor/temperature
- Process – Dual MCU
- Interface – 4-20 mA, HART, MAU
- Analyse – predictive maintenance/fault reporting

Common features of IOT for this valve controller application:

Measurement of a variety of inputs depending on what is being controlled, such as:

- Motor phase/frequency;
- Pressure;
- Temperature;
- Voltages on board.

Processing:

- On board microcontroller for feedback loop operation and status monitoring.

Interfaces:

- MAU – common media access unit (MAU) for Profibus or Fieldbus interfaces
- HART – Two tone FSK modulation with 4-20 mA current loop at 1200 baud rate.
- 4-20 mA supply current with valve position controlled based on current supplied.

Analyse:

- Monitoring motor vibrations to predict when maintenance may be necessary;
- Customer claims 70% reduction in Opex through predictive maintenance rather than just maintenance on fixed schedule.

IIOT –Valve Controller Topics

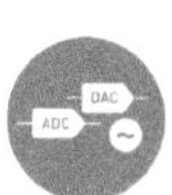

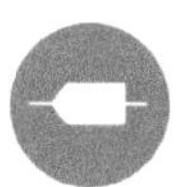

Architecture Circuitry Measurement Energy Interface Economics

- System Requirements
- RFIC Architecture
- Cost impact
- RF Circuit details and measurements

Valve controller topics to be further discussed.

IIOT – Cloud

- Analysis of data can lead to valuable information
- Reduce maintenance costs by as much as 30%
- Find misplaced assets
- Leverage information from across plant to see higher level correlations

Reducing maintenance costs is a hot topic.

Keeping records of actual usage patterns of pieces of equipment, vibration from pump and valves, can be useful. These records can be correlated to show when a device is beginning to show wear and ear and when a maintenance service should be done.

Servicing costs can be substantially reduced by not servicing equipment that does not need it – 30% reduction in service cost which for a large plant can correspond to several millions of dollars per year.

Locating plant using GPS and beacons can also save significant costs. Oddly not knowing where plant is located is a common problem in big companies, e.g., communications companies that do not know where specific routers/switches are located – that information is somewhere on a spreadsheet that never gets updated.

IIOT – Valve Controller System 1

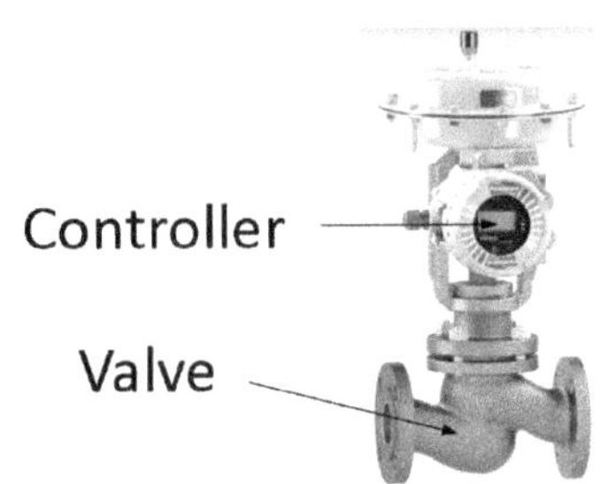

- Need to *accurately* control the position of a valve
- Need to be intrinsically safe
- Need to reduce costs from current BOM

The position of the gas pipeline valve must be accurately controlled and must involve less than an amount of electrical energy that could cause an explosion.

38

IIOT – Valve Controller System 2

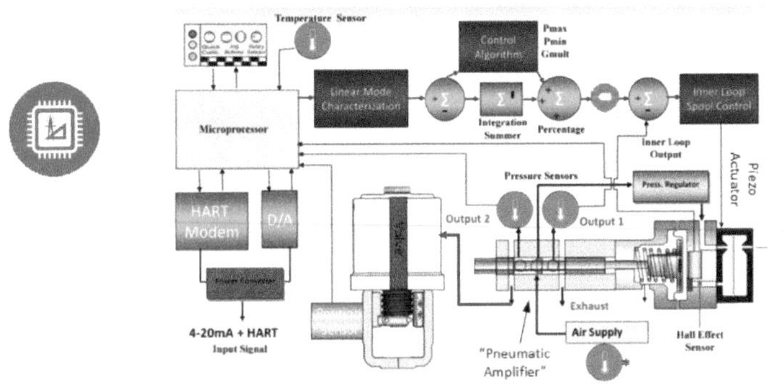

- Piezo membrane controls pneumatic system
- Need to monitor a variety of pressures and temperatures

Valve controlled via pneumatic power.

Piston assembly is used to control how much pneumatic power is sent to the valve. Piston position is in turn controlled by smaller pneumatic flow regulated using a piezo membrane.

The voltage applied to the piezo membrane is controlled from the microcontroller.

Measurements of pressure are taken at various points in piston assembly. Position sensor is also used to measure how open the valve is at the moment. All these measurements serve as input to the control algorithm used to control the valve to the desired set point.

IIOT – Process Choice

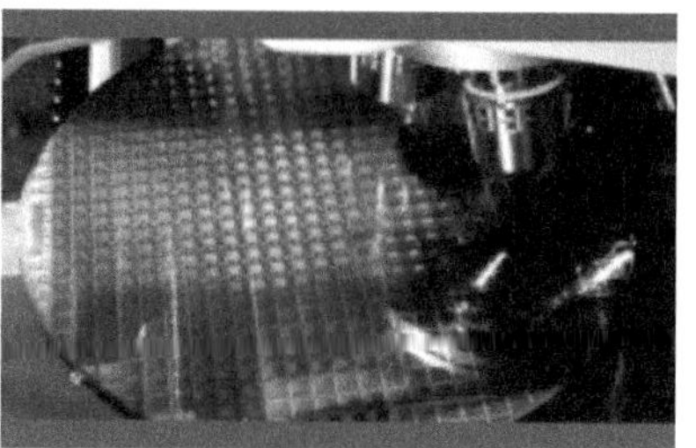

- Process choice
  - Power
    - 0.18 ell – 98% reduction in leakage power from 0.18 G
  - Flash
  - Cost

The process choice is based on the following principles on the lowest cost while maintaining a power footprint within the limits imposed by the 4-20 mA current loop power supply. Lowest cost is meant per chip when including the NRE for manufacture (mask costs) and also other extra costs if the process used does not include an on board flash.

To simplify the decision, it can be decided early on to only go with a process that could support flash and to limit only to a foundry, e. g., TSMC.

The 180 ell process node is a good compromise for cost (particularly in a lower volume design) while maintaining good leakage current performance. It also supports flash and is a popular node that has a lot of pre-existing IP (costs of porting or reinventing IP can be reduced for this node).

IIOT – System Architecture Valve

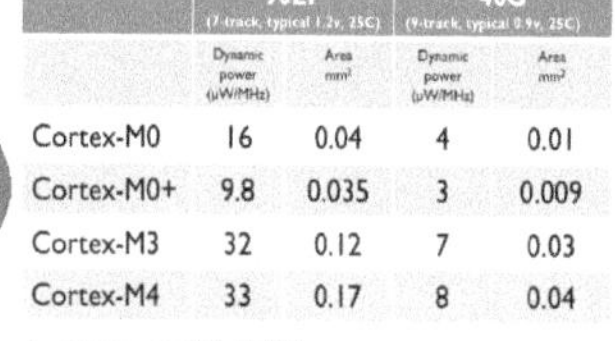

	90LP (7-track, typical 1.2v, 25C)		40G (9-track, typical 0.9v, 25C)	
	Dynamic power (µW/MHz)	Area mm²	Dynamic power (µW/MHz)	Area mm²
Cortex-M0	16	0.04	4	0.01
Cortex-M0+	9.8	0.035	3	0.009
Cortex-M3	32	0.12	7	0.03
Cortex-M4	33	0.17	8	0.04

- Static power <0.7 µW/MHz

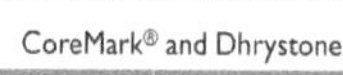

CoreMark® and Dhrystone

Dhrystone (official)	Dhrystone (max options)	CoreMark
DMIPS/MHz	DMIPS/MHz	CoreMark/MHz
0.84	1.21	2.33
0.94	1.31	2.42
1.25	1.89	3.32
1.25	1.95	3.40

* CoreMark data from ARM website & CoreMark.org website

Cortex-M0 Base usable configuration includes 1 IRQ + NMI, excludes debug
Cortex-M0+ Base usable configuration includes 1 IRQ + NMI, excludes debug
Cortex-M3 Base usable configuration includes 1 IRQ + NMI, excludes ETM, MPU and debug
Cortex-M4 Base usable configuration includes DSP extensions, 1 IRQ + NMI, excludes ETM, MPU, FPU and debug

ARM® Cortex®-M4: Nested Vectored Interrupt Controller; Wake Up Interrupt Controller Interface; CPU (with DSP Extensions); FPU; Code Interface; Memory Protection Unit; SRAM & Peripheral Interface; Bus Matrix; Data Watchpoint Flash Patch & Breakpoint; Debug Access Port; ITM Trace; ETM Trace; Serial Wire Viewer, Trace Port

- MCU Choice
  - A R M
    - Application
    - Realtime
    - Microcontroller
- ARM Cortex M4 Selected

Many different features are preponderant in the choice of a processor but the key ones for this particular case are summarized: Ecosystem: could they get support particularly around software development in the case of the provider is no longer available. It is also needed to meet a certain level of performance (MIPs).

Security might have swayed some decision making, but in this case is not a big concern for the product since it is to be used in a closed ecosystem and therefore (in theory) it needs no security.

(ARM's announced last year their new microcontroller architecture that builds in security from the ground up).

Power is the big driver for many choices in the chip, but power for CPU is not much worse than power from SRAMs and Flash and other peripherals, and so even big increases in power efficiency would not necessarily mean big power improvements over all.

IIOT – Memory

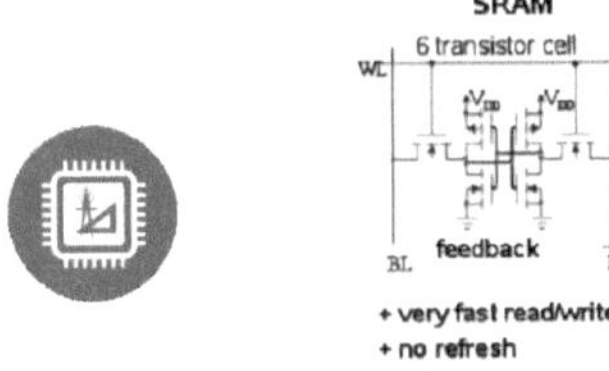

DRAM
1 transistor- / 1 capacitor cell
Stored charge
WL
BL
+ fast read/write
+ small cell
- volatile
- complicated technology

FLASH
1 transistor cell
Stored charge
WL
BL
+ nonvolatile
+ very small cell
- slow write/erase
- only block erase
- limited endurance

- Memory Model
  - On-chip Flash
  - SRAM

It can have memory on chip, off chip or a combination of both.

Key decisions are around:

- Size of memory - has a cost associated to put it on chip;
- Availability of memory types in chosen process (Flash);
- Memory bandwidth - off chip memory is often slower;
- Memory is often defined by the processing power of processor;
- Memory that is as fast as possible (as close to single cycle as possible).

Some memory as working memory is necessary (SRAM). Other non-volatile program memory is also necessary, such as:

- SRAM loaded from an external non volatile memory store;
- Flash co-packaged with the chip; On chip Flash.

IIOT – Architecture

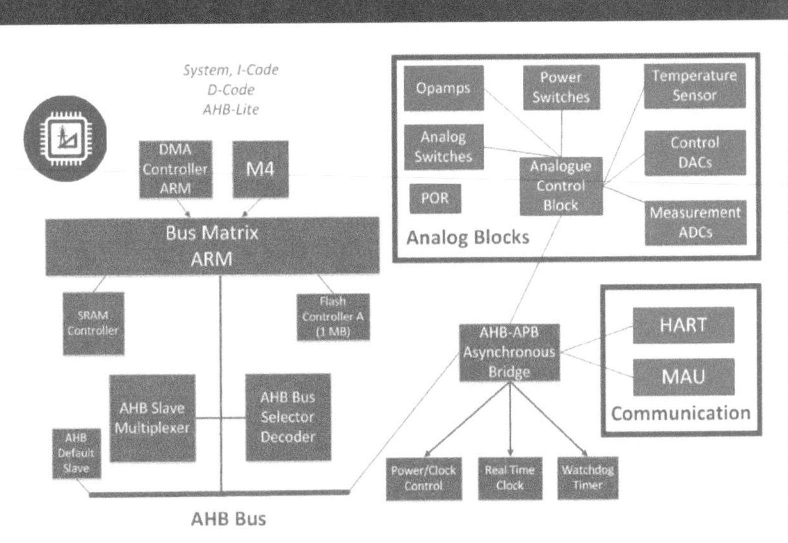

Simple overview of the ARM architecture.

ARM uses a high performance bus system (AHB).

It offers a variety of analog blocks, and HART and MAU interfaces.

IIOT – Measurement Channel

43

1. External sensors to be measured;
2. These can either be measured directly by the ADC or through a programmable gain amplifier (with chopping to reduce 1/f noise);
3. Low Noise Voltage Reference;
4. ADC

SAR ADCs are more advantageous than Sigma-Delta converters:

- Smaller latency
- Multiplexing of sensors is less complex

A typical measurement channel is shown which consists of:

IIOT - ADC Specifications

44

- Resolution: 14 bits
- Sampling Frequency: 200kHz
- Single-Ended or Differential Input
- Vref Range = [1.25V, 3.0V]
- Input Range: Vref pp (Single-Ended and Diff.)
- 16:1 Analog Input MUX
- DNL < 0.9 LSB, INL < 4.0 LSB
- ENOB: 11.9 bit (@Vref=2.4V)
- Sleep mode w/ fast wake-up time

ADC is used to digitize the output of various sensors. Sensors output signals are in the range of DC – 500 Hz bandwidth.

ADC is also used as a 12-bit ADC for HART Rx path, with input signal at 26.4 kHz. Some sensors output signals are single-ended, others are differential.

ADC specifications are driven by the customer.

14-bit resolution / 12-bit ENOB:

Defines the precision of the measurement, and therefore of the control over the valve position.

DNL < 0.9 LSB, INL < 4.0 LSB:

With DNL < 1 LSB, monotonicity of the ADC transfer function is guaranteed, which is desirable to avoid oscillation of the control loop.

Fs = 200kS/s, Single-Ended or Differential Inputs:

Vref range = [1.25V, 3.0V]:

Some sensors requires the same Vref to be applied to the sensor and to the ADC in order to cancel out gain/offset error in the measurement. Various types of sensors usually means having various reference voltages.

Internal reference voltages generated on-chip = 1.25V and 2.4V. Other reference voltages are generated off-chip. Vdd=3.0V can also be used as reference voltage.

45 IIOT - SAR ADC Architecture (1)

- DAC non-linearity calibration: sizing for kT/C noise rather than capacitor mismatch.
- Comparator offset calibration: needed for DAC calibration, reduce overall ADC offset error
- DAC redundancy: dynamic errors during first bit decisions can be digitally corrected

Capacitive-DAC non-linearity calibration:

This allows to size DAC for kT/C noise rather than mismatch. Total DAC capacitance can be significantly reduced which in turn reduces the area.

This also leads to significant reduction in power consumption. DAC switches can be scaled down (logic power), the reference buffer and ADC input buffer power can also be reduced.

DAC layout design effort is significantly reduced because parasitic and mismatch effect are calibrated, and therefore no specific matching layout technique is needed.

Comparator offset calibration:

DAC non-linearity error is measured using the comparator, therefore the comparator offset needs to be calibrated prior to DAC calibration.

Comparator can be sized for noise and speed rather than for low-offset, resulting in power and area reduction.

Both calibrations are performed in foreground (at start-up) and are robust to voltage supply and temperature variation.

DAC redundancy:

- Dynamic errors (e.g., DAC settling error, comparator decision error) during first bit decisions can be digitally corrected by subsequent bit decisions.
- Relax DAC settling error requirement and comparator noise during first bit decision.

IIOT - SAR ADC Architecture (2)

46

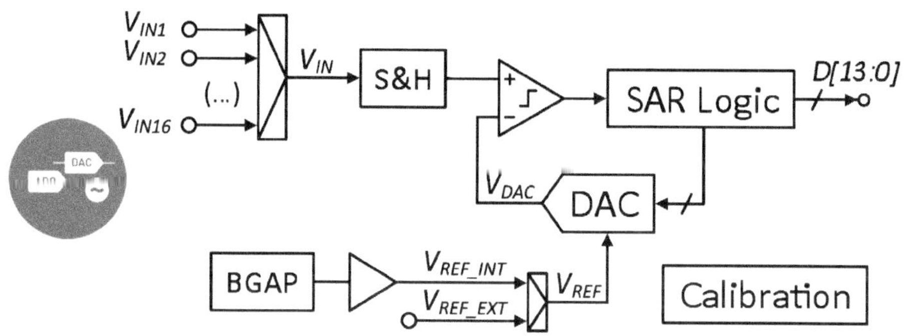

- Configurable Single-Ended or Diff. Inputs
- 16:1 input MUX: using CMOS switches
- Reference circuit requirements:
  - low-noise, low-power w/ fast wake-up time
  - settling error is relaxed by DAC redundancy

Configurable single-ended or differential inputs:

Single-ended mode is implemented using unipolar pseudo-differential sampling scheme(i.e., VINP = signal and VINN = signal ground reference).

Requires small changes to DAC switching sequence and calibration logic.

16:1 input MUX:

Low-frequency input signal (BW=500Hz) can be considered as a DC input, switches on-resistance variation over Vin range is not an issue as long as there is a sufficient time to settle Vin onto the sampling capacitor.

Number of settling time-constants required to settle to within 0.5 LSB :

Ton / = ln(2^(N+1)), where Ton is the time during which the sampling switch is on.

= R*C where R is the total series resistance from the ADC input source to the ADC sampling capacitor C, N is the ADC resolution.

Reference circuit requirements:

Low-noise: 14-bit ADC working at Fs=200kS/s, so both flicker and thermal noise need to be reduced.

Low-power with fast wake-up time: in customer's application, ADC is to be used intermittently (i.e., ADC is off most of the time), and from time-to-time, ADC is turned-on and convert a few samples, then back to power- down mode. To save power, ADC and reference voltages must have a quick wake up from power-down mode.

Low-noise and low-power contradicts with fast wake up time, i.e., a external capacitor can not be simply connected to Vref.

IIOT - ADC Noise & Power Breakdown

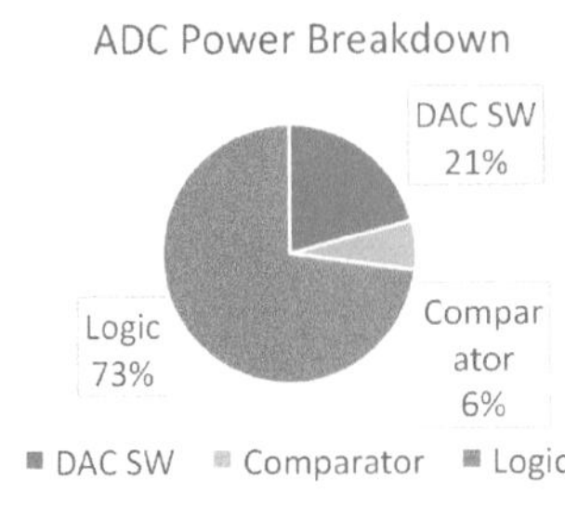

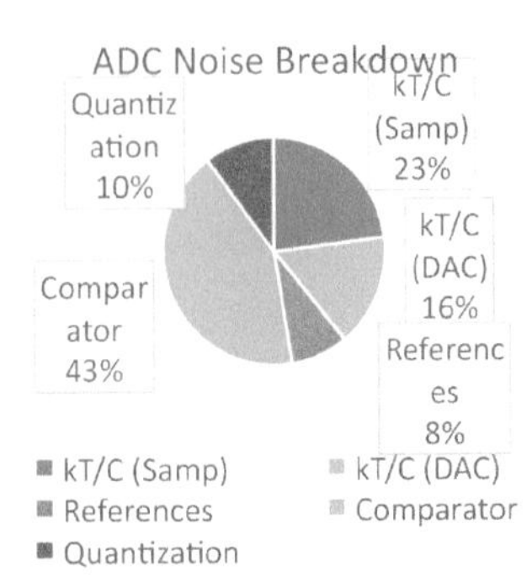

- Total Power = 160uW at 200kS/s (wo/ ref. power)
- Total Noise = 133uVrms

ADC logic power dominates the power consumption for following reasons:

Logic was implemented using standard cells to reduce design effort and facilitate future porting, and not optimized for power.

Logic is working with 1.8 V, and DAC switches with 3.0 V, therefore, level shifters are needed on each DAC.

Technology - CMOS 0.18um.

Total Power from all supplies (3.3V and 1.8V) is 160uW at 200kS/s. Power scales linearly with sampling rate.

Total Noise of 133 uVrms yields 12.18-bit ENOB considering 2.4 Vpp input range. Adding distortion plus some margin gives 11.9-bit ENOB (as in specification).

48

IIOT - ADC Layout

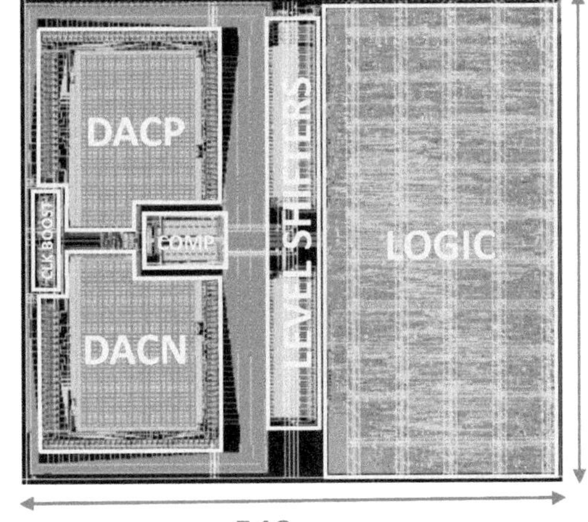

ADC area is 0.243mm2 (540um x 450um).

Not shown in the picture are:

- The bandgap - area is 0.044mm2 (177um x 245um).
- The reference buffer - area is 0.07mm2 (410um x 170um).
- The 16:1 analog input multiplexer - area is 1.6mm2 (84um x 960um x 2). Has a large area due to ESD protection.

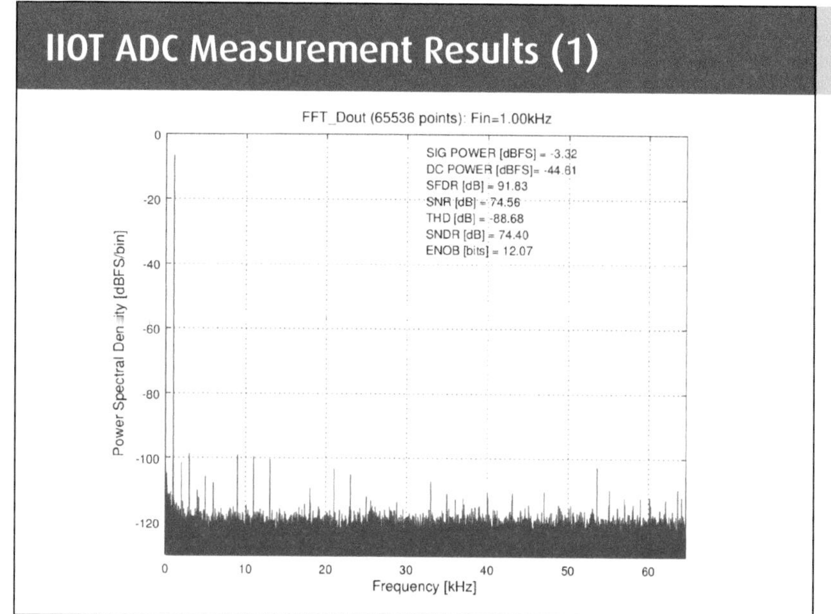

The following measurement results presented (FFT and INL/DNL) are for single-ended input mode.

Fs is aproximately 130kS/s in the FFT plot due to test setup, but internally the ADC is working at full speed, i.e., 200kS/s.

Input signal amplitude is -0.3 dBFS but appears at -3.3 dBFS in the FFT plot. This is due to blackman harris window function applied to the ADC output before computing FFT. Blackman harris window was used because Fin is not coherent to Fs due to limitations of the test setup.

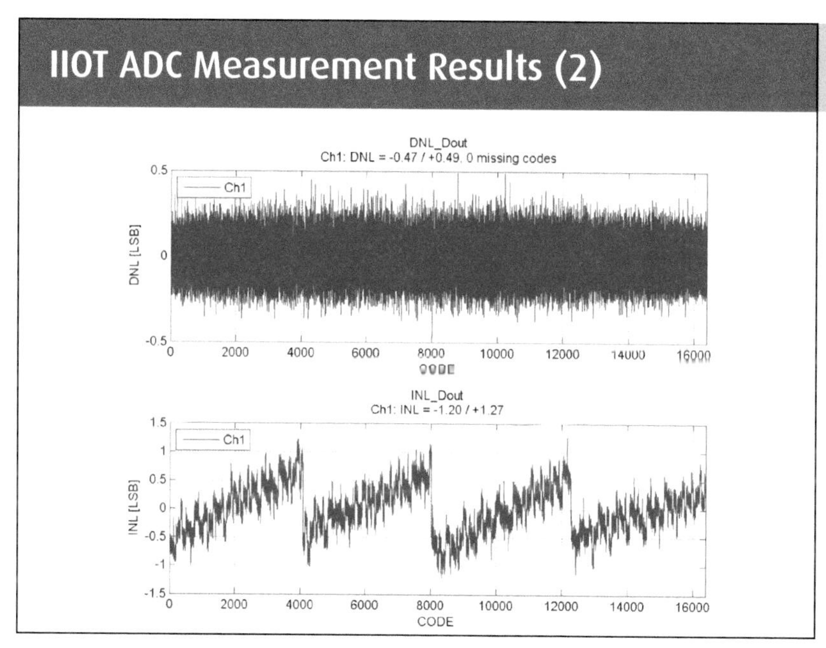

ADC DNL and INL measurements.

IIOT – ADC Low Noise Reference

The reference circuit has a lower power mode for normal operation and a higher power mode for adc operation. The transition from lower to higher power mode is achieved quickly.

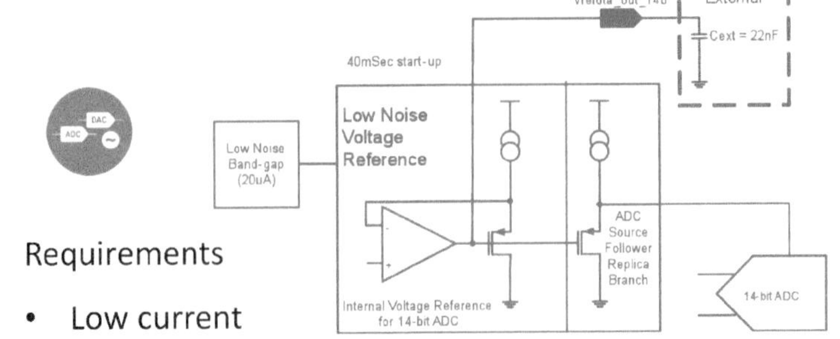

Requirements

- Low current
- Band gap flicker noise < 60uVrms
- Thermal noise <100uVrms
- External filter cap, slow start-up

IIOT – PGA

Programmable gain amplifier with chopping.

It is shown a representation of the spectrum at different circuit nodes, showing the chopping operation.

The flicker noise noise is up-converted to the chopping frequency and then attenuated by an output filter.

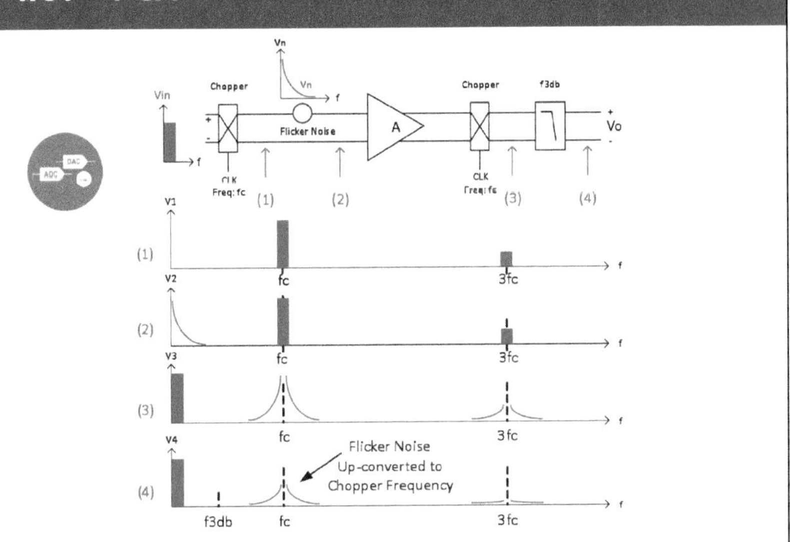

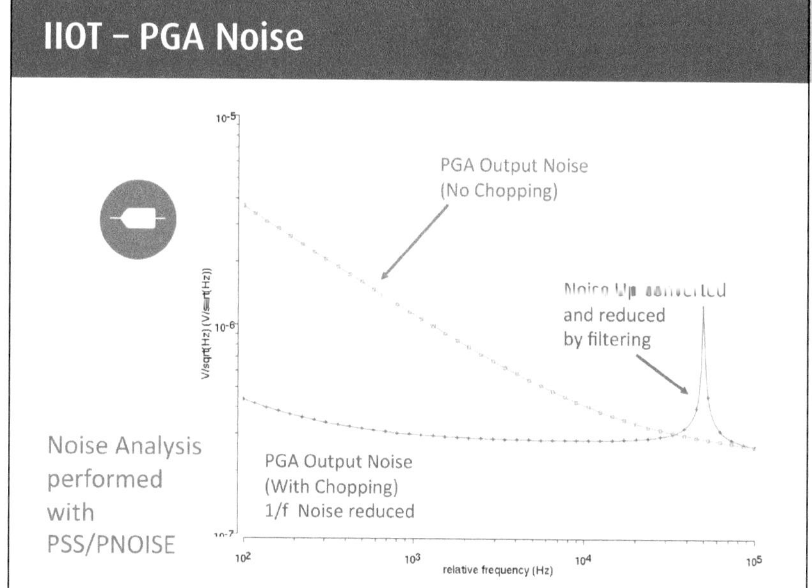

53

Simulation of PGA noise using PSS/PNOISE analysis.

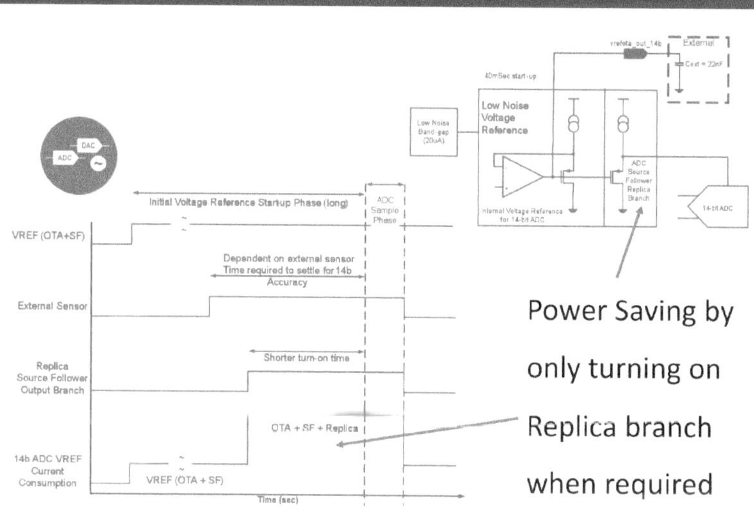

54

Operation of the reference circuit in lower and higher power modes.

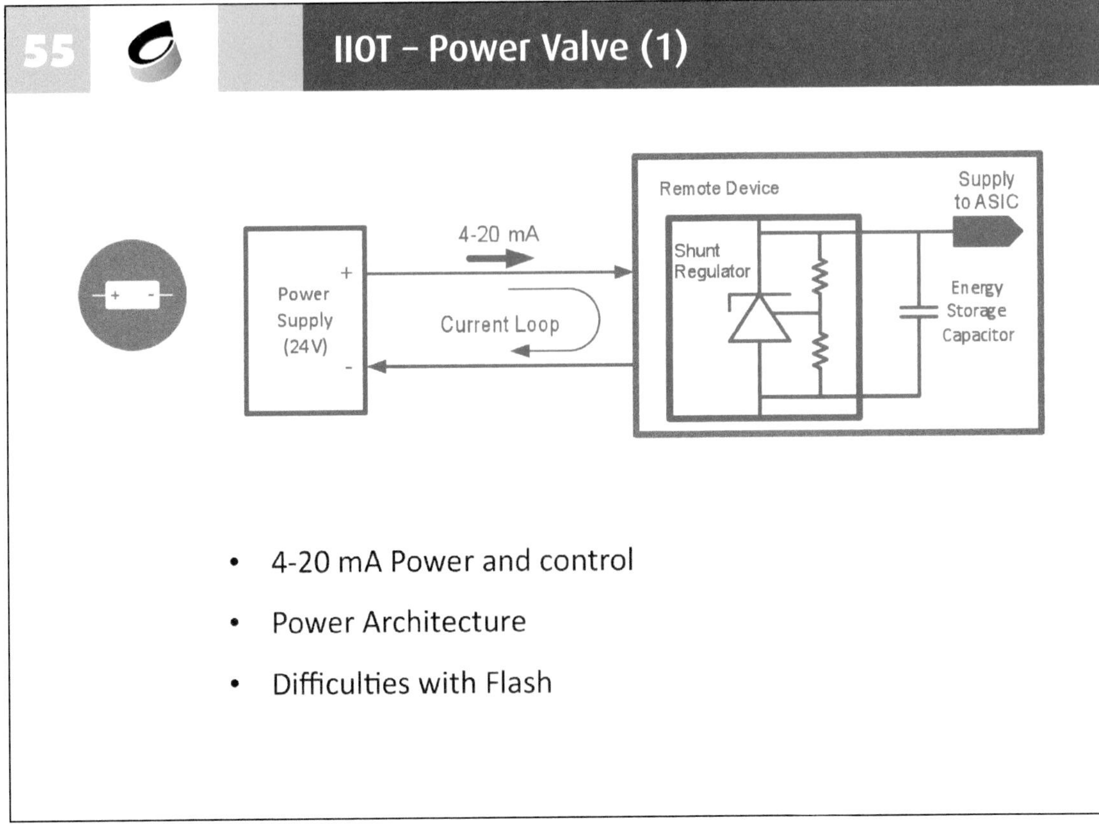

4-20 mA current used to control the valve position. Small power footprint for entire product

Some tasks take more instantaneous power than is available. However, measurements are not done frequently, so it is imperative to store power in a battery such an array of capacitors and then burst to high power for short periods to do measurements.

The peak power and static power consumption need to be reduced so that there is enough energy available in the capacitors when needed.

Power architecture details:

- Current from HART bus is used to charge capacitors. Stored energy is used to power local devices.
- The overall amount of charge is limited to what can be stored (intrinsic safety).
- Robust power-n-reset circuitry is necessary.

Flash memory:

- Flash memory needs a large current to program.
- Uses internal charge pumps to get the high voltage necessary for operation, but still draws a high current.
- Need enough energy to sustain communications to get programmed and also to program to shadow
- Flash which is switched to when download is complete and verified. Reprogramming the device can therefore take many hours.

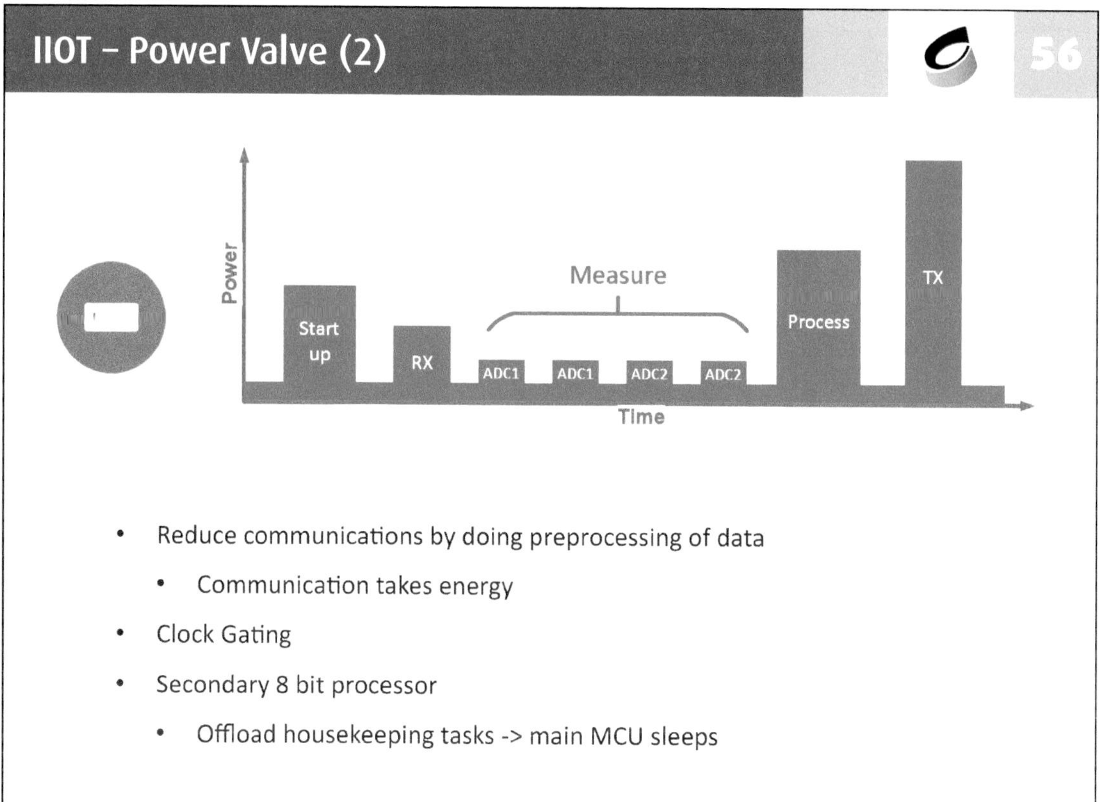

IIOT – Power Valve (2)
56
Power
Measure
Start up
RX
ADC1
ADC1
ADC2
ADC2
Process
TX
Time
Reduce communications by doing preprocessing of data
Communication takes energy
Clock Gating
Secondary 8 bit processor
Offload housekeeping tasks -> main MCU sleeps

Clock Gating:

- SoC is divided into many different clock domains.
- Allows fine grained control for power consumption.
- Clock speed is also configurable between 1 MHz and 40 MHz.
- Can be switched on the fly so during sleep mode clocks can be run slower.
- SRAM and Flash divided into blocks that can be individually power gated to allow the control of power footprint.

Secondary 8-bit processor:

- Some status monitoring and low level measurements are always running in background.
- Power footprint is minimized by putting the main processor and most of logic into sleep mode and allowing small secondary processor with dedicated (small) resources (memory) to do the tasks.

57 

IIOT – Valve Interfaces – HART

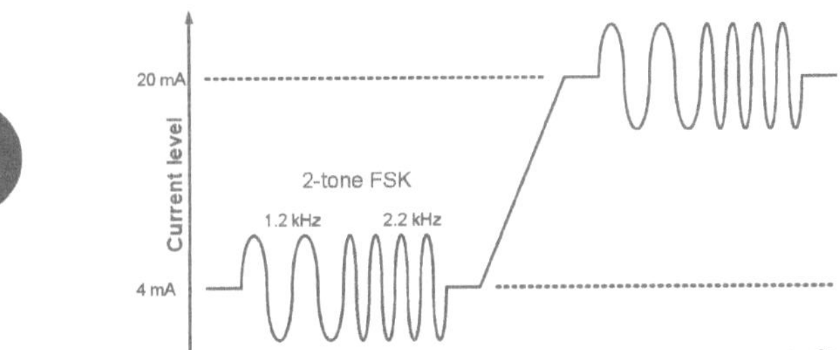

- 4-20 mA
- HART
  - 2-tone FSK communications protocol (1.2k Baud)
  - 1.2 kHz and 2.2 kHz modulated onto line

4-20 mA:
- Used to power the device.
- Current level is also used to control the position of the valve. Very simple but there is no return path.
- Give information if something is wrong with the device.

HART:
- Two-tone FSK communications protocol.
- Operates tones at 1.2 KHz and 2.2 KHz on 4-20 mA line. Half-Duplex (only sending or receiving).
- Baud Rate is 1.2 Kbaud.
- Sufficient for monitoring the device but is slow to do software updates.
- Not very well defined standard.
- Many products in marketplace claim to be HART compliant but cannot inter-operate.

Industrial protocols use current signalling as it is very immune to interferences. The overall average current per bit of information should be zero.

58 

IIOT – Valve Interfaces - MAU

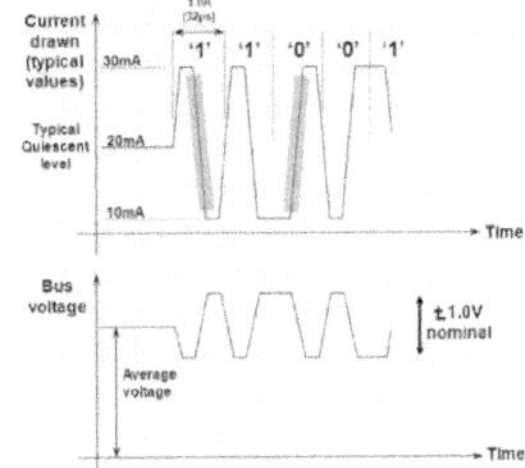

- MAU (Foundation ™ Fieldbus)
  - Provides communication and power
  - 2 wire bus connection
  - 31.25 kbit/s data rate, Manchester encoded

MAU (Media attachment unit):

Much faster protocols are used to control the device (no use of 4-20 mA for control) and also to get better communications with device.

As 4-20 mA is not used for control, it is turned up to full current so that power is not so much of an issue with products that use these interfaces.

IIOT – Economics (1)

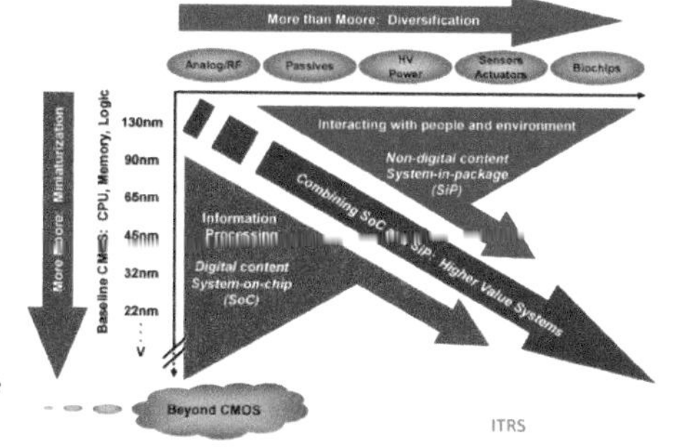

ITRS

- 180 nm is relatively inexpensive
- Chip price can be as low as $10
- Save ~ $100 of COTs chips
    - Integrate as much as possible
    - Including Op-Amps, switches, etc.
- Reduce test and board level costs

International Technology Roadmap for Semiconductors (ITRS) diagram for IOT:

180 nm is relatively inexpensive. Chip price can be as low as $10.

To save costs on chips:

- Integrate as much as possible
- Including Op-Amps, switches, etc.

Reduce test and board level costs.

IIOT – Economics (2)

- Ownership of IC on balance sheet
- An asset as distinct from an expense
- Control of supply chain

Other economical aspects:

Owning chip design means IP on balance sheet.

An asset as distinct from an expense.

Control of supply chain.

CHAPTER 5

Microprocessors/ MCUs for Internet of Things

Leonel Sousa

IST/UL, INESC-ID

Lisbon, Portugal

In the "Microprocessors/MCUs for IoT" chapter, the state of the art architectures of the current MPUs and MCUs will be analyzed, as well as the benchmarks for evaluating their performance and efficiency. Furthermore, the support for interconnection and communication with users, things and cloud services will also be discussed. Examples of commercial MPUs and MCUs will be provided, and the main investigation paths for developing the future processing devices for the IoT will be underlined.

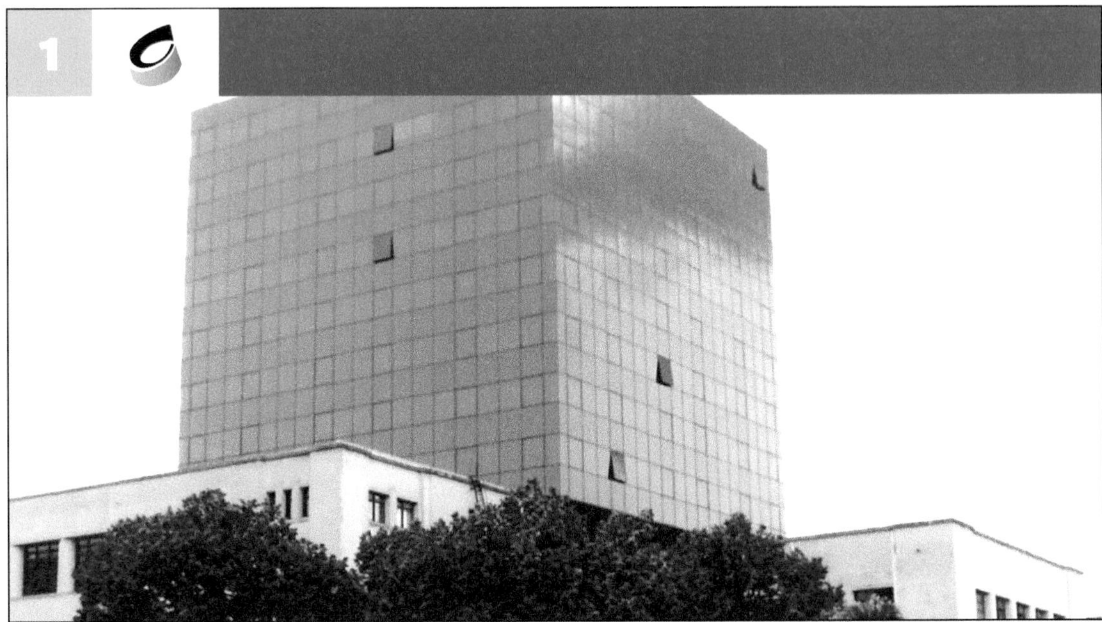

This presentation talks about microprocessors (MPUs) and microcontrollers (MCUs) that can be applied in IoT domain and applications. This is quite challenging because for the last 20 years these devices were designed by focusing on their performance, while "neglecting" power/energy consumption, which is a very important aspect in IoT. This talk will present some devices, most of them available in the market and can be used in IoT, by showing their different characteristics and application domains.

2 

What is Internet of Things (IoT)?

- No single, universal definition
  - *IoT generally refers to scenarios where network connectivity and computing capability extends to objects, sensors and everyday items*
    - exchange and consume data with minimal human intervention
  - Machine-to-Machine (M2M), IoT
    - Billions of interconnected devices - Everybody is connected!

Internet of Things (IoT) is a very hot topic in research and it has no universal definition. However, it generally refers to scenarios where different machines, such as objects, sensors and everyday items, are connected and require minimal human intervention – machine-to-machine interface (M2M).

IoT has many application domains: health, automotive, home, telecommunications, etc. It also connects all these different applications, thus allowing to integrate everything in a huge system.

Projections and Impact

- 2019
  - Cisco: 24 billion Internet–connected objects
- 2020
  - Morgan Stanley: 75 billion networked devices
- 2025 (looking out further)
  - Huawei: 100 billion IoT connections
- and also by 2025
  - McKinsey Global Institute: economical impact of IoT \$3.9 to \$11.1 trillion

Variability in predictions but paint a picture of significant growth and influence

The projections regarding the IoT dissemination are very interesting, revealing that in the next four years there will be about 75 billion devices connected, having a huge impact in our society. These projections are attracting the industry big players (e.g. Intel) to design MCUs for IoT. The predictions for 2025 are even more optimistic, with 100 billions of IoT connects and an economic impact between \$3.9 Trillion to \$11.1 Trillion.

Projections and Impact

- * "Cloud and Mobile Network Traffic Forecast - Visual Networking Index (VNI)." Cisco, 2015. http://cisco.com/c/en/us/solutions/serviceprovider/visual-networking-index-vni/index.html
- ** Danova, Tony. "Morgan Stanley: 75 Billion Devices Will Be Connected To The Internet Of Things By 2020." Business Insider, October 2, 2013. http://www.businessinsider.com/75-billion-devices-will-be-connected-to-the-internet-by-2020-2013-10
- *** "Global Connectivity Index." Huawei Technologies Co., Ltd., 2015. Web. 6 Sept. 2015. http://www.huawei.com/minisite/gci/en/index.html
- **** Manyika, James, Michael Chui, Peter Bisson, Jonathan Woetzel, Richard Dobbs, Jacques Bughin, and Dan Aharon. "The Internet of Things: Mapping the Value Beyond the Hype." McKinsey Global Institute, June 2015.

Some references of projections and predictions of the IoT impact in the world.

Outline

- Requirements and Energy Sources
- Microprocessor/MCU Architectures
- SOC: System on Chips
- Security
- Applications and Future Directions

Talk's outline. A question can be posed: "How to interconnect microprocessors? There are no more standalone microprocessors!"

Outline

7

- Requirements and Energy Sources
- Microprocessor/MCU Architectures
- SOC: System on Chips
- Security
- Applications and Future Directions

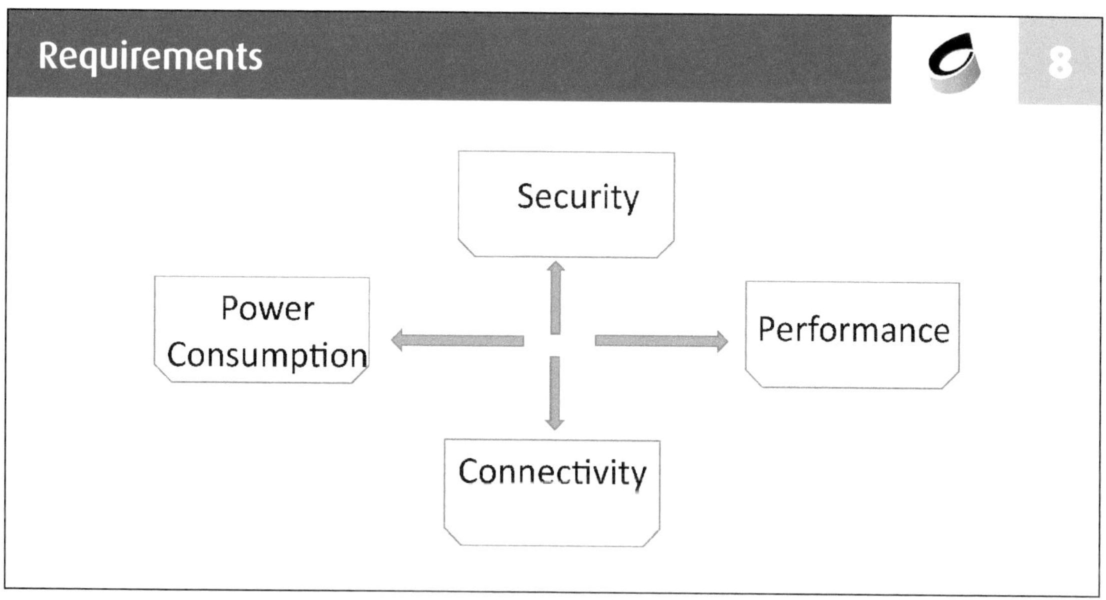

One of the main difficulties when designing MCUs for IoT is that there are four different "forces" that are moving toward different directions: Security, Performance, Connectivity, and Power Consumption. IoT devices must have ultra-low power consumption, the security must be high to avoid possible information leakage (very sensitive data may be sent by the devices, e.g. health data), the devices should be able to connect to the network and send/receive data to/from other devices. Finally, they should be able to process data (e.g. from sensor readings) and to run the stack for wireless communication.

Requirements and Energy Sources

- At heart of IoT systems
  - a processor or microcontroller unit (MCU) processes data (from sensors) and runs software stack interfaced to a wireless device for connectivity

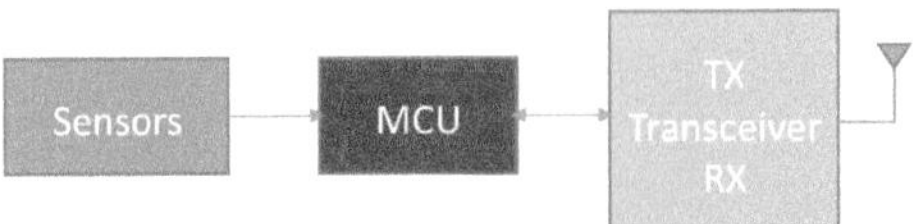

- Transceiver consumes significant amount of the energy used by the system for a single sensor reading (typically 50%!)

There are many different techniques to reduce CU power consumption. However, considering that a device reads data from a single sensor, processes that data with an MCU, and communicates to a network through a transceiver, the transceiver is responsible for about 50% of the power consumption.

10

Requirements and Energy Sources

- Microcontrollers
  - **Integration**: implement a whole design in a chip
  - **Operating Frequency**: lower than CPUS (typically < 200MHz)
  - **Memory**: limited, less than 1MB (typically 16kB)
  - **Power consumption**: Low, battery operations and energy harvesting

Regarding microcontrollers (MCUs), all different blocks, such as sensors, transceiver, memory, should be completely integrated in the same chip (System on Chip – SoC). The operating frequency should be low (tens to hundreds of MHz) to minimize power consumption. The memory is very limited, also to reduce power consumption, which means that the data processing should not be complex. MCUs have a very lean program that just acquires the data and sends it outside. Some adjustments will be detailed.

Requirements and Energy Sources

- Current handheld devices
  - Around 3V lithium-ion (Li-ion) battery, typically 2200mAh
- Energy Harvesting
  - thermoelectric generators convert heat to electricity
  - piezo elements convert mechanical vibration
  - photovoltaics convert sunlight (or any photon source)...

Current handheld devices are usually powered by 3V/2200mAh batteries. The most used energy harvesting sources are thermoelectric (heat to electricity), piezoelectric (mechanical vibration to electricity) and photovoltaic (light to electricity).

Energy Sources

12

ENERGY SOURCES		
Small solar panels	Hundreds of mW/cm^2	Handled electronic devices
Piezoelectric devices	Hundreds of mW/cm^2	Handled electronic devices or remote wireless actuators
Seedbeck devices (body heat)	Tens of mW/cm^2	Remote wireless sensors
RF energy from an antena	Hundreds of pW/cm^2	Remote wireless sensors

- IC characteristics required for inclusion/use in the alternative energy
  - Low standby quiescent currents, typically less than **6 µA** and as low as **450 nA**
  - Low startup voltages, as low as **20 mV**
  - Multiple output capability and autonomous system power management

To use alternative energy sources, it is required to assure that systems have ultra-low power consumption; otherwise it will be impossible to power them with energy harvesting sources. It is especially important to assure low standby currents and low startup voltages, because ideally these systems will be sleeping for most of the time.

13 

Outline

- Requirements and Energy Sources
- Microprocessor/MCU Architectures
- SOC: System on Chips
- Security
- Applications and Future Directions

14 

Microprocessor/MCU Architectures

- Microcontrollers
  - **Integration**: implement a whole design in a chip
  - **Operating Frequency**: lower than CPUS (typically < 200MHz)
  - **Memory**: limited, less than 1MB (typically 16kB)
  - **Power consumption**: Low, battery operations and energy harvesting

The used MCUs architectures can range from 8-bit to 32-bit, and should run a small RF stack. To reduce power consumption, the devices should be powered by a battery and connected to gateways data. Thus, it is possible to significantly reduce these devices' power consumption and improve the overall system's autonomy.

Microprocessor/MCU Architectures

- Current handheld devices
  - Around 3V lithium-ion (Li-ion) battery, typically 2200mAh
- Energy Harvesting
  - thermoelectric generators convert heat to electricity
  - piezo elements convert mechanical vibration
  - photovoltaics convert sunlight (or any photon source)...

The IoT brings new requirements and challenges that are not present in conventional systems. The most important challenges will be discussed in the next slides.

Low Power Techniques for MCUs

ENERGY SOURCES		
Small solar panels	Hundreds of mW/cm^2	Handled electronic devices
Piezoelectric devices	Hundreds of mW/cm^2	Handled electronic devices or remote wireless actuators
Seedbeck devices (body heat)	Tens of mW/cm^2	Remote wireless sensors
RF energy from an antena	Hundreds of pW/cm^2	Remote wireless sensors

- IC characteristics required for inclusion/use in the alternative energy
  - Low standby quiescent currents, typically less than **6 µA** and as low as **450 nA**
  - Low startup voltages, as low as **20 mV**
  - Multiple output capability and autonomous system power management

The largest portion of power consumption of an MCU is dynamic, i.e. due to the switches' transitions, and it is proportional to the operating frequency and to the square of the supply voltage. One possible solution to reduce power consumption is to use clock gating, i.e. disable the clock of specific blocks when they are not performing any operation, thus reducing the overall switches' transitions of the system. By reducing supply voltage, it is also possible to significantly reduce power consumption. However, below certain values, this is not trivial task (e.g. in CMOS).

17 

Low Power Techniques for MCUs

- Several levels of clock gating:
  - **Idle mode** gates the CPU clock domain and the Flash clock domain while the peripherals and interrupt system continue operating.
  - **Power Down mode** gates all clock domains, only enabling asynchronous operation; external oscillator is also stopped
  - **Standby mode** is the same as Power Down mode except that the main oscillator is kept running

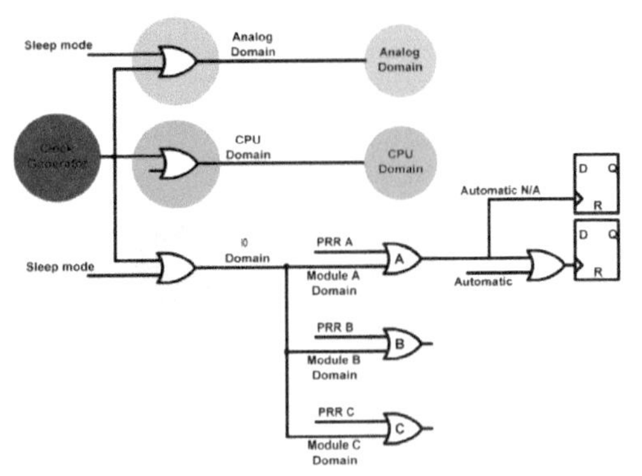

It is possible to have different clock gating levels in a MCU, according to its operating mode.

18 

Low Power Techniques for MCUs

- Automatic clock gating (or multi-level clock gating)
  - clock enabled only when an update is required (i.e. when a value is changed)
  - Rather than updating the register with the previous state, the clock to the register is gated.

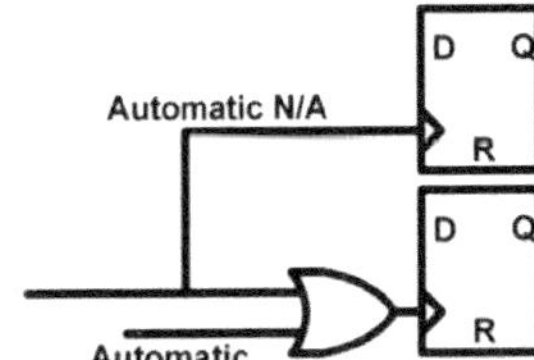

Low Power Techniques for MCUs

19

- The only circuits needed to be on during a **power down** mode are the *reset* and the *wake up* circuits.
  - This should reduce the power consumption close to zero during the power off mode, since there are no oscillators and clock sources on, just some simple logic blocks.

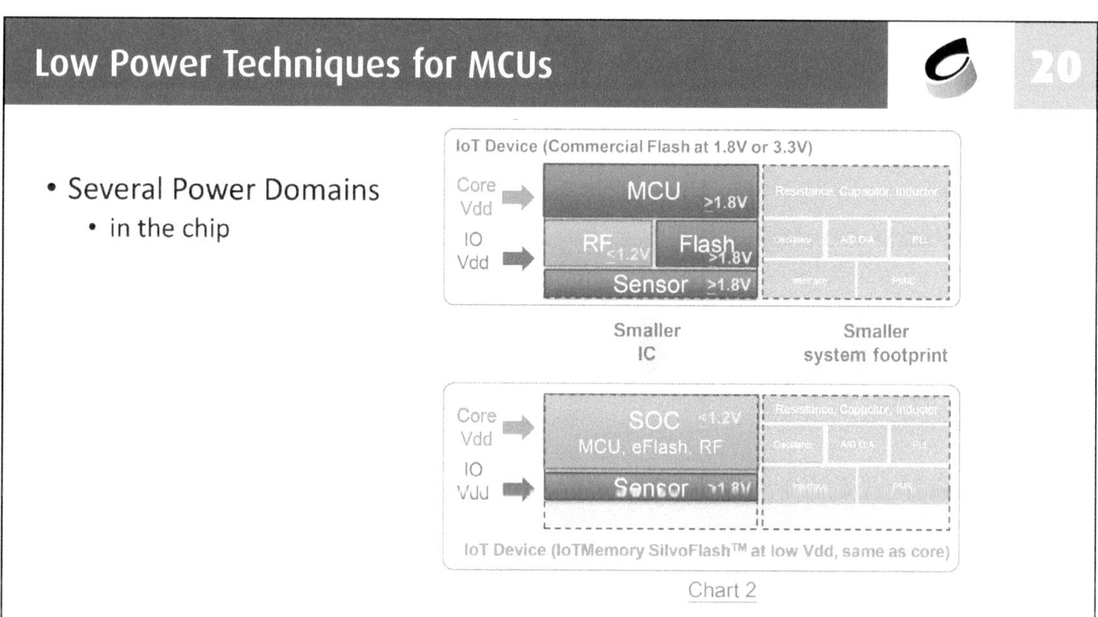

Chart 2

In typical SoCs for IoT there are different power domains. For instance, the MCU, flash and sensor blocks are supplied by 1.8V and the RF block is supplied by 1.2V. By making the IC more compact and designing all blocks to be in the same power domain, e.g. all blocks supplied by 1.2V, it is possible to reduce the system supply voltage and to improve the clock gating techniques, further decreasing power consumption.

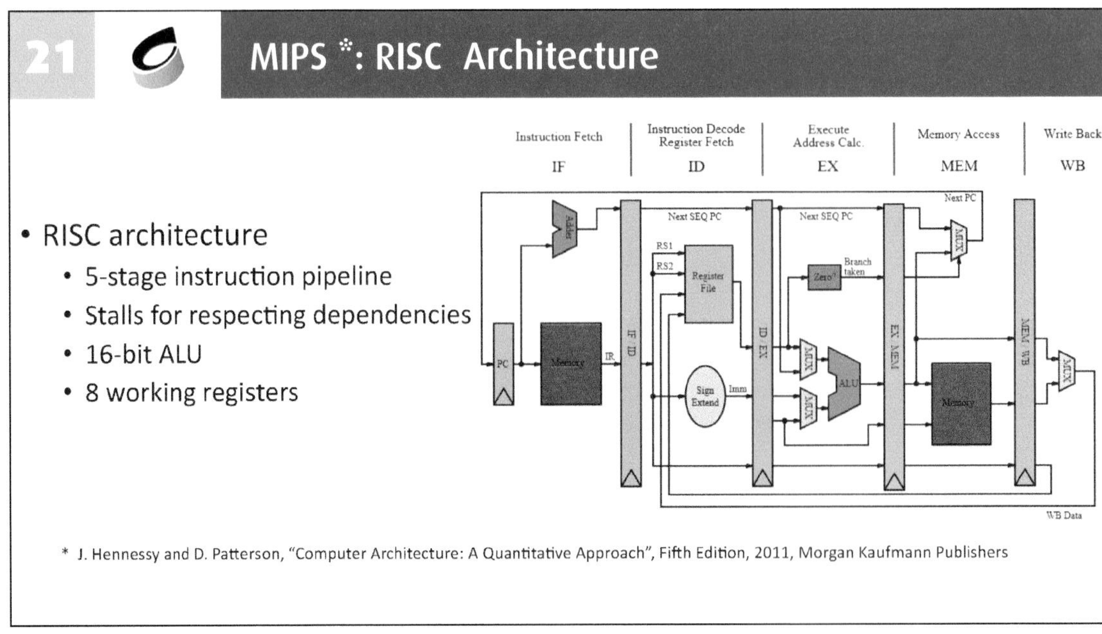

In early 90's the MIPS architecture became very popular. Although this Reduced Instruction Set Computer (RISC) highly simplifies the architecture and hardware of processors regarding the Complex Instruction Set Computer (CISC), this architecture has significant power consumption due to data and control hazards that arise from pipeline's operation.

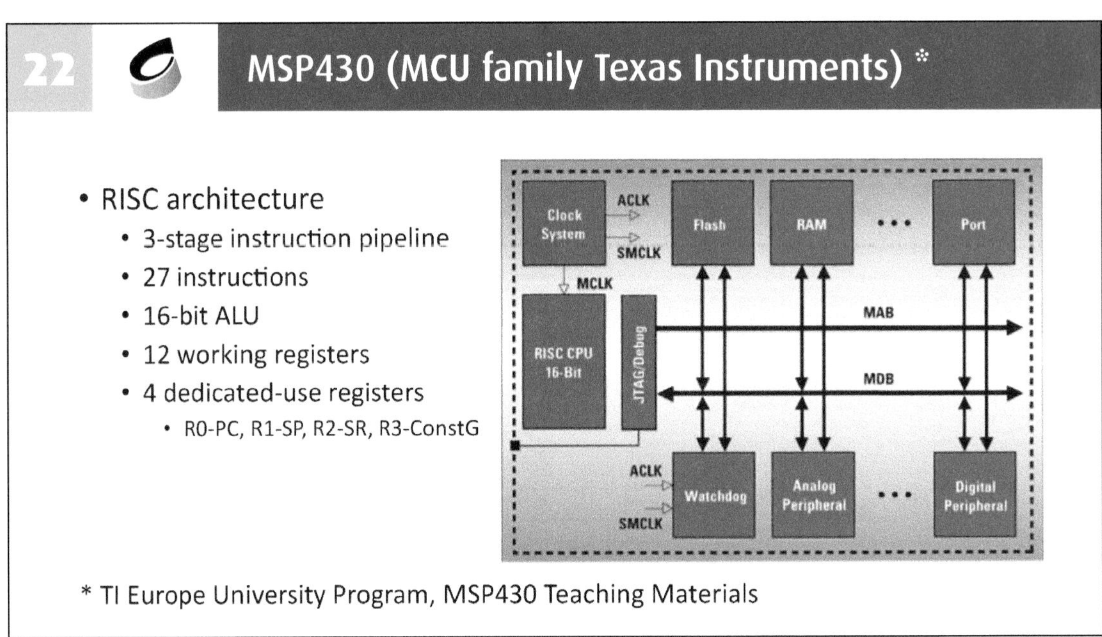

22
MSP430 (MCU family Texas Instruments) *
• RISC architecture
• Stalls for respecting dependencies
• 3-stage instruction pipeline
• 27 instructions
• 16-bit ALU
• 12 working registers
• 4 dedicated-use registers
• R0-PC, R1-SP, R2-SR, R3-ConstG
Clock System
ACLK
SMCLK
MCLK
Flash
RAM
Port
RISC CPU 16-Bit
JTAG/Debug
MAB
MDB
Watchdog
Analog Peripheral
Digital Peripheral
* TI Europe University Program, MSP430 Teaching Materials

Instead of the 5-stage instruction pipeline, the MSP430 has a 3-stage pipeline, which decreases the MCU performance, while also reducing the impact of dependencies in the code. Furthermore, the architecture is simpler and it only has 27 instructions.

MSP430 (MCU family Texas Instruments) 23

- Datapath
  - MSP430X – 20 bits, but buses of 16 bits
  - ALU: addition, subtraction, comparison and logical (AND, OR,XOR)
  - Multiplier (optional)

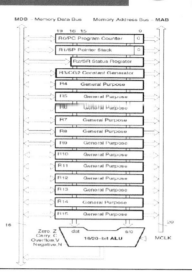

Operation cycle is much simpler (decode, fetch, execute). Simple execution step: get from register, execute, store in register (Load/store architecture).

MSP430 (MCU family Texas Instruments)

24

- Power consumption
  - **0.1 µA** for RAM data Retention
  - **0.8 µA** for real-time clock mode operation
  - **250 µA /MIPS** dynamic run mode
- Wake-up time: **< 1 µs**
- Port leakage: **< 50 nA**
- Low operation voltage: **1.8V-3.6V**
- Maximum frequency: **25MHz**
  - Drhystone MIPS (DMIPS)@8 MHz: 2.3 (#iterations/s divided by 1757)
- dynamic voltage scaling, low-power peripherals (LP UART, LP timers)
  - OP Amps, comparator gated timers, LCD, Supply Voltage Supervisor, 10/12/16-bit SAR ADCs, 12-bit dual DAC

The MCU MSP430 has low power consumption, about 250/MIPS (million instructions per second), fast wake-up time, which is a very important aspect as IoT devices are usually in sleep mode for most of the time. Low voltage and current operation allow the MCU to be supplied by simple batteries or by alternative energy sources. It has many peripherals to communicate with other devices (e.g. UART, LCD drivers) and to read sensors (e.g. op-amps, ADCs).

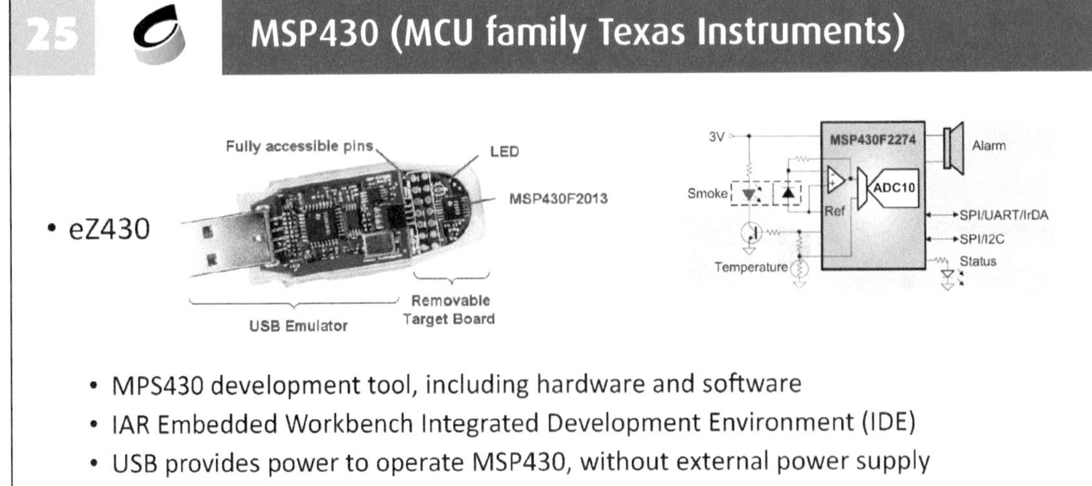

• IAR Embedded Workbench Integrated Development Environment (IDE)
• USB provides power to operate MSP430, without external power supply
• Price around $20

Low cost development tools, including the required software and hardware. The image on the right shows a smoke sensor designed with a MSP430 and few external devices.

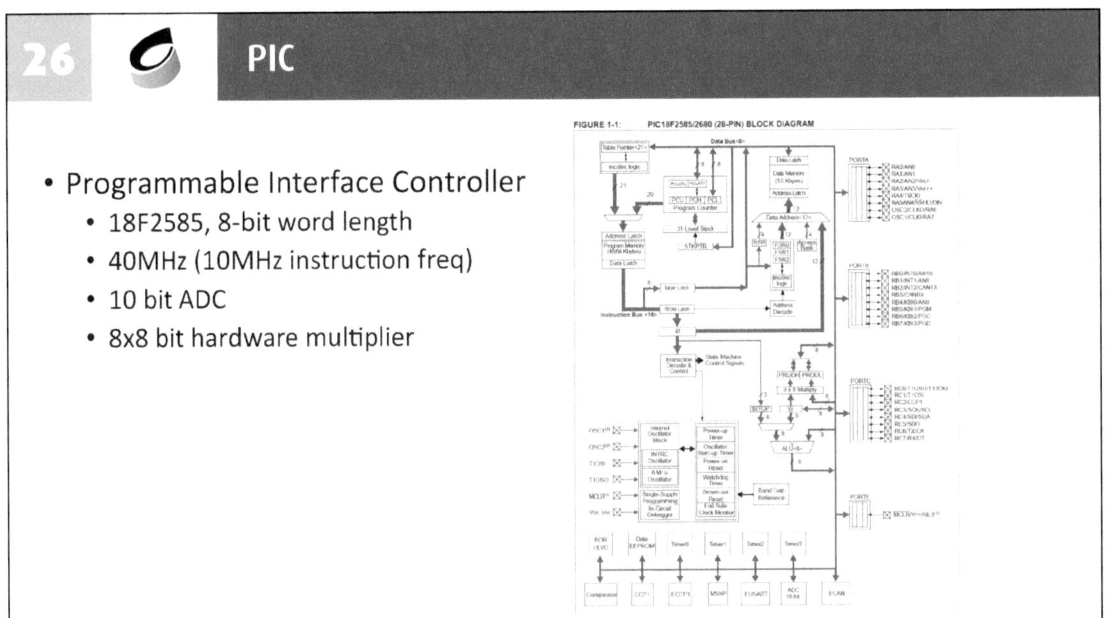

26
PIC
• Programmable Interface Controller
• 18F2585, 8-bit word length
• 40MHz (10MHz instruction freq)
• 10 bit ADC
• 8x8 bit hardware multiplier
FIGURE 1-1: PIC18F2585/2680 (28-PIN) BLOCK DIAGRAM

The PIC architecture is similar to the MSP but the ALU has only 8 bits instead of 16. In this specific MCU the program memory has 48kB. When choosing a MCU, the user should be aware of the memory requirements and choose a device that fits all its needs, because adding external memory will significantly reduce the system performance. MCUs usually have a lot of integrated peripherals such as timers, comparators and ADCs.

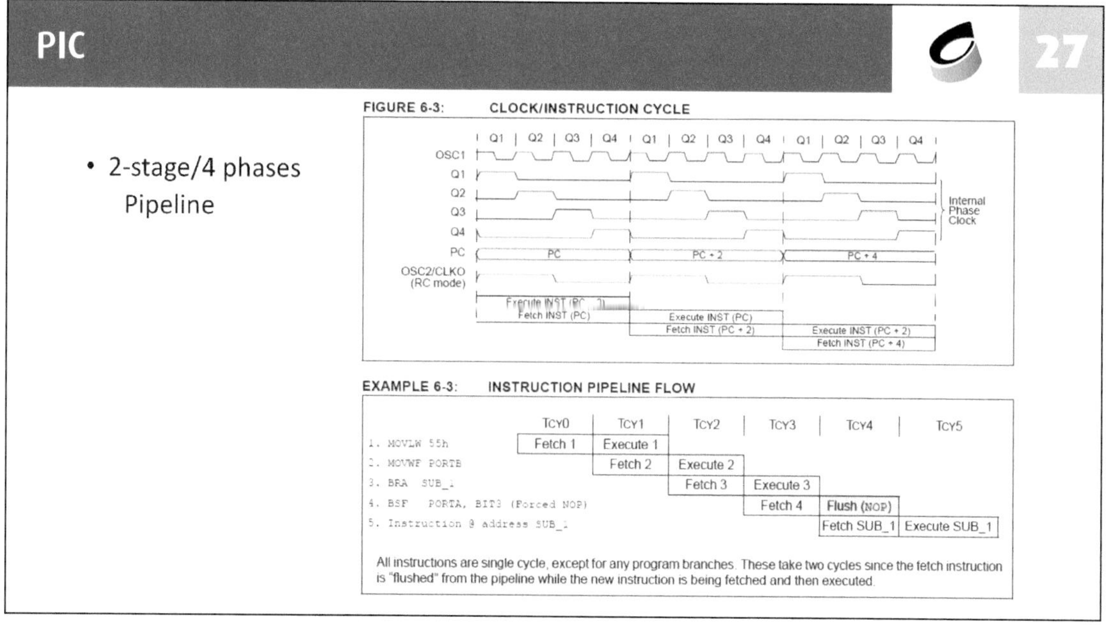

In the PIC architecture one instruction requires 4 clock cycles to be executed, i.e. the instruction period is four times longer than the clock period. In the same instruction cycle it is possible to fetch an instruction while executing the previous instruction.

PIC (for IoT)

28

- eXtreme Low Power (XLP) PIC technology (1.8V to 3.6V)
- 8- and 16- bit Low Power MCU that enable energy harvesting
- Ultra low power consumption
  - Sleep currents as low as **9 nA**
  - Watch-dog timer (WDT) down to **200nA**
  - Real time clock/calendar (RTCC) down to **400nA**
  - Run currents down to **35 μA/MHz**
- Integration with wireless 802.15.4 RF connectivity
  - 2.4 GHz

PICs that are specifically designed for IoT exhibit extreme low power consumption and integrated wireless connectivity (standard 802.15.4).

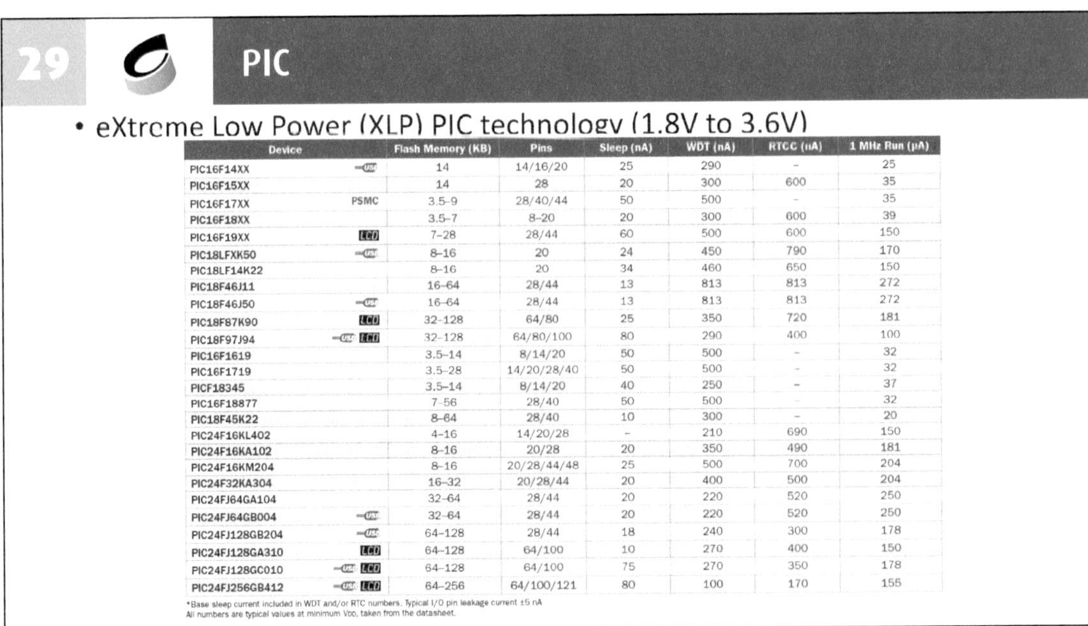

29 PIC

- eXtreme Low Power (XLP) PIC technology (1.8V to 3.6V)

Device		Flash Memory (KB)	Pins	Sleep (nA)	WDT (nA)	RTCC (nA)	1 MHz Run (µA)
PIC16F14XX	USB	14	14/16/20	25	290	-	25
PIC16F15XX		14	28	20	300	600	35
PIC16F17XX	PSMC	3.5-9	28/40/44	50	500	-	35
PIC16F18XX		3.5-7	8-20	20	300	600	39
PIC16F19XX	LCD	7-28	28/44	60	500	600	150
PIC18LFXK50	USB	8-16	20	24	450	790	170
PIC18LF14K22		8-16	20	34	460	650	150
PIC18F46J11		16-64	28/44	13	813	813	272
PIC18F46J50	USB	16-64	28/44	13	813	813	272
PIC18F87K90	LCD	32-128	64/80	25	350	720	181
PIC18F97J94	USB LCD	32-128	64/80/100	80	290	400	100
PIC16F1619		3.5-14	8/14/20	50	500	-	32
PIC16F1719		3.5-28	14/20/28/40	50	500	-	32
PICF18345		3.5-14	8/14/20	40	250	-	37
PIC16F18877		7-56	28/40	50	500	-	32
PIC18F45K22		8-64	28/40	10	300	-	20
PIC24F16KL402		4-16	14/20/28	-	210	690	150
PIC24F16KA102		8-16	20/28	20	350	490	181
PIC24F16KM204		8-16	20/28/44/48	25	500	700	204
PIC24F32KA304		16-32	20/28/44	20	400	500	204
PIC24FJ64GA104		32-64	28/44	20	220	520	250
PIC24FJ64GB004	USB	32-64	28/44	20	220	520	250
PIC24FJ128GB204	USB	64-128	28/44	18	240	300	178
PIC24FJ128GA310	LCD	64-128	64/100	10	270	400	150
PIC24FJ128GC010	USB LCD	64-128	64/100	75	270	350	178
PIC24FJ256GB412	USB LCD	64-256	64/100/121	80	100	170	155

*Base sleep current included in WDT and/or RTC numbers. Typical I/O pin leakage current ±5 nA
All numbers are typical values at minimum VDD, taken from the datasheet.

The table shows the program memory, number of pins and power consumption of multiple PIC devices. Users can choose the most convenient device according to the system that they are designing.

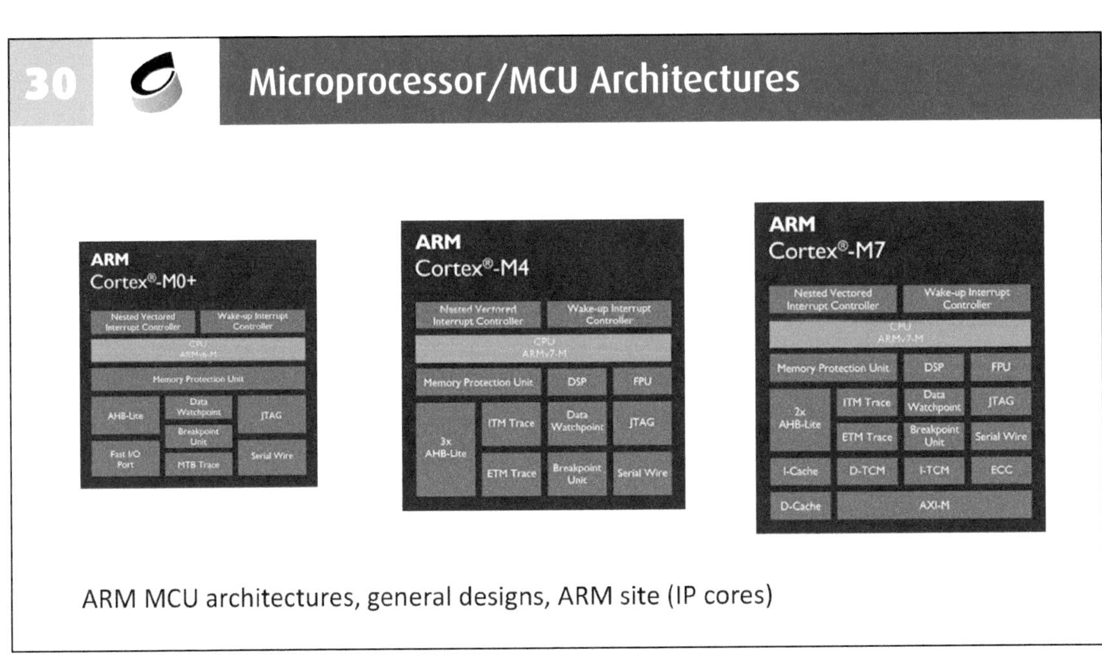

ARM MCU architectures, general designs, ARM site (IP cores)

A more recent architecture of microprocessors used for IoT is the ARM, which is being used in a wide number of MCUs from multiple vendors. The cores designed by ARM also include a large number of peripherals such as DSPs, FPUs, advanced high-performance buses (AHB) with AMBA (Advanced Microcontroller Bus Architecture), etc. This architecture permits to have more processing power at the IoT nodes. There are a lot of different MCU options, with varying amount of peripherals, that allow the user to choose the right MCU according to the system's requirements.

Microprocessor/MCU Architectures

31

- MCU ARM Cortex-M0, ..., Cortex M7
  - 32 bit RISC load/store processors
  - 58 ARM instructions
  - High code density with ARM Thumb® and Thumb®-2 technology (mixed 16-bit and 32-bit instructions)
  - Low cost ($1)
  - low power consumption and high energy efficiency (Cortex-M0+ achieves a power consumption of just **9.4μW/MHz**)

More instructions when compared with MSP and PIC, which are compressed in memory and allow to mix 16-bit and 32-bit instructions. These devices are also more efficient than the previous ones.

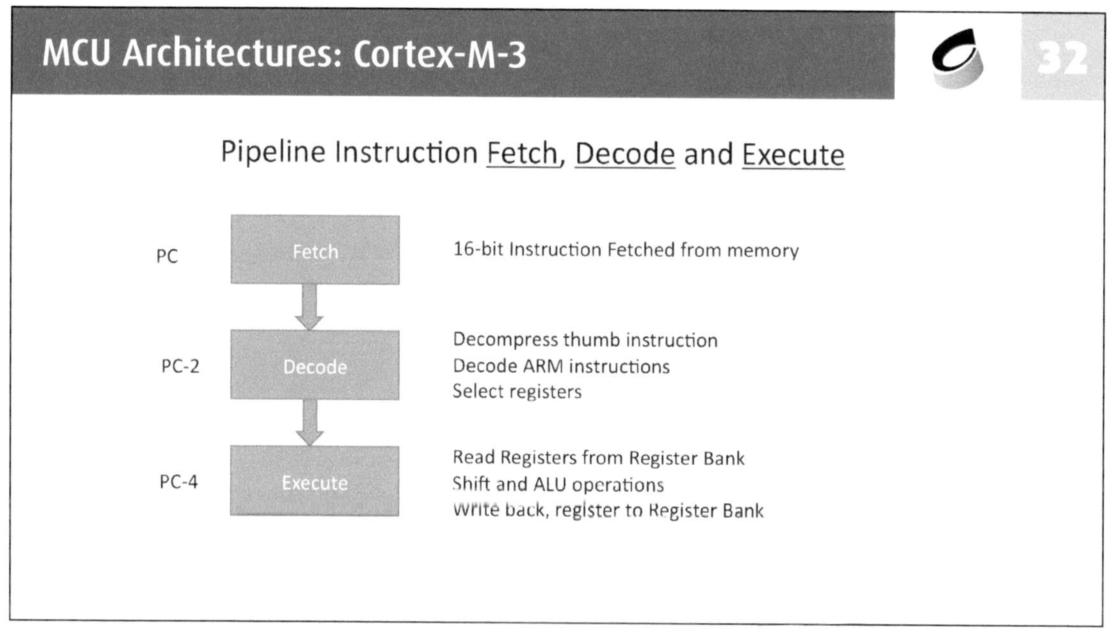

MCU Architectures: Cortex-M-3
32
Pipeline Instruction Fetch, Decode and Execute
PC
Fetch
16-bit Instruction Fetched from memory
PC-2
Decode
Decompress thumb instruction
Decode ARM instructions
Select registers
PC-4
Execute
Read Registers from Register Bank
Shift and ALU operations
Write back, register to Register Bank

The instruction pipeline of the ARM Cortex-M3 is similar to MSP and PIC (fetch, decode, execute), but the instructions need to be decompressed before their execution.

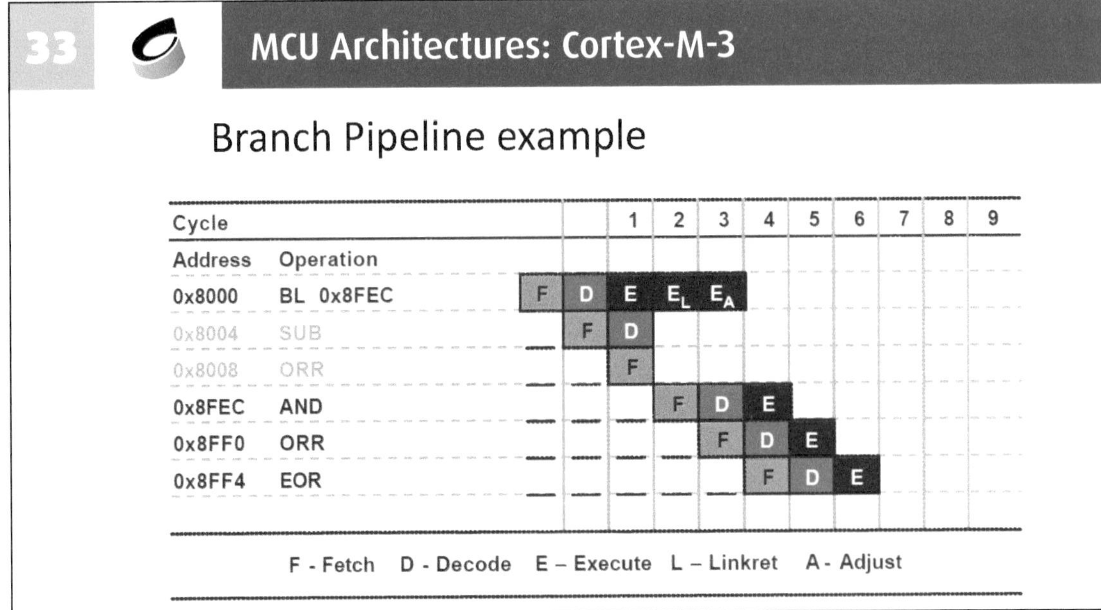

33
MCU Architectures: Cortex-M-3
Branch Pipeline example
Cycle 1 2 3 4 5 6 7 8 9
Address Operation
0x8000 BL 0x8FEC
0x8004 SUB
0x8008 ORR
0x8FEC AND
0x8FF0 ORR
0x8FF4 EOR
F D E E_L E_A
F - Fetch D - Decode E – Execute L – Linkret A - Adjust

When there is a branch, it takes three cycles to execute the instruction.

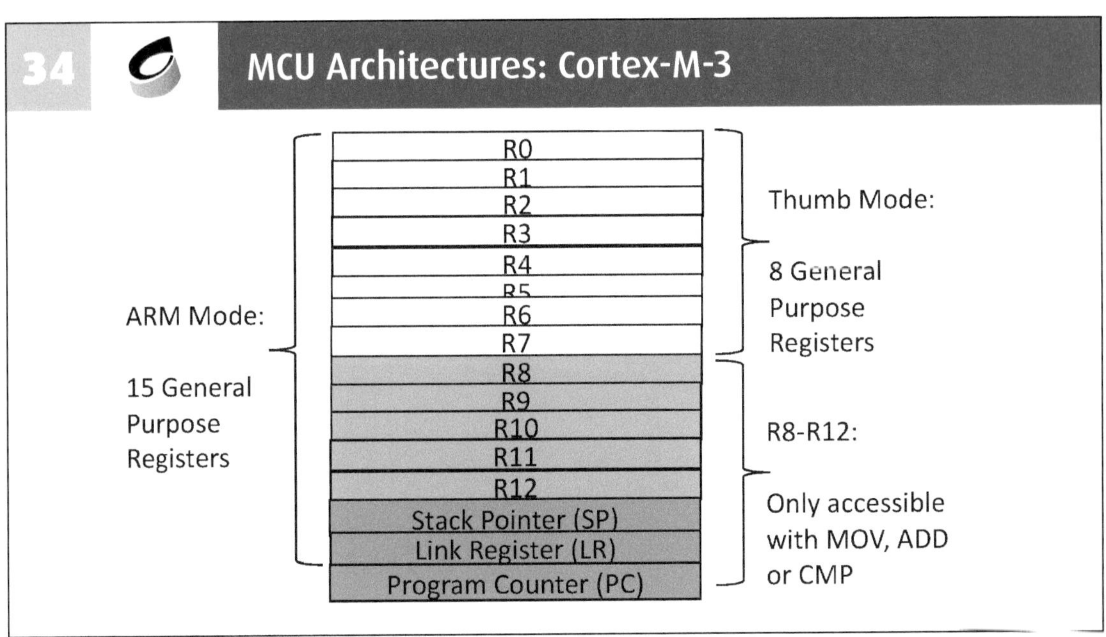

The ARM Cortex-M3 MCU has 16 registers, which can have different functions according to its operating mode (ARM mode or thumb mode).

MCU Architectures: Cortex-M3 Register File

35

- Simple LD/ST microarchitecture with three pipeline stages (IF/DEC/EXE) and a simple datapath

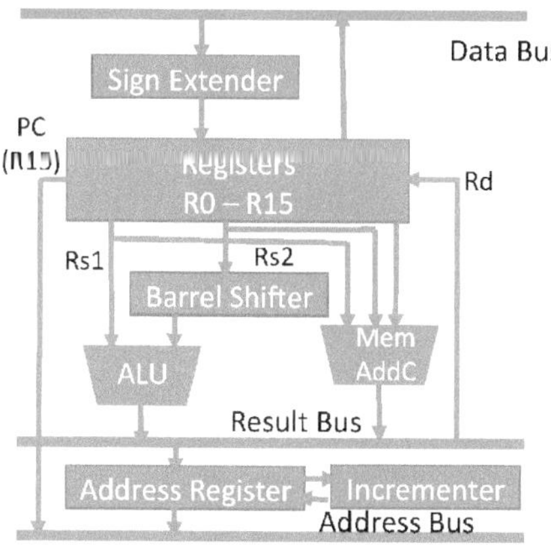

Shown in the figure, the datapath of the ARM architecture is simple, but very beneficial in terms of power-consumption.

Power Management

36

- Multiple sleep modes supported
  - Controlled by Nested Vectored Interrupt Controller (NVIC)
  - Sleep Now – Wait for Interrupt/Event Instructions
  - Sleep on Exit – Sleep immediately on return from last ISR
  - Deep sleep
    - Long duration sleep, so PLL can be stopped
    - Exports additional output signal SLEEPDEEP
- Cortex-M3 system is clock gated in all sleep modes
  - Sleep signal is exported allowing to clock gate also the external system
  - NVIC interrupt interface stays awake
- Wake-up Interrupt Controller (WIC)
  - External wake-up detector allows Cortex-M3 to be fully power down
  - Effective State-Retention / Power Gating (SRPG) methodology

"ARM Cortex-M3 Introduction", ARM University Relations, The Architecture for the Digital World, ARM

ARM based MCUs have multiple sleep modes, controlled by the Nested Vectored Interrupt Controller (NVIC), and it employs clock gating to reduce power consumption.

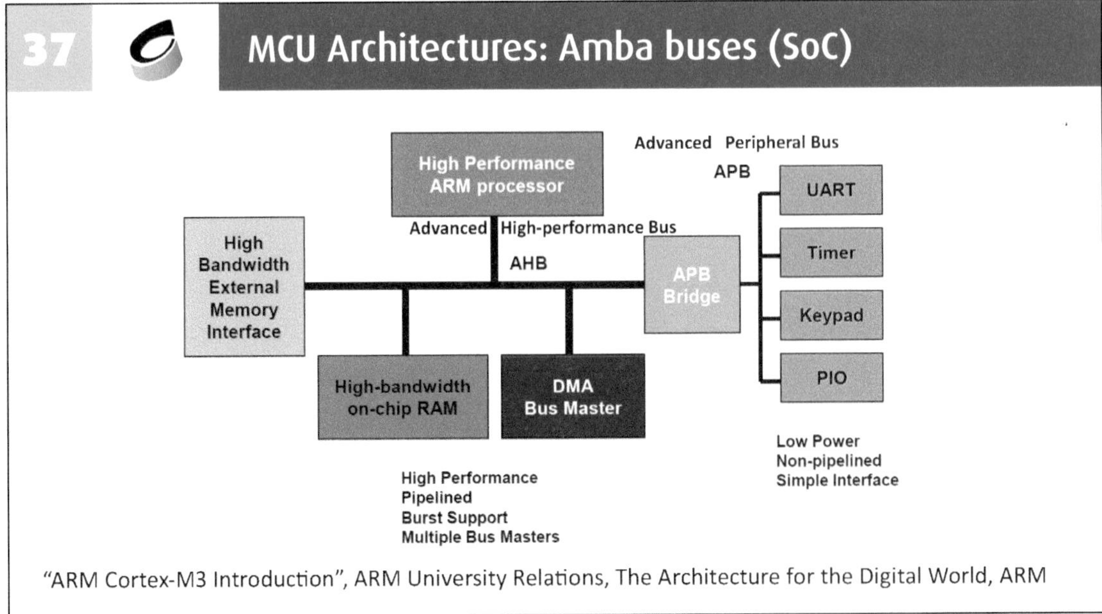

The MCU includes an interface to connect external memory, if the included RAM is not sufficient, and it allows to connect many different peripherals (USART, timers, etc.) using the advanced peripheral bus.

38 Ultra Low Power MCU

- STM32L4 Series (STMicroelectronics)
  - based on its ARM® Cortex®-M4 core with FPU
  - Drhystone MIPS (DMIPS)@80 MHz: **100**
    - number of program iteration completions per second divided by 1757 (Vax 11/780)
- Dynamic voltage scaling, low-power peripherals (LP UART, LP timers)
  - OP Amps, compar., LCD, 12-bit DACs and 16-bit ADCs (hardware oversampling)
- Ultra-low-power mode, backup registers without RTC: **8 nA**
- Ultra-low-power mode + 16 Kbytes of RAM + RTC: **450 nA**
- Dynamic run mode: down to **84 μA/MHz**
- Wake-up time: **5 μs**

Example of a MCU from the STM32L4 series, from STMicroelectronics, ARM Cortex-M4 core, which is used for IoT.

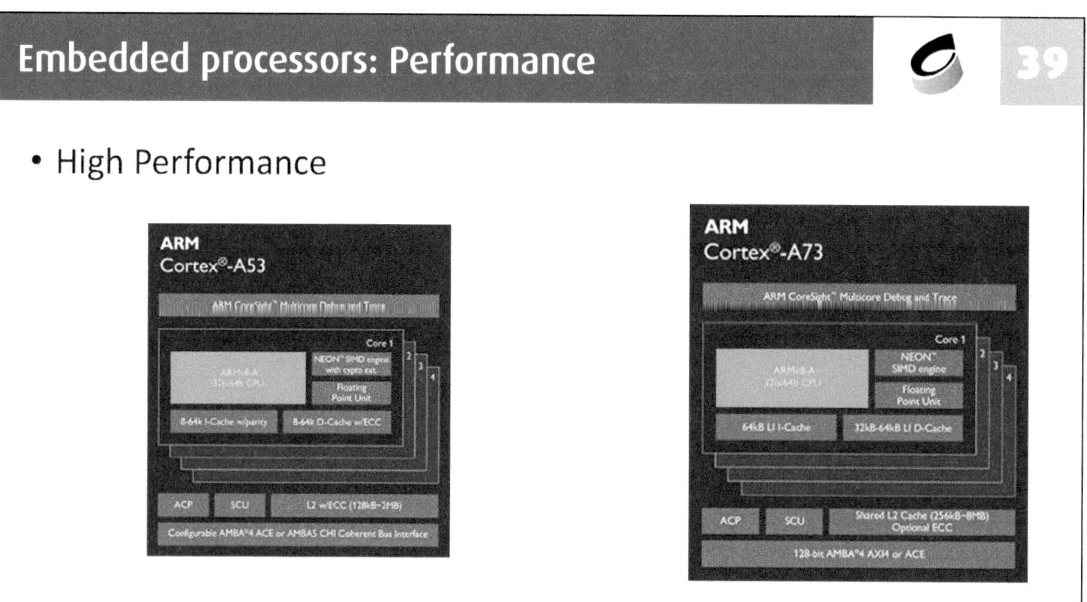

While the ARM M series provide many different MCUs, for high performance systems there is, for instance, the series, which are aimed for embedded systems that include several CPU cores and several levels of memory, among other features.

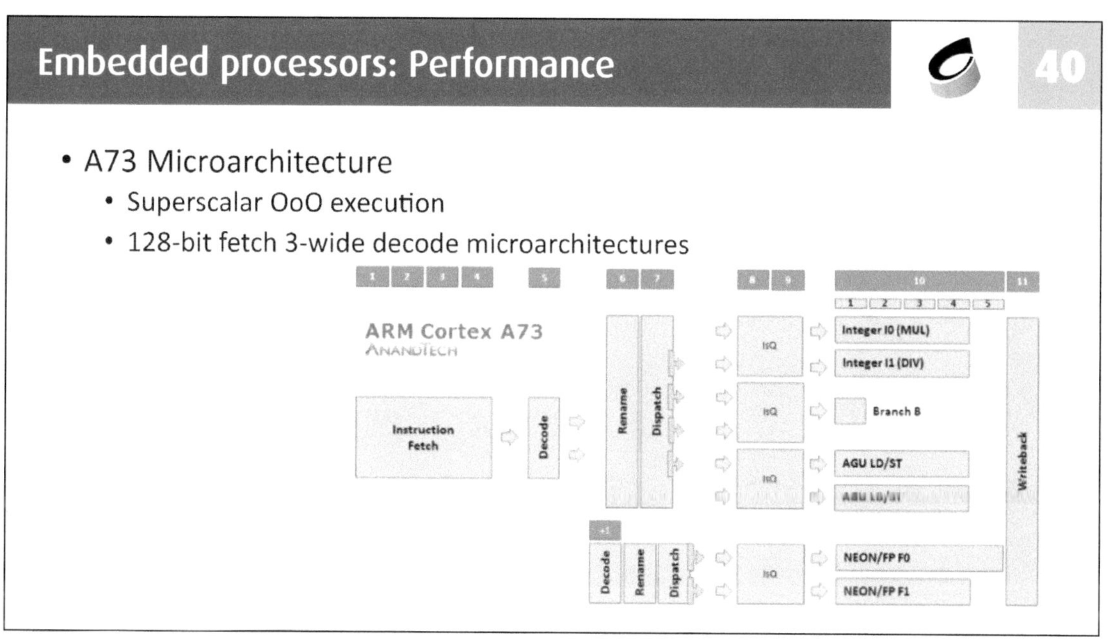

These systems have many more functions than the ARM MCUs. They have a 64-bit architecture, and the instruction pipeline is more complex, which may provide Out-of-Order execution (e.g. A73 model).

41 

Embedded processors: NEON

- 75% higher performance for multimedia processing in embedded devices
- With very low increase in power consumption
- Simultaneous computation of 8x16-bit or 4x32-bit
- Exploited by the compiler or manually in the code, transparent to the OS

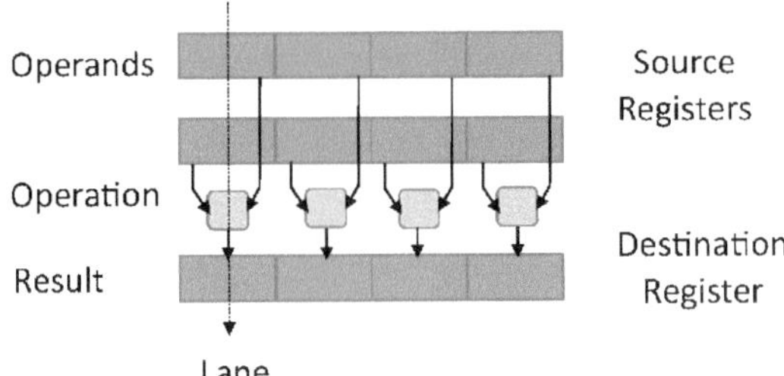

NEON Single Instruction Multiple Data (SIMD) Instruction Set Architecture (ISA) extensions allow to achieve significant performance improvements with very low increase both in cost and power consumption. This kind of techniques and ISA extensions have been exploited in the last years, also for IoT, due to above-mentioned advantages.

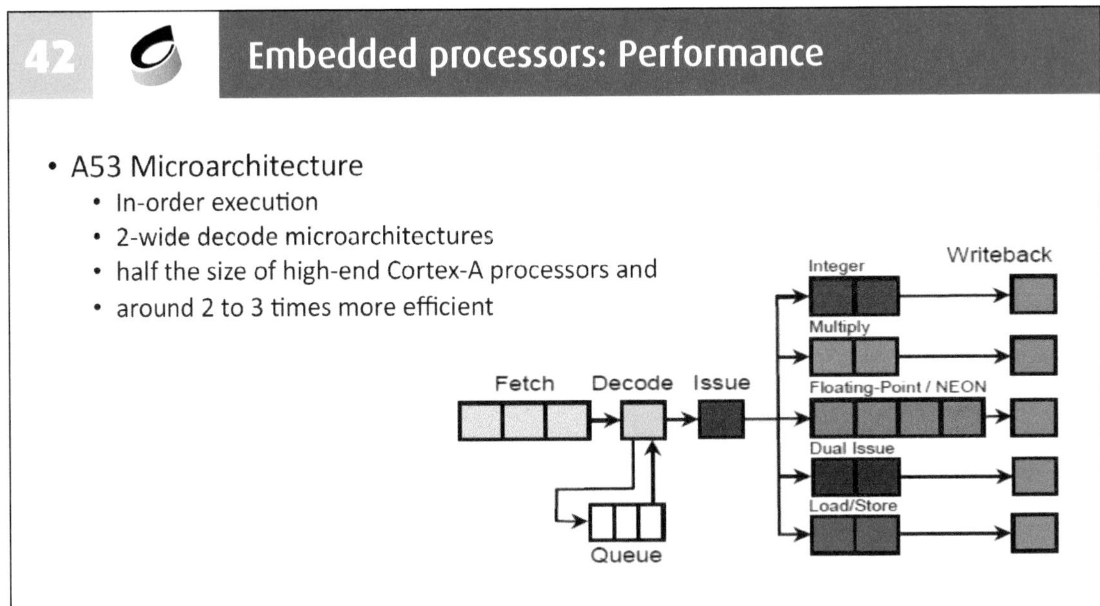

42
Embedded processors: Performance
• A53 Microarchitecture
• In-order execution
• 2-wide decode microarchitectures
• half the size of high-end Cortex-A processors and
• around 2 to 3 times more efficient
Fetch
Decode
Issue
Queue
Integer
Multiply
Floating-Point / NEON
Dual Issue
Load/Store
Writeback

Between the ARM M series and the top-tier A processors, such as the previously referred A73, there are other superscalar architectures that perform in-order processing (e.g. A53).

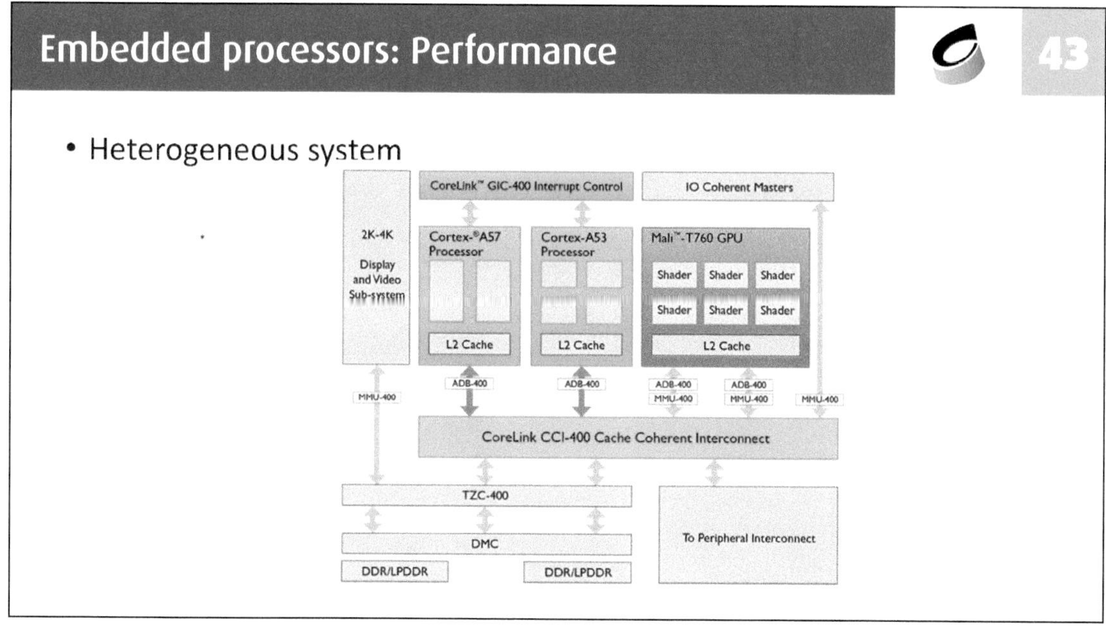

Include more than one microarchitecture in the same chip and use these heterogeneous processors according to processing power or consumption requirements. In this example, the A57 and A53 processors are used in the same SoC. It is also possible to include processors with different characteristics and ISAs, e.g. CPUs and GPUs. The main problem of heterogeneous systems is that the level of complexity is significantly increased and the programming of these devices becomes more challenging.

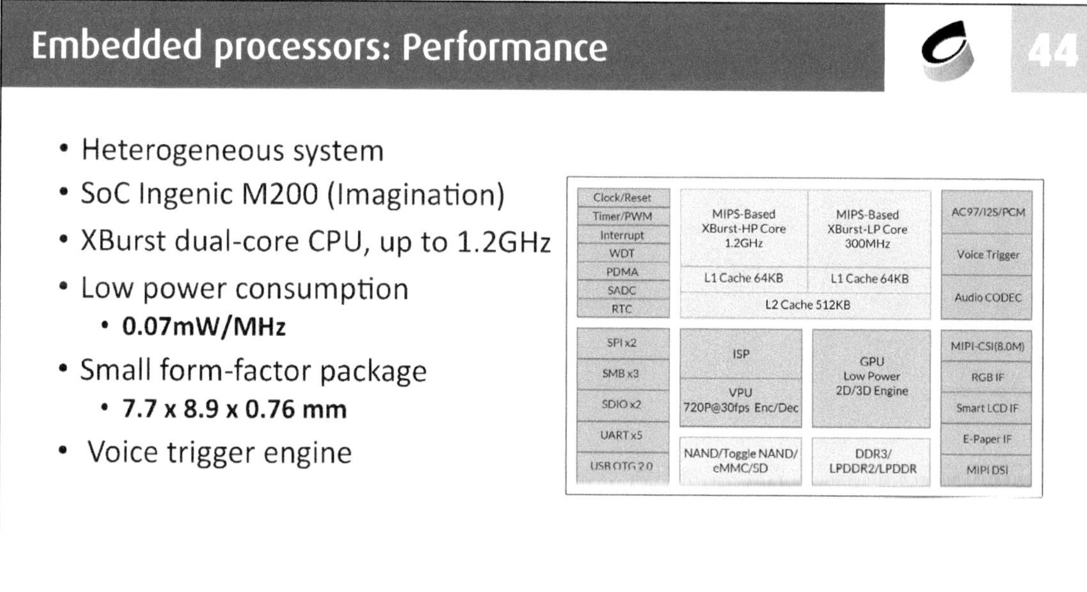

In addition to ARM, there are other companies/academia that are investing in these techniques. These systems are not designed for IoT nodes, but for applications that require much more processing (e.g. smartphones). In these architectures, the instruction set tends to be the same for both cores, allowing to easily perform multithreading for parallel processing.

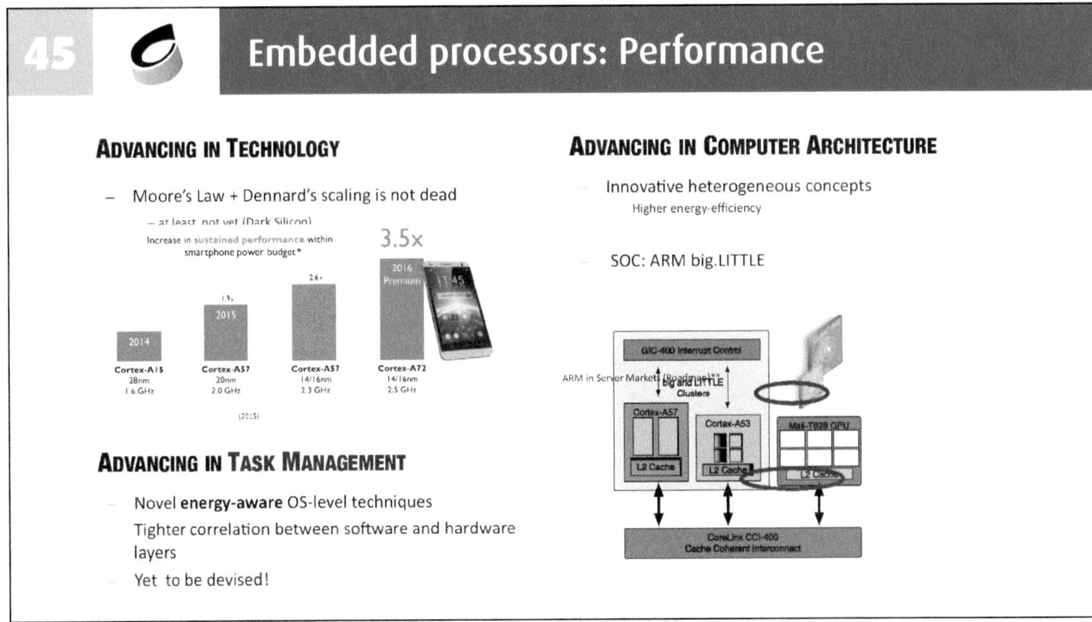

Several aspects are important to improve the efficiency of embedded systems, not only the technology and the architecture but also the software, namely the operating systems.

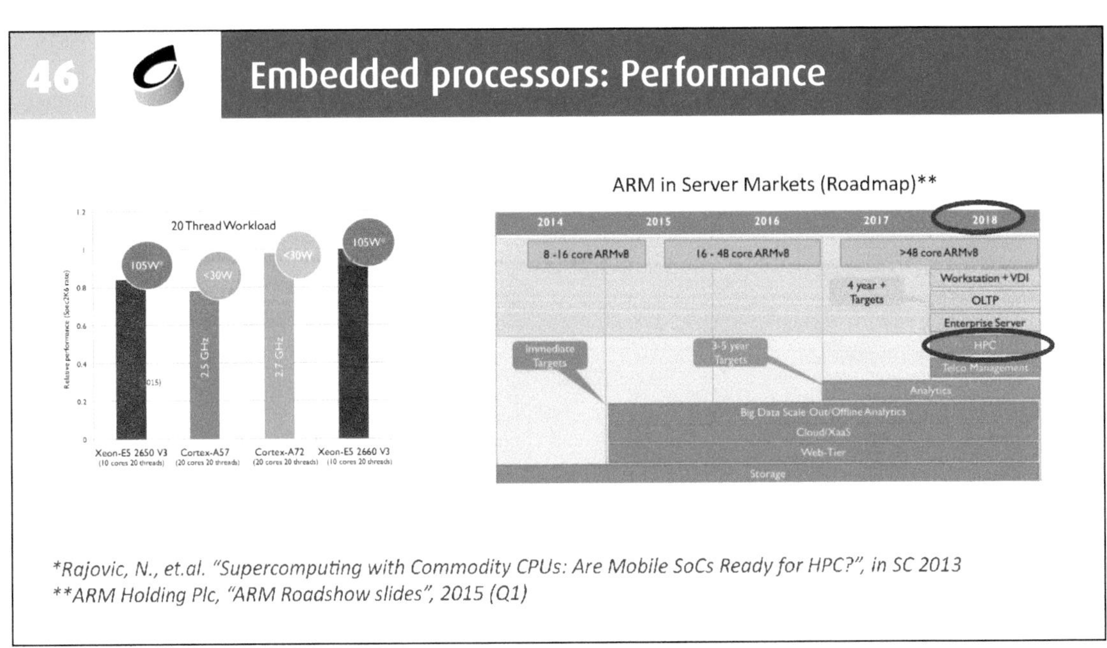

Figure shows that the performance achieved with ARM processors is comparable with Intel Xeon processors, but with much less power consumption. Since ARM processors are very efficient, in the next years it will be possible to have systems with a high number of cores (more than 48) with relative low power consumption and low cost.

Power/Performance Optimization of a SoC

47

- Integrate different ASICs
- Customize processors
- Reduce memory bandwidth and frequency
- Mixing with V_t low transistors
- Power gating/Clock gating/DVFS
  - Different modes: Run, Sleep, Shutdown...
  - Fine-grained pipeline shutdown
  - Faster register save and restore
  - Power domains and voltage domains

To optimize performance and power consumption of a SOC, it is necessary to consider several methods and techniques.

SoC for Connecting

48

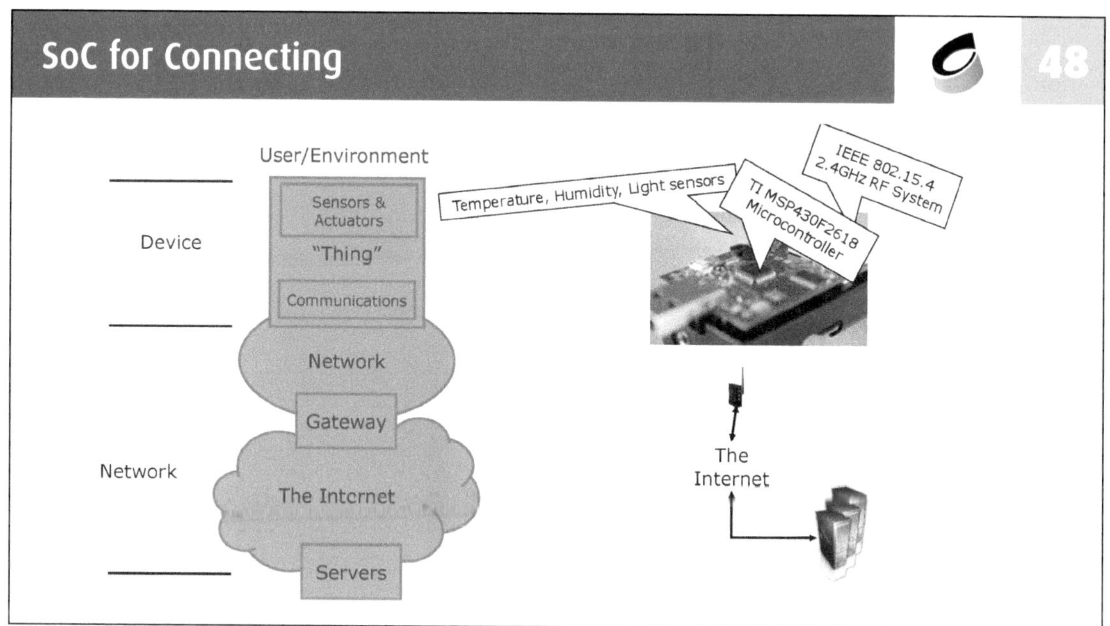

By connecting IoT devices to the cloud it is possible to perform the computation in the cloud, reducing the nodes' complexity and power consumption. However, this significantly increases communication and internet usage.

AWS IoT - Amazon Web Services

- MQTT protocol
  - machine-to-machine (M2M)/"Internet of Things" connectivity protocol
  - Designed as an extremely lightweight publish/subscribe messaging transport
  - Useful for connections with remote locations where a small code footprint is required and/or network bandwidth is at a premium
- X.509 standard
  - For a public key infrastructure (PKI) to manage digital certificates and public-key encryption
  - Key part of the Transport Layer Security Reduce memory bandwidth and frequency

Example of protocols and standards for the IoT, the MQTT protocol and the X.509 standard.

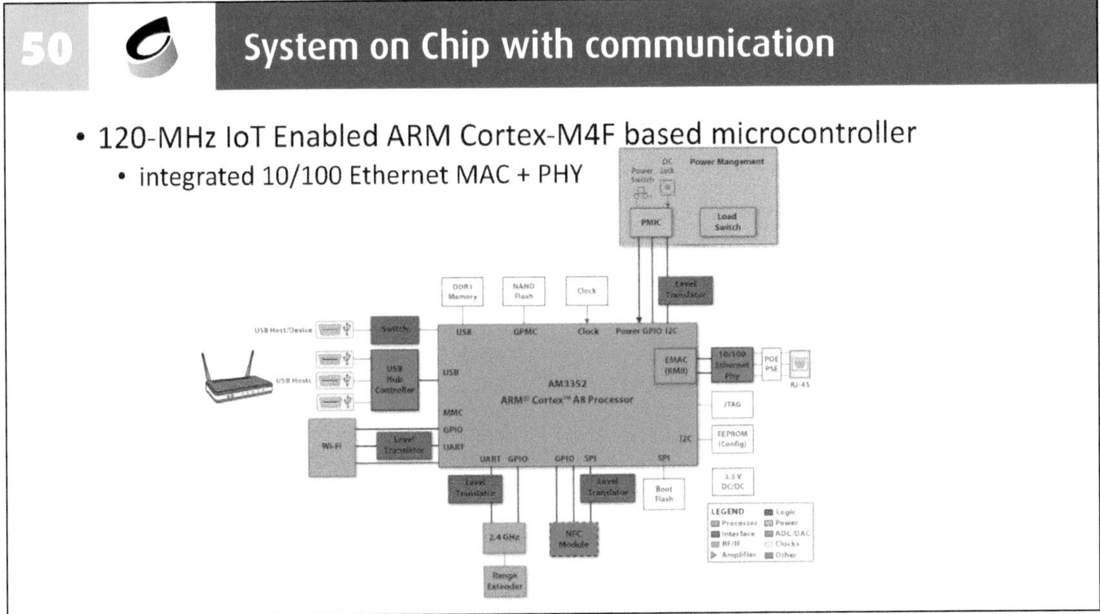

Example of an ARM Cortex-M4F based microcontroller.

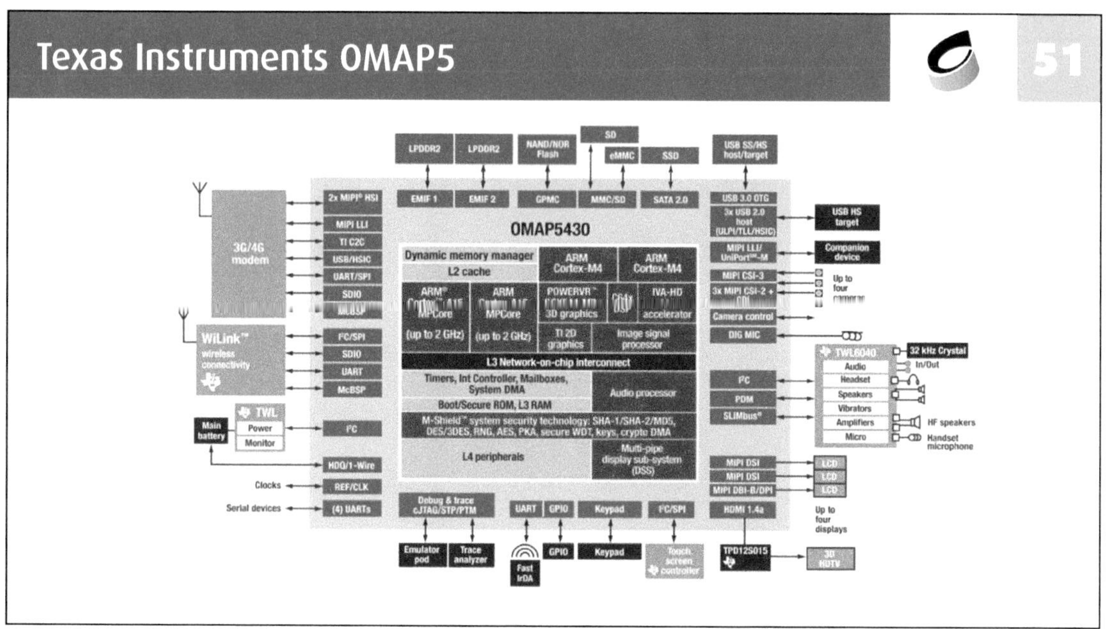

Example of a TI OMAP5430 SoC.

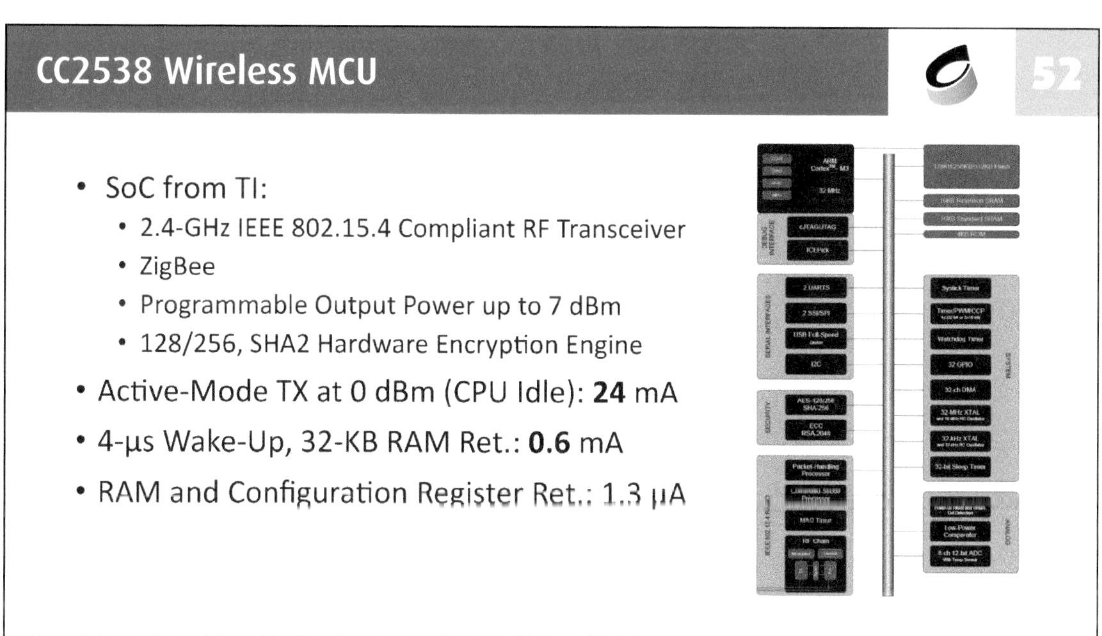

Example, CC2538 Wireless MCU from TI.

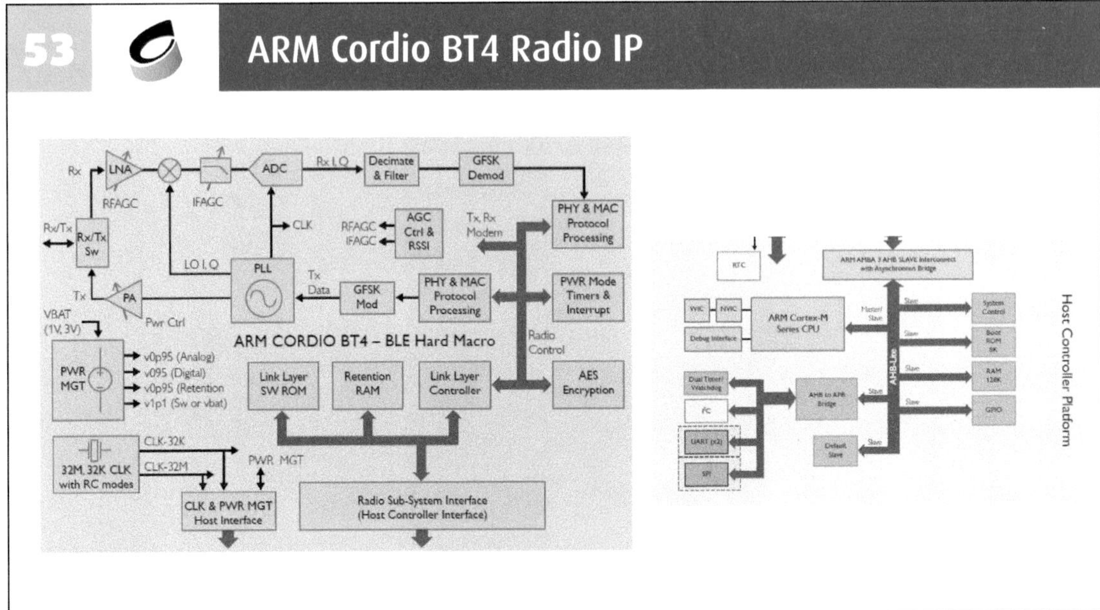

Another example, the ARM Cordio BT4 Radio IP. It is impressive the variety of implemented SoC and IP, with different characteristics and for different purposes.

54 ARM Cordio BT4 Radio IP

- IP for TSMC 55nm
  - Incorporates Bluetooth Core Specification version 4.2
  - Sub-volt (950mv nominal) low-power core designed for optimal efficiency with AMD Cortex (M)
  - Total power consumption during Tx and Rx: **6.5mA @1V**
  - Receive sensitivity of **-94dBm**
- **2400 – 2484** MHz
- GFSK modulation, Frequency-hopping spread spectrum (FHSS)
- AES Encryption Engine

The ARM Cordio BT4 allows different signal modulations and provides AES Encryption to assure secure data transmission/reception.

Raspberry Pi

55

- "The Raspberry Pi is a series of credit card-sized single-board computers developed in the United Kingdom by the Raspberry Pi Foundation to promote the teaching of basic computer science in schools and developing countries"
  - Cost: $35

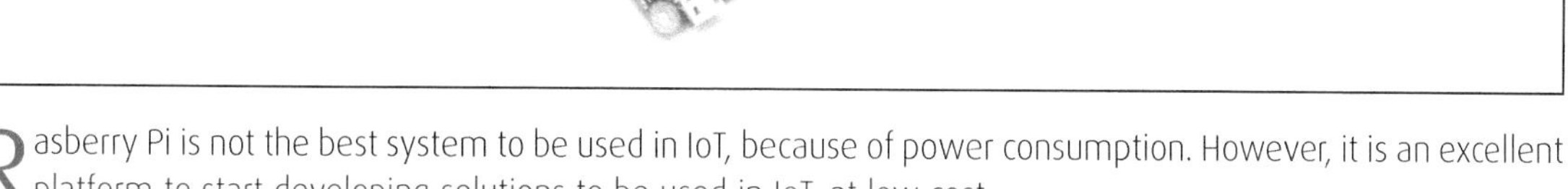

Rasberry Pi is not the best system to be used in IoT, because of power consumption. However, it is an excellent platform to start developing solutions to be used in IoT, at low cost.

Raspberry Pi

56

- Processor: 1.2GHz Quad Core BCM2837 ARMv8 64bit Processor
- Connectivity: Built-in WiFi (802.11B/G/N) and Bluetooth Low Energy (BLE)
- RAM: 1GB
- GPU: VideoCore IV
- 4 x USB 2.0 ports and 10/100 Ethernet Port
- Full-Size HDMI and Composite, and Sound L/R Stereo Line-out
- Operating System: microSD card to load and store OS
- Digital Interfaces: 1 x CSI Camera Port and 1 x DSI Display Port
- GPIO: 40 General Purpose Input / Output Pins
- Power: Requires **5V 2.4A** USB Power Supply

Raspberry Pi main specifications.

57  Cryptography and Security

- Engines for **symmetric-key** encryption have been provided my MCUs/ processors
  - Advanced Encryption Standard (AES), block-cypher that supersedes the Data Encryption Standard (DES) for symmetric-key encryption
  - National Institute of Standards and Technology (NIST) in 2001
  - block size of 128 bits, but three different key lengths: 128, 192 and 256 bits
- Hardware support for True Random Number Generator (RNG)
  - NIST SP 800-90 method to implement Pseudo Random Generators (PRNG)
  - Typical PRNG implementation requires a True Random Number to seed PRNG

Security is one of the main aspects of the IoT, mainly when communicating sensitive data and information. Many MCUs provide symmetric-key encryption, for which the encryption and decryption keys are the same, which means that both sender and receiver have access to that key. These devices also have (pseudo-)random number generators (RNG) which are useful for cryptography, namely to generate the keys.

58  AES Encryption

- Each round consists of four similar but different sets
  - including one that depends on the encryption key itself

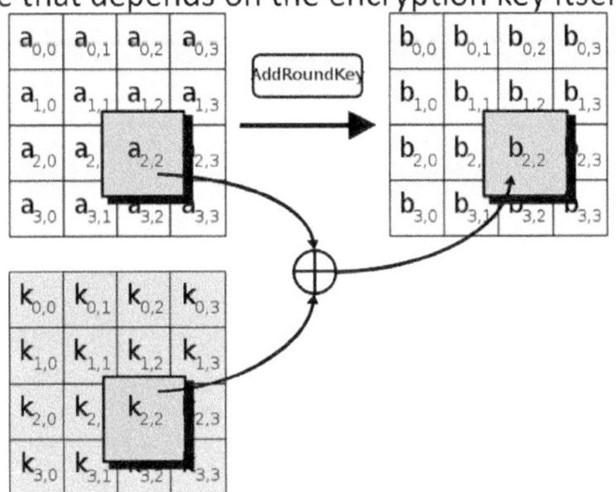

Basic operation of Advanced Encryption Standard (AES) Encryption.

PIC 24 family: Cryptographic engine *

- DES and Triple DES (3DES) Encryption and Decryption (64-bit block size):
  - Supports 64-bit keys and 2-key or 3-key Triple DES
- AES Encryption and Decryption (128-bit block size):
  - Supports key sizes of 128, 192 or 256 bits
- Supports ECB, CBC, modes for Both DES and AES Standards
- 512-bit OTP array for key storage, not readable from other memory spaces
- Hardware Support for TRNG
- Support for PRNG, NIST SP800-90 Compliant

* "Cryptographic Engine", Microchip, DS70005133B, 2013

Most of the IoT processors provide hardware encryption engines, which is more effective than software encryption and uses fewer resources from the MCU. These devices offer different encryption methods, namely for symmetric-key encryption, such as AES and Data Encryption Standard (DES). One example is the Microchip PIC24 family.

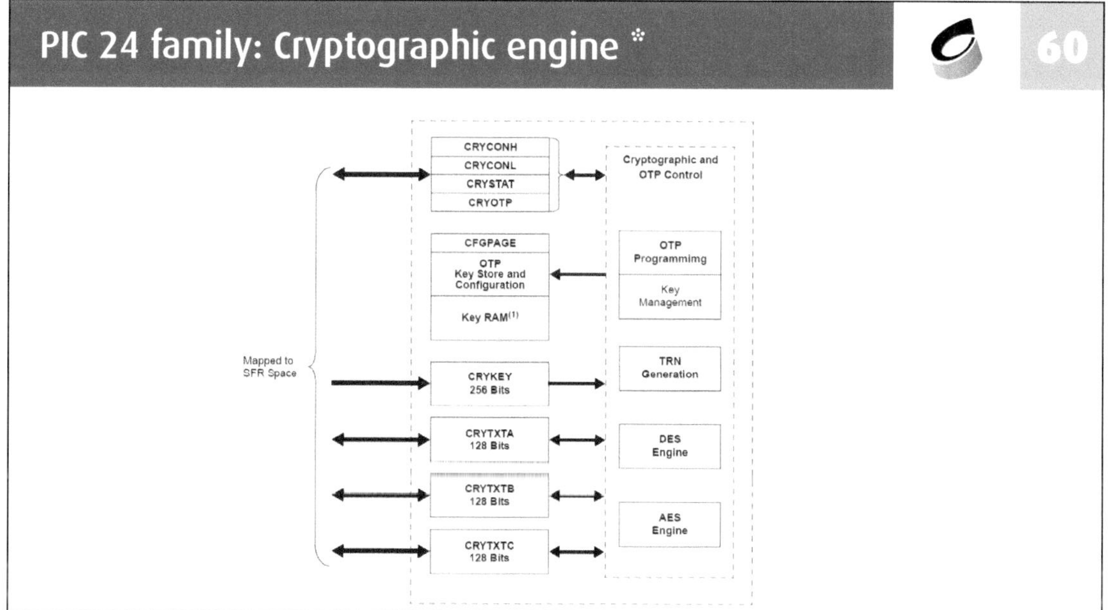

Architecture of the cryptographic engine of the PIC24.

61 PIC 24 family: Cryptographic engine

Mode	Clock Cycles (approximate)	
	Per Block	Additional Load/Unload
DES Encryption/Decryption (Block 64 bit)	10	2
128-bit AES Encryption/Decryption (Block 128 bit)	219	32
192-bit AES Encryption/Decryption (Block 128 bit)	275	32
256-bit AES Encryption/Decryption (Block 128 bit)	299	32

Evaluation of the cryptographic engine of the PIC24.

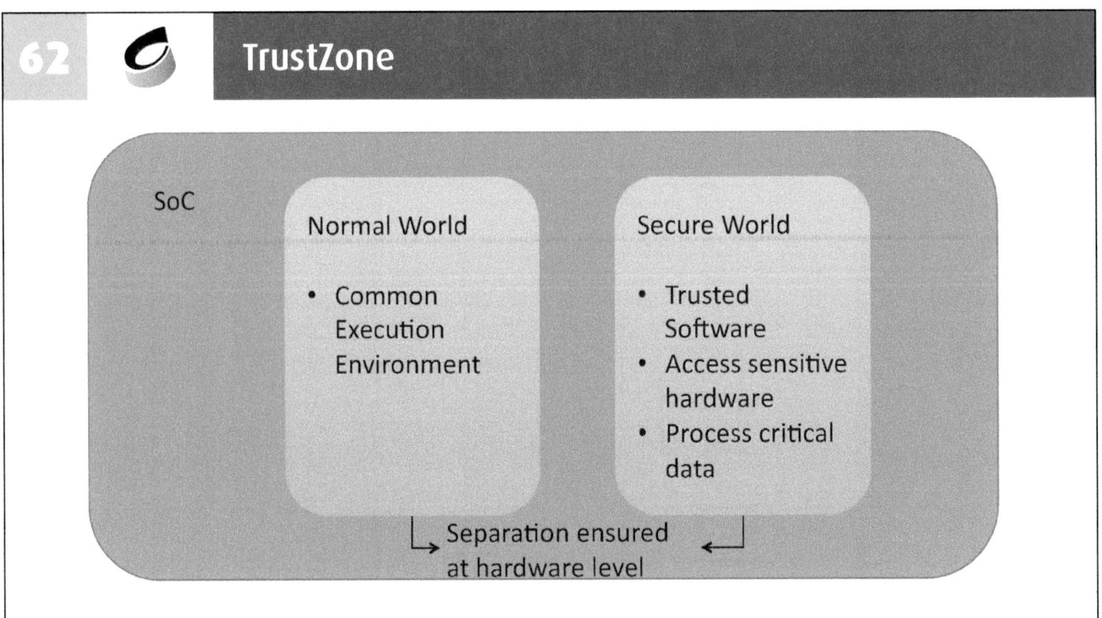

ARM created the TrustZone technology for improving security. Two "execution worlds", which are hardware separated, are created: Normal World, for general execution; Secure World for sensitive execution. According to the operating world, the device has different security mechanisms. The memory address space is different for both worlds.(Slides 62 to 66).

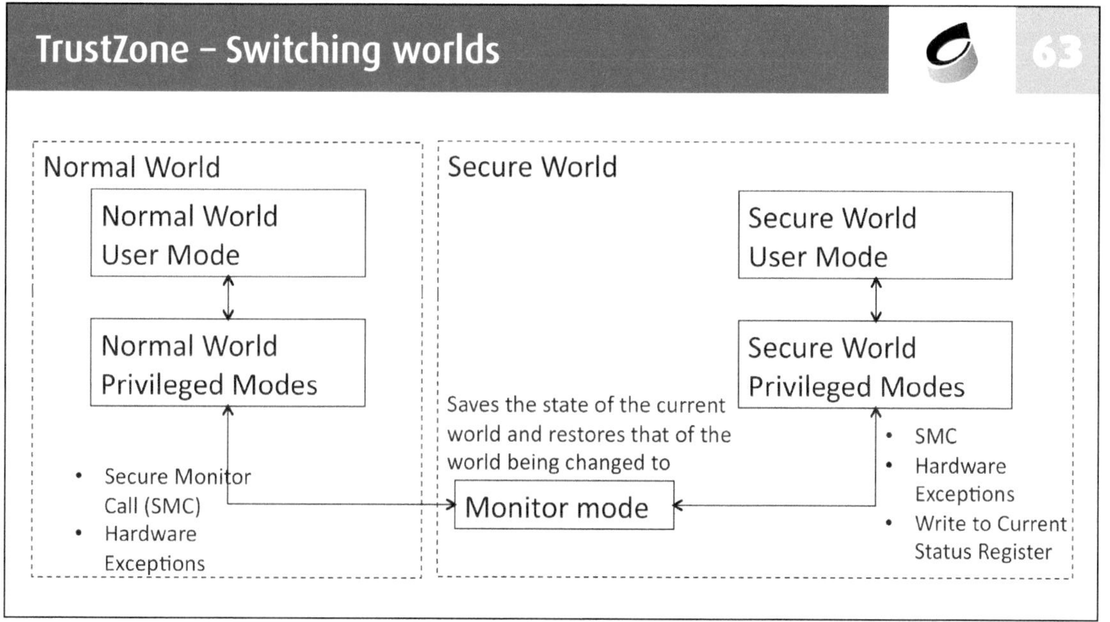
TrustZone – Switching worlds
63
Normal World
Normal World User Mode
Normal World Privileged Modes
• Secure Monitor Call (SMC)
• Hardware Exceptions
Secure World
Secure World User Mode
Secure World Privileged Modes
Saves the state of the current world and restores that of the world being changed to
Monitor mode
• SMC
• Hardware Exceptions
• Write to Current Status Register

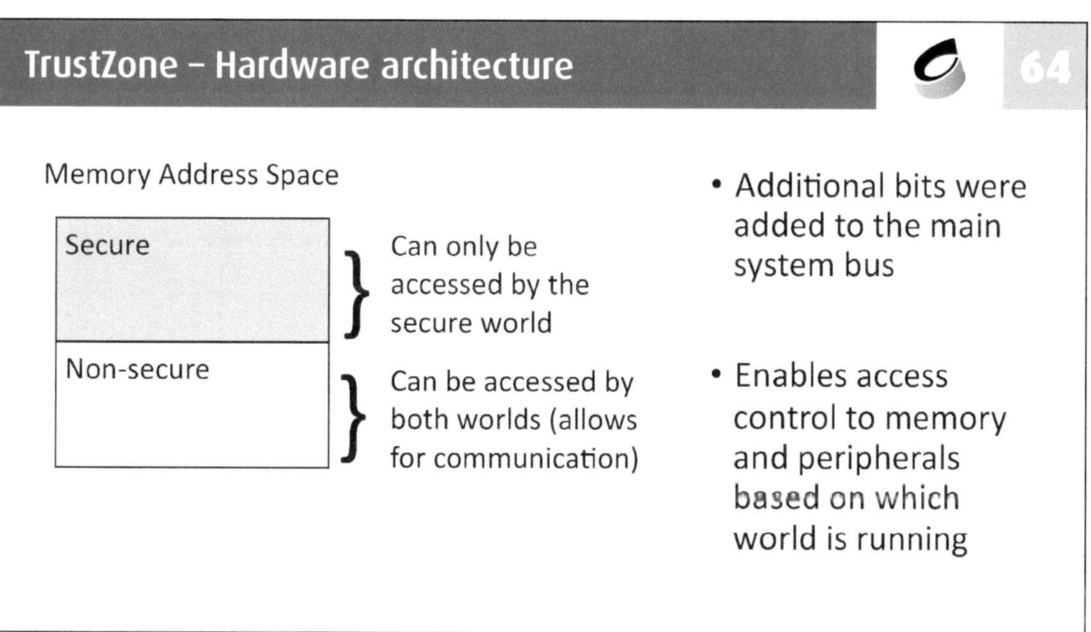
TrustZone – Hardware architecture
64
Memory Address Space
Secure
Can only be accessed by the secure world
Non-secure
Can be accessed by both worlds (allows for communication)
• Additional bits were added to the main system bus
• Enables access control to memory and peripherals based on which world is running

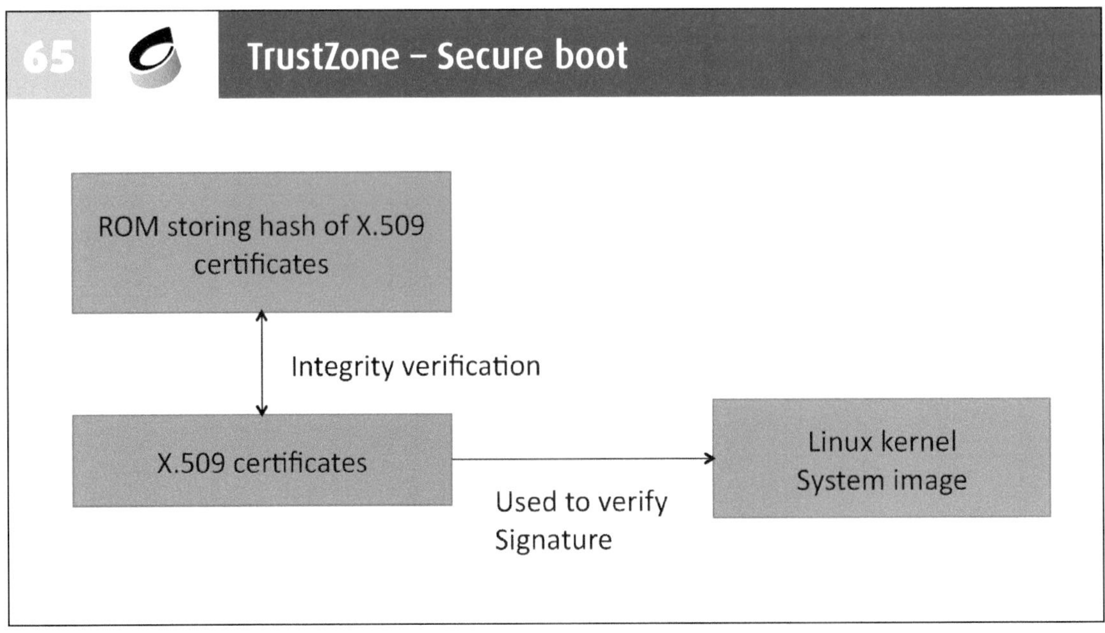
65
TrustZone – Secure boot
ROM storing hash of X.509 certificates
Integrity verification
X.509 certificates
Used to verify Signature
Linux kernel
System image

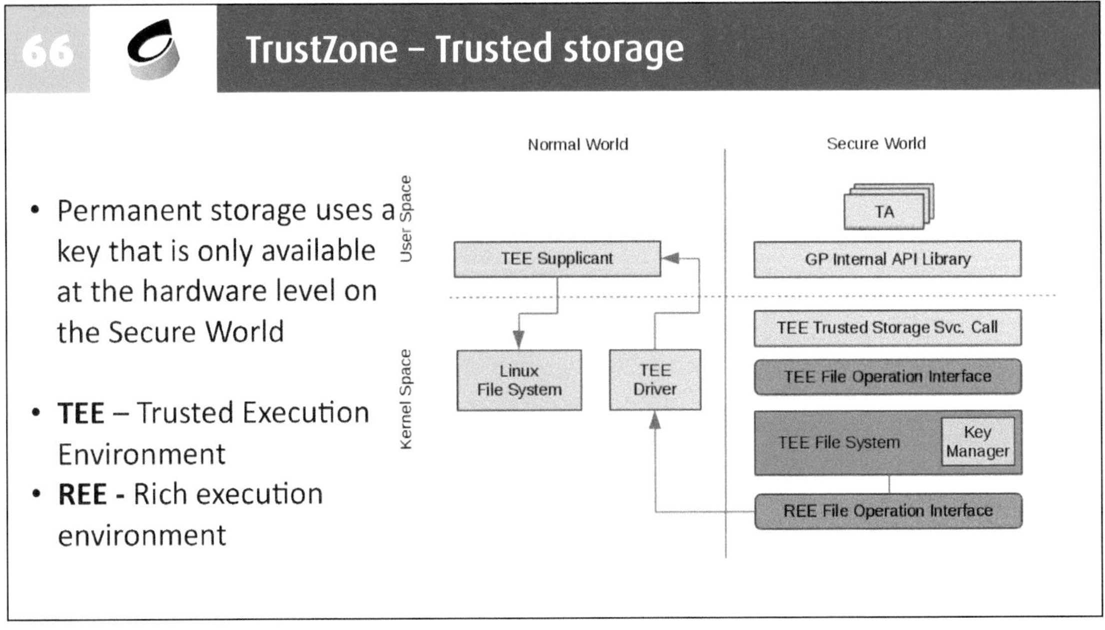
66
TrustZone – Trusted storage
• Permanent storage uses a key that is only available at the hardware level on the Secure World
• TEE – Trusted Execution Environment
• REE - Rich execution environment
Normal World
Secure World
User Space
Kernel Space
TA
TEE Supplicant
GP Internal API Library
TEE Trusted Storage Svc. Call
Linux File System
TEE Driver
TEE File Operation Interface
TEE File System
Key Manager
REE File Operation Interface

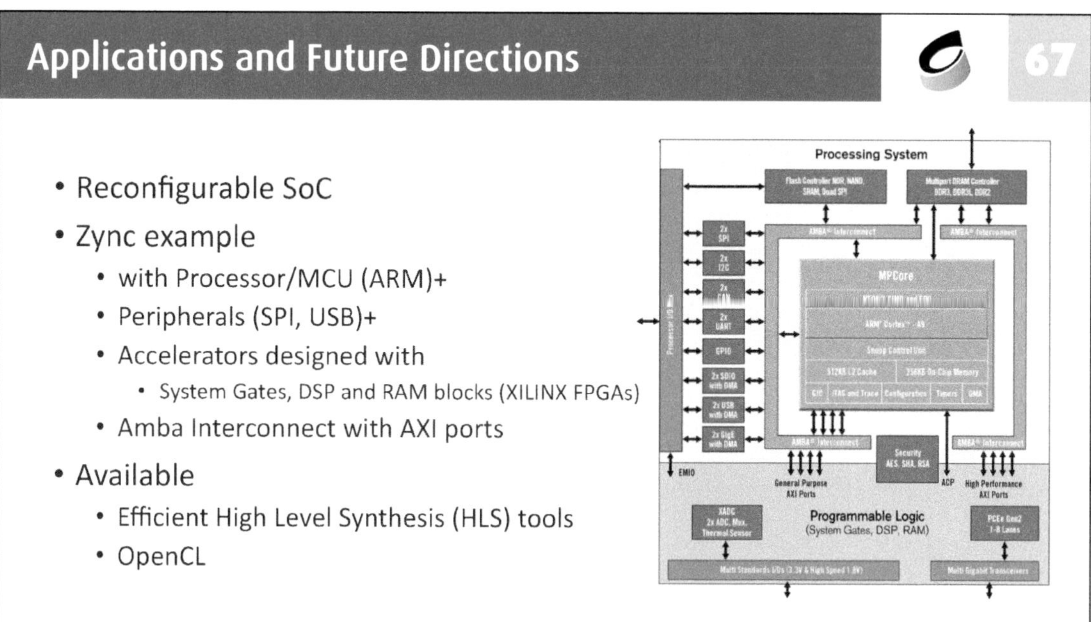

Applications and Future Directions
67
• Reconfigurable SoC
• Zync example
• with Processor/MCU (ARM)+
• Peripherals (SPI, USB)+
• Accelerators designed with
• System Gates, DSP and RAM blocks (XILINX FPGAs)
• Amba Interconnect with AXI ports
• Available
• Efficient High Level Synthesis (HLS) tools
• OpenCL
Processing System
MPCore
EMIO
General Purpose AXI Ports
ACP
High Performance AXI Ports
Programmable Logic
(System Gates, DSP, RAM)

The future is to have reconfigurable SoCs, with many different processors, peripherals, FPGAs, DSPs. By using FPGAs, it is possible to hardware instead of software, making faster, simpler and more efficient systems. These systems will also be more reconfigurable and heterogeneous. We can observe nowadays the example of the Zync system.

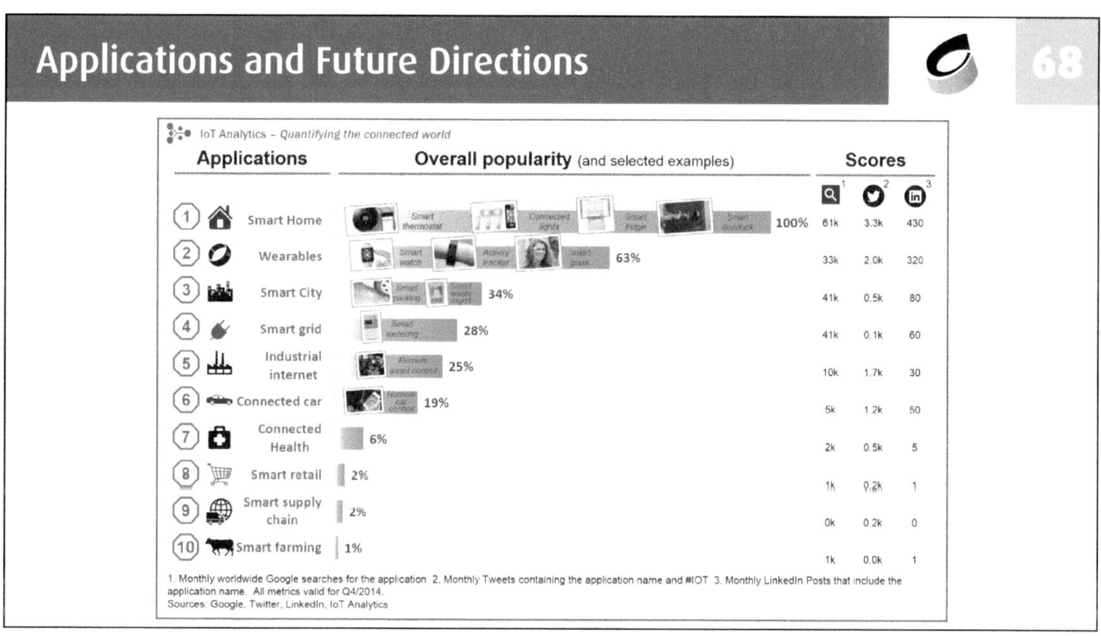

Smart home applications are becoming more popular and will be the main field for IoT systems, since it is an area that has the highest impact in the world population. Wearables are also very popular, especially due to smart watches. Health care devices are more difficult to accomplish, mainly because they are more expensive and require more security mechanisms.

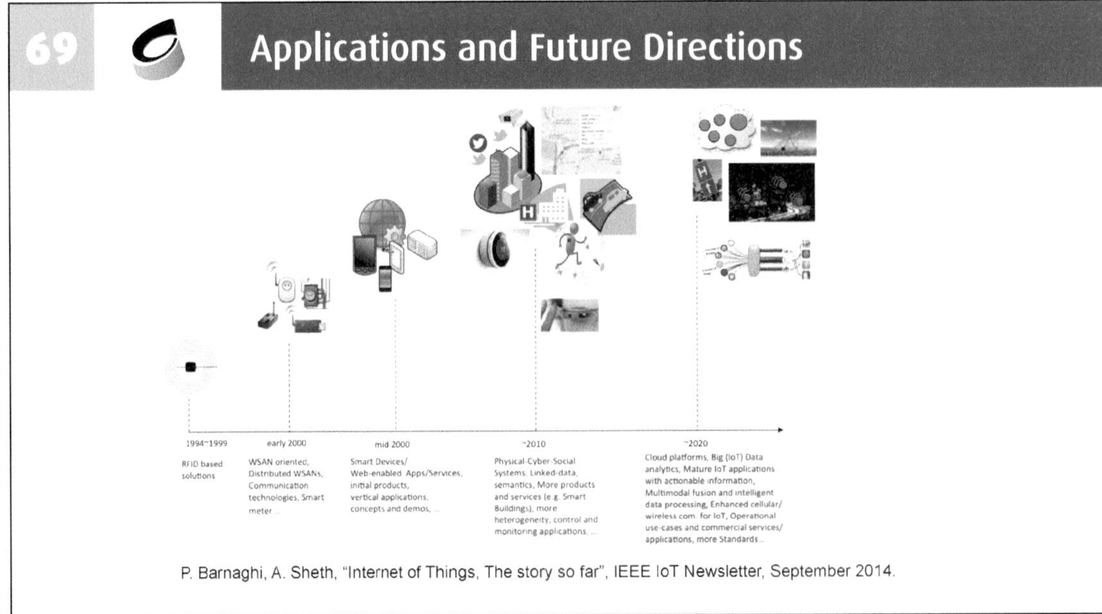

Applications and future directions of IoT.

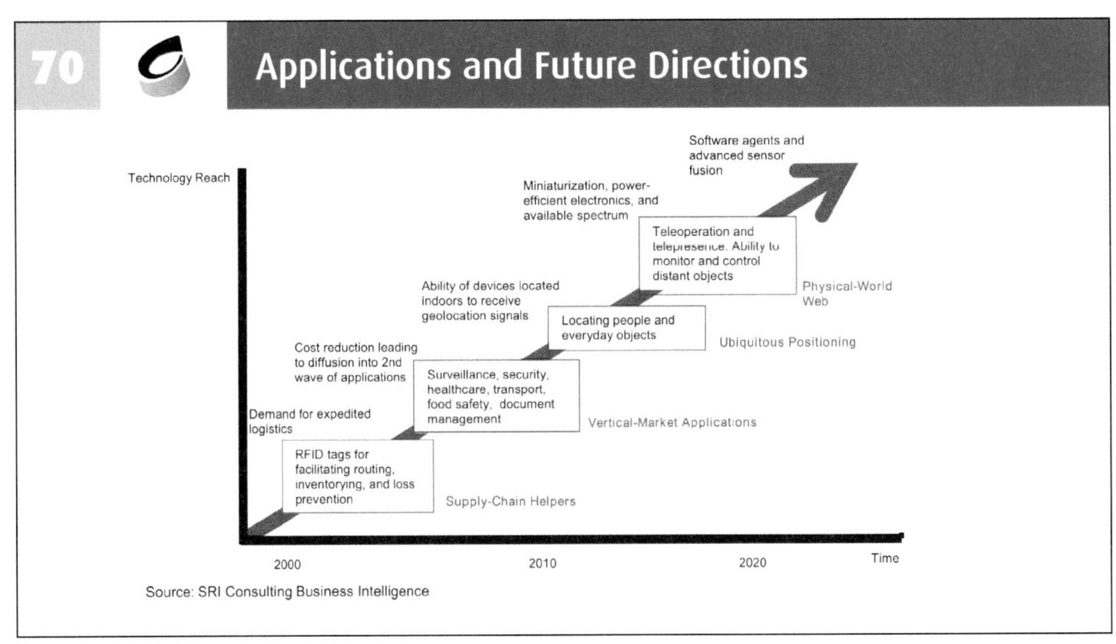

Bibliography

- "Cloud and Mobile Network Traffic Forecast - Visual Networking Index (VNI)." Cisco, 2015. http://cisco.com/c/en/us/solutions/serviceprovider/visual-networking-index-vni/index.html
- Danova, Tony. "Morgan Stanley: 75 Billion Devices Will Be Connected To The Internet Of Things By 2020." Business Insider, October 2, 2013. http://www.businessinsider.com/75-billion-devices-will-be-connected-to-the-internet-by-2020-2013-10
- "Global Connectivity Index." Huawei Technologies Co., Ltd., 2015. Web. 6 Sept. 2015. http://www.huawei.com/minisite/gci/en/index.html
- Manyika, James, Michael Chui, Peter Bisson, Jonathan Woetzel, Richard Dobbs, Jacques Bughin, and Dan Aharon. "The Internet ofThings: Mapping the Value Beyond the Hype." McKinsey Global Institute, June 2015.
- Karen Rose, Scott Eldridge, Lyman Chapin. "An Overview: Understanding the Issues and Challenges of a More Connected World", WWW.INTERNETSOCIETY.ORG
- MQTT - http://mqtt.org/
- ITU-T Recommendation X.509 : Information technology - Open Systems Interconnection - The Directory: Public-key and attribute certificate frameworks - http://www.itu.int/rec/T-REC-X.509
- J. Hennessy and D. Patterson, "Computer Architecture: A Quantitative Approach", Fifth Edition, 2011, Morgan Kaufmann Publishers
- TI Europe University Program, MSP430 Teaching Materials
- "ARM Cortex-M3 Introduction", ARM University Relations, The Architecture for the Digital World, ARM
- "eXtreme Low Power (XLP) PIC Microcontrollers", Microchip
- Arne Martin Holberg, Asmund Saetre. "Innovative Techniques for Extremely Low Power Consumption
- with 8-bit Microcontrollers", Atmel White Paper
- Rajovic, N., et.al. "Supercomputing with Commodity CPUs: Are Mobile SoCs Ready for HPC?", in SC 2013
- ARM Holding Plc, "ARM Roadshow slides", 2015 (Q1)
- "ARM Cotex-A9", Technical Reference Manual
- "ARM Cotex-A9 MPCore", Technical Reference Manual
- A. Lamber et al., "Keys to Silicon Realization of GigaHertz Performance and Low Power Cortex-A15", ARM Technology Conference
- Jörg Henkel, Sri Parameswaran (Editors), "Designing Embedded Processors: A Low Power Perspective", Springer, 2007th Edition
- "The Untapped Potential of Trusted Execution Environments on Mobile Devices", IEEE Security & Privacy, July/August, 2014, pp. 29-37

CHAPTER 6

Power & Energy Management for Internet of Things

Marcelino Santos

CTO SiliconGate, IST/UL, INESC-ID

Lisbon, Portugal

Traditional Power Management Units (PMUs) generate reference voltages and currents, produce power on reset signals and control the system start up sequence, sequencing the enable of voltage regulators. The new generation of PMUs, targeting IoT applications, is capable of entering different power gated modes, achieving nano-Amp range operation, being controlled by the Real Time Clock (RTC). Why and how are RTCs being used as brains of the IoT PMUs is going to be analyzed in the "Power-and-Energy Management for IoT" chapter.

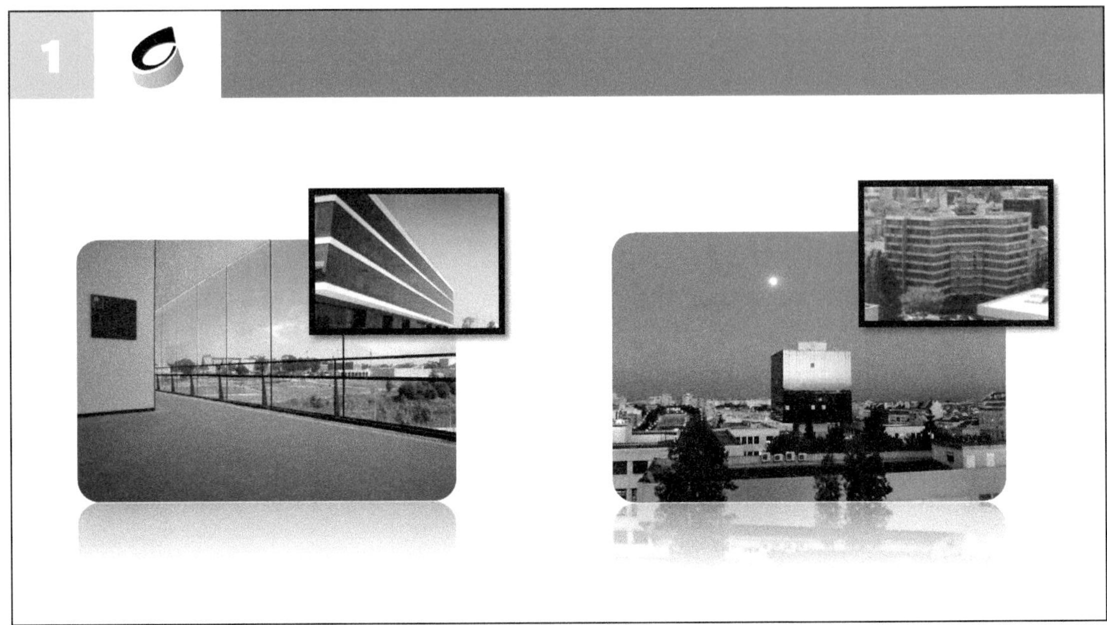

Devices of Internet of Things (IoT) typically present challenging specifications for their Power Management Unit (PMU) as most of them are small gadgets meant to stay operational, autonomously, for a long period of time. The PMU is responsible for not only efficiently powering the system but must also be able to manage the energy storage and harvesting to ensure a long lifetime of the device with the lowest external maintenance.

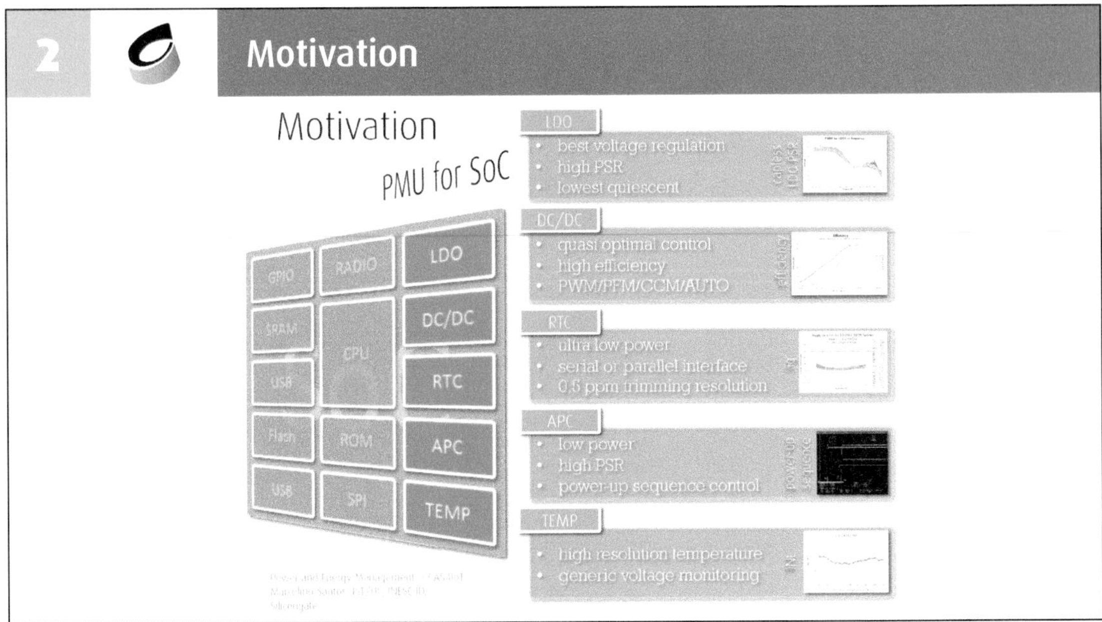

A power management unit is composed by several building blocks, such as voltage regulators, low drop out regulators (LDO) and DC-DC converters, a real time clock (RTC), an advanced power controller (APC) and might also have additional functionalities like defining power modes based on special signals and/or sensors.

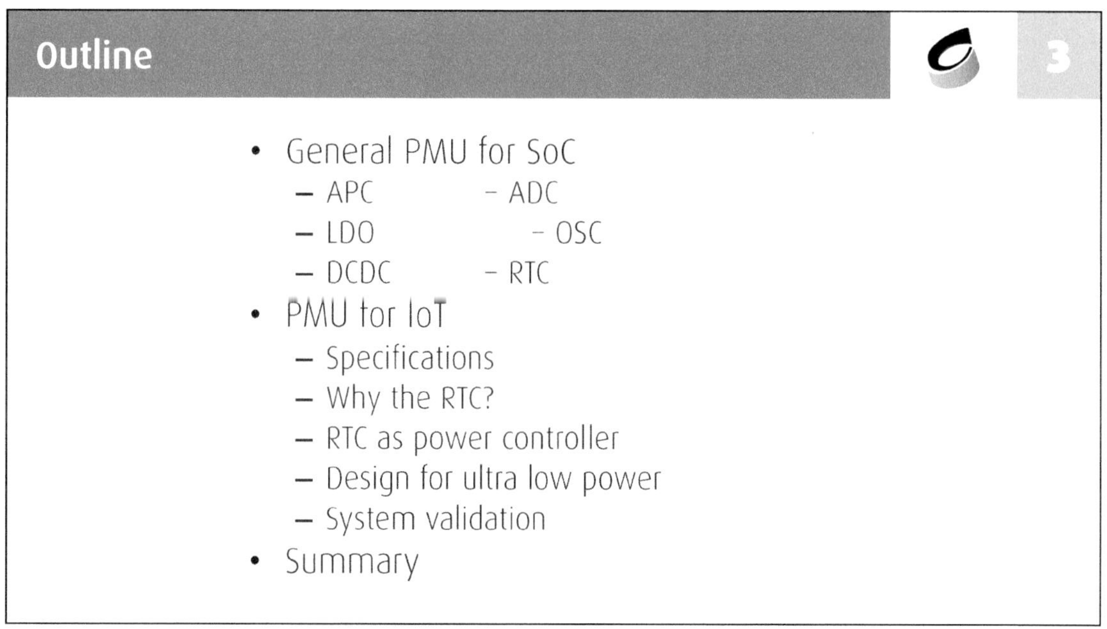

In this tutorial, a general overview of the main building blocks of a PMU for System-on-Chip (SoC) are going to be described, as well as the specifications and requirements imposed by IoT applications to the PMU.

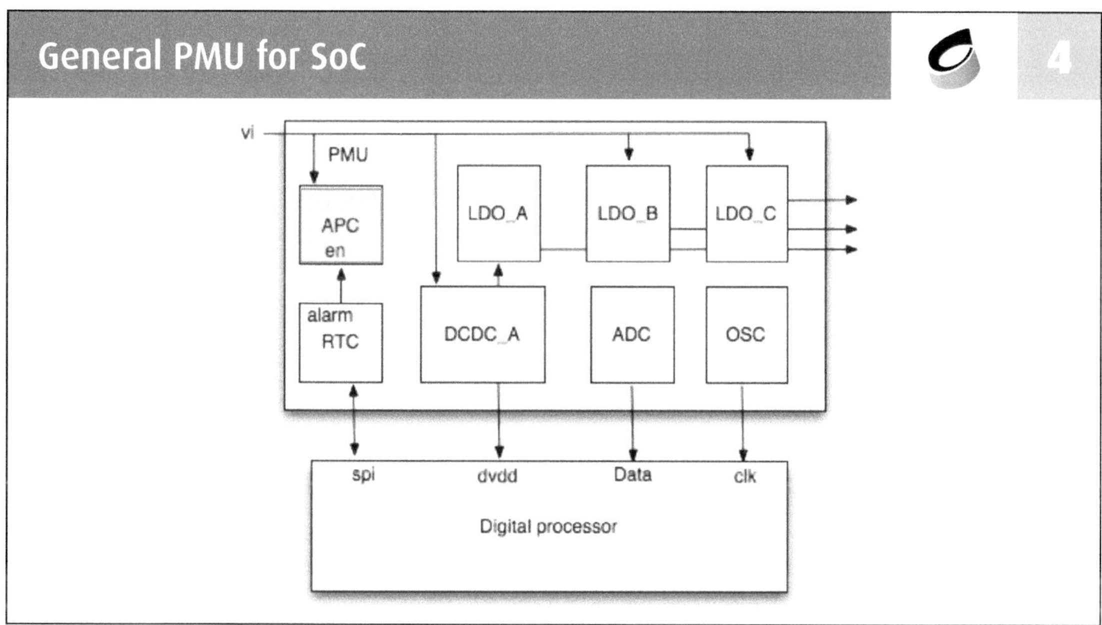

A general architecture of a PMU can be seen in this slide, with each block performing a specific function. Each of these blocks will be analyzed further on. We'll look first into the Advanced Power Controller (APC), which can also be known as Internal Supply Module (ISM).

5 

General PMU for SoC - APC

Main functions:
- Provides reference voltages
- Provides bias currents
- Generates power-on-reset (por)
- Controls power up sequence (multiple voltage regulators)
- Detects undervoltage-lockout

The APC is the backbone of the PMU. It is responsible for enabling/disabling the reference voltage (used in LDOs and DC-DCs), bias currents, providing a power-on-reset for the digital logic, and it includes a undervoltage-lockout detector, so that the PMU is turned OFF whenever the voltage is not enough to ensure its proper behavior. This is done using hysteresis to avoid oscillation.

General PMU for SoC - APC

Auxiliary functions:
- Provides clock signals
- Detects brownout
- Provides cascode protection voltages
- Regulates voltage for other PMU low power cores
- Provides the control signals for the digital interface of all cores
- Allows trimming for process and temperature compensation

More than the basic functions, the APC has some auxiliary functions that depend on the specific application of the System on Chip (SoC). It can be used to:

- Provide clock signals, for example for the DC-DC converters;
- Detect brownout, i.e. generates a warning when the voltage is getting close to the lowest allowable value;
- Generate cascode protection voltages, which are mandatory for a voltage supply higher than 3.6 V to ensure a good reliability of the PMU;
- Control the voltage regulators of internal cores of the PMU;
- Provide control signals for the digital interface and also for voltage regulators that are programmable;
- Trim reference voltages/currents in order to compensate for process and temperature variations.

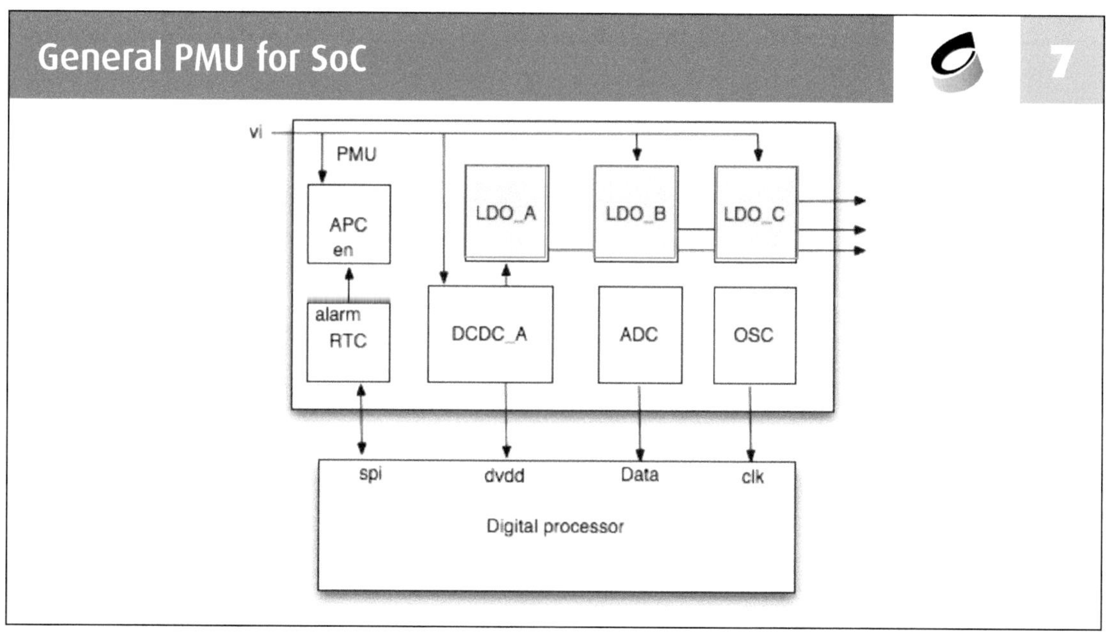
General PMU for SoC
7
vi
PMU
APC
en
alarm
RTC
LDO_A
LDO_B
LDO_C
DCDC_A
ADC
OSC
spi
dvdd
Data
clk
Digital processor

The focus now shifts to the voltage reference circuits. These can be implemented using either LDO regulators or DC-DC converters. Here, 3 LDOs are shown just as a mere example.

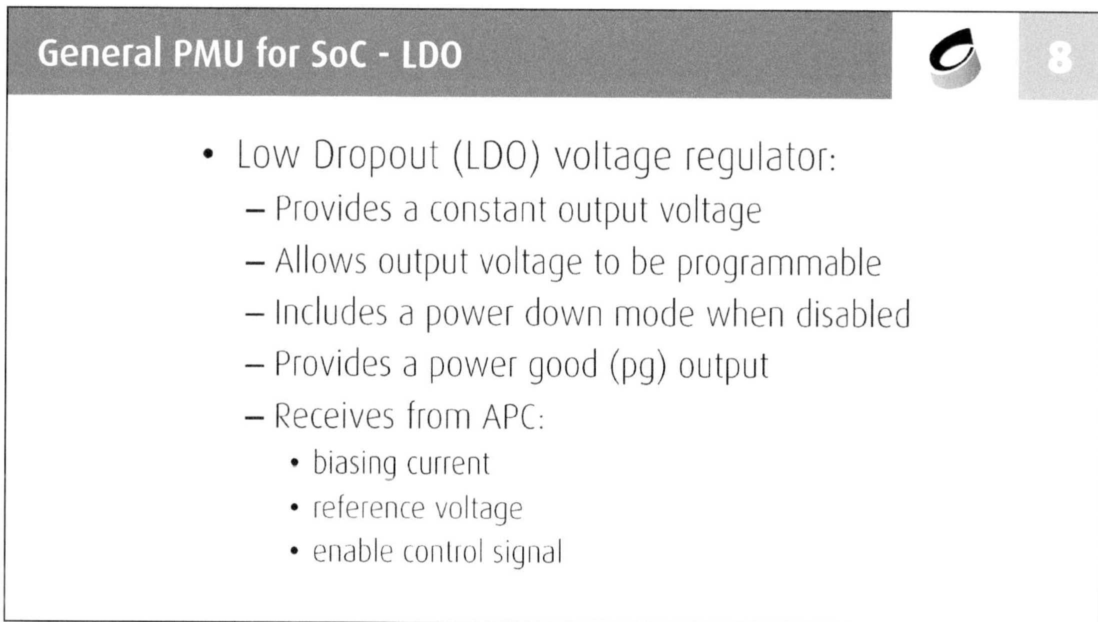
General PMU for SoC - LDO
8
• Low Dropout (LDO) voltage regulator:
– Provides a constant output voltage
– Allows output voltage to be programmable
– Includes a power down mode when disabled
– Provides a power good (pg) output
– Receives from APC:
• biasing current
• reference voltage
• enable control signal

The LDO regulators can provide a constant and programmable output voltage as long as it is lower than the supply voltage. It can be easily disabled if it is not needed. These circuits require a biasing current, a reference voltage for the feedback regulation, and the enable signal, which are supplied by the APC.

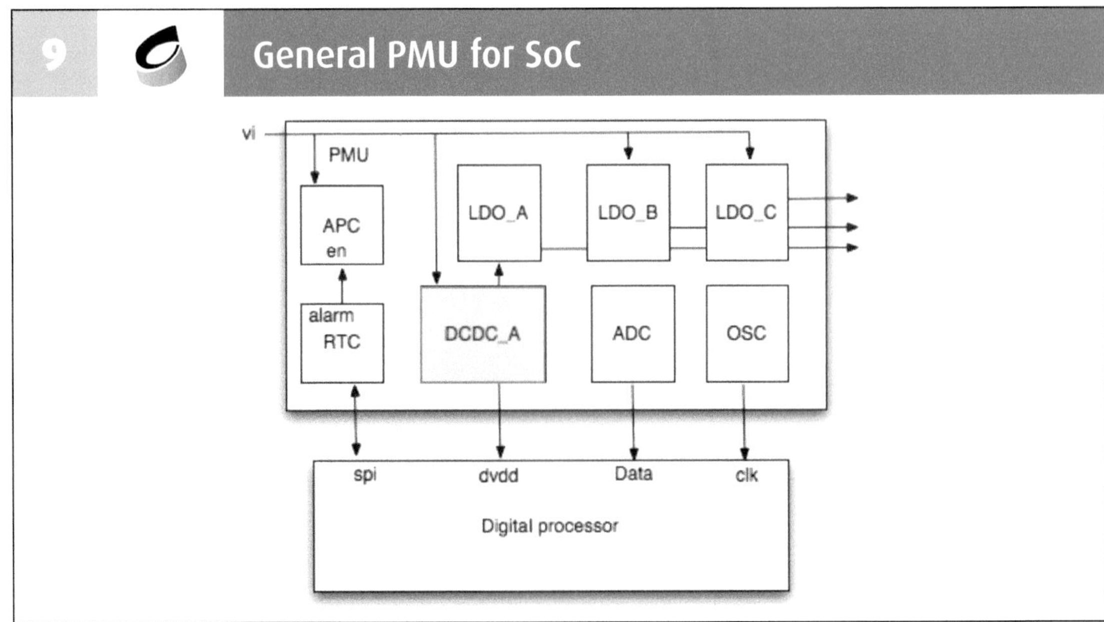

The main disadvantage of an LDO regulator is that high efficiency is only achieved when the regulated output voltage has a value close to the supply voltage. For voltages both lower and higher than the supply voltage, an inductor or capacitor based DC-DC converter is required.

10

General PMU for SoC – DCDC

- DCDC voltage regulator:
  - Provides a constant output voltage
  - Allows output voltage to be programmable
  - Includes a power down mode when disabled
  - Provides a power good (pg) output
  - Receives from APC:
    - biasing current
    - reference voltage
    - enable control signal
    - clock

These converters can provide a stable output voltage with high efficiency (above 90%). Similar to LDO regulators, the output voltage of a DC-DC convert can also be programmable, and can be powered down. They also require a biasing current, reference voltage, clock signal, and an enable signal to trigger them when needed.

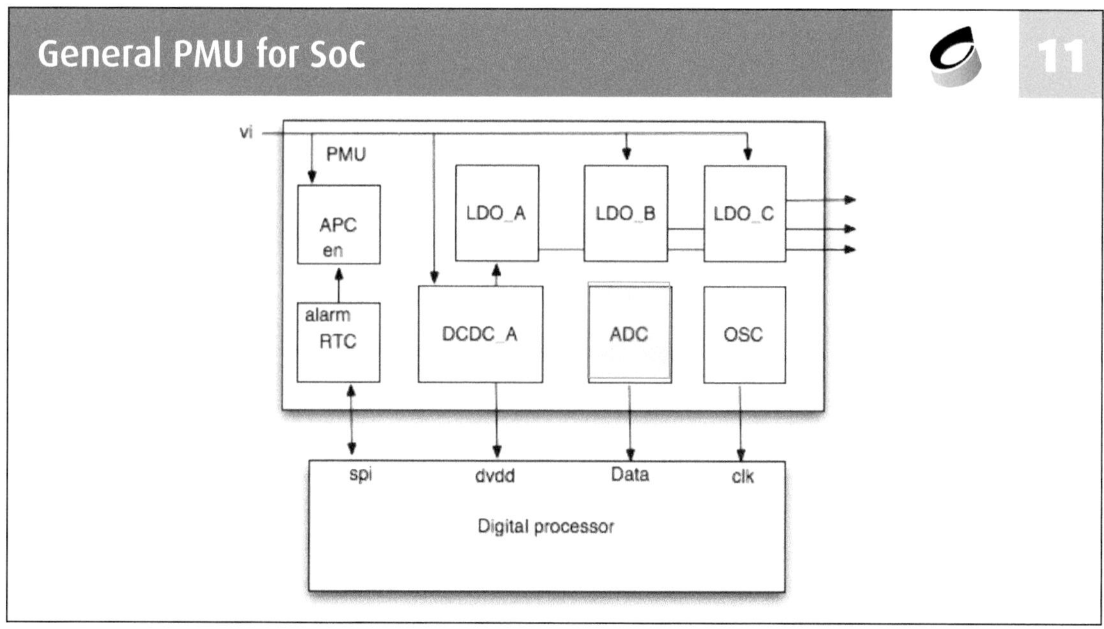

An Advanced Digital Controller (ADC) block can sometimes be included in the PMU, with the main function being the monitoring of its temperatures and voltages.

General PMU for SoC – ADC

12

- Temperature and voltage monitoring :
  - Measures temperature (diodes) and/or voltage
  - Multiplexes different input sources
  - Outputs the digital value and end-of-conversion (eoc)
  - Receives from APC:
    - biasing current
    - reference voltage
  - Receives from digital control
    - Start of conversion (soc) command
    - clock

It recurs to different input sources to monitor such information and also provides both a digital value and an end-of-conversion (EOC) signal at its output. In order to provide this information, it requires both the biasing current and reference voltages from the APC, as well as a clock signal and a start of conversion (SOC) command from the digital control.

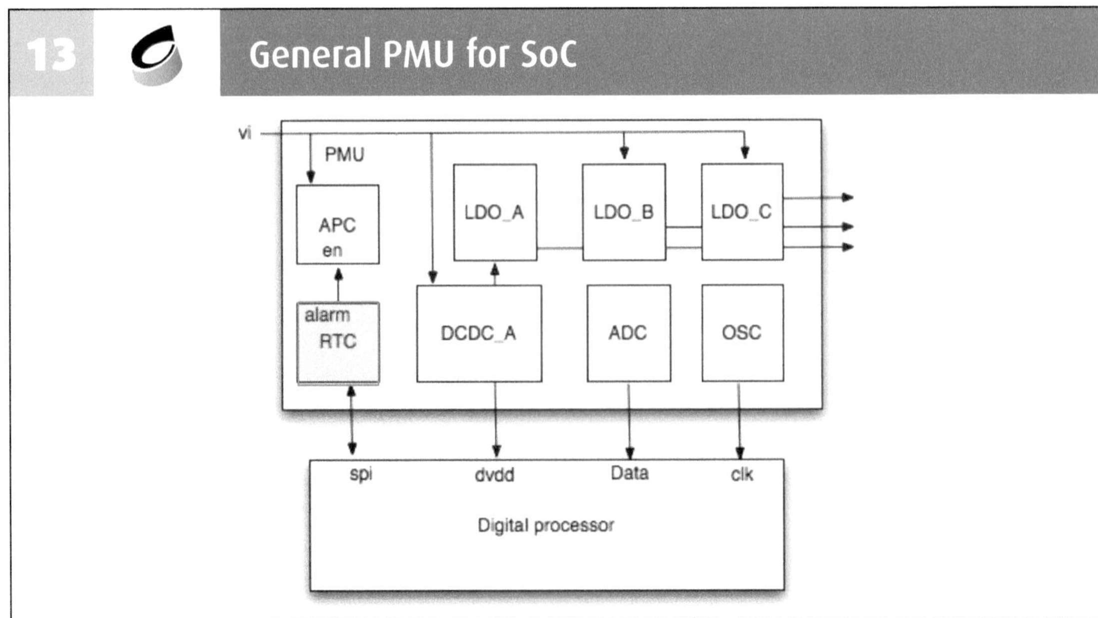

We now move to the The Real Time Clock (RTC) module, which tracks time and is a fundamental block within the PMU.

14 General PMU for SoC – RTC

- Real Time Clock (RTC):
  - Includes a 32768 Hz crystal oscillator
  - Divides crystal frequency to obtain 1s clock
  - Increments 32bits time_reg with 1s clock
  - Activates alarm output if time_reg == alarm_reg
  - Includes a digital serial interface
  - Can be trimmed to compensate for crystal dependence on temperature (0.5 ppm)

It is implemented by a crystal oscillator ($FCLK \approx 32768Hz$) that generates a clock signal that, after divided by 2^{16}, provides a 1 Hz (T=1s) signal that is used to increment a 32-bit register to increment time (typically 32 bits is enough to keep a record of several years). Alarm signals can be triggered by comparing the RTC signal to a fixed value. For example, it can be used to power up the system regularly. The crystal oscillator depends on the temperature. Trimming can be achieved by stealing or adding a specific number of clock cycles after 64 seconds, achieving a precision of 0.5ppm. This trimming process can be used, together with a temperature sensor, to compensate for the frequency dependency on temperature.

General PMU for SoC – RTC

- Real Time Clock (RTC):
  - Uses ultra-low power
  - Is usually permanently powered (for time keeping)
  - Alarm output is the enable of *High Power* mode - enables the APC that commands the power-up sequence:
    - PMU is enabled when time_reg==alarm_reg
    - PMU is disabled when the digital control resets alarm
  - Digital information can be kept in general purpose registers (GPRs) in the RTC – the only core supplied during *Low Power* operation

The power consumption of the RTC is very low, usually never higher than 200 nA, and thus the RTC is the only block in the PMU that is always enabled. The main reason behind it is to keep track of time. Consequently, it is used also to store data that needs to be kept even after all possible power down states. Since the RTC is always powered on, it can be used as a permanent memory even when the processor is powered down.

General PMU for SoC – RTC

- Real Time Clock (RTC):
  - An RC oscillator can be used alternatively or complementarily to the crystal oscillator. Reduces the BOM or allows anti-tempering: detection if the frequency is changed at the crystal pins
  - The RC oscillator has 1% accuracy in the range of temperature -40°C to 125°C

In order to decrease the bill of materials (BOM), an RC oscillator can be used in applications where the crystal accuracy is not required. In applications where it is important to detect attacks, the RC oscillator can be used to monitor the crystal frequency since its operation does not depend on external pins and, therefore, it is more robust.

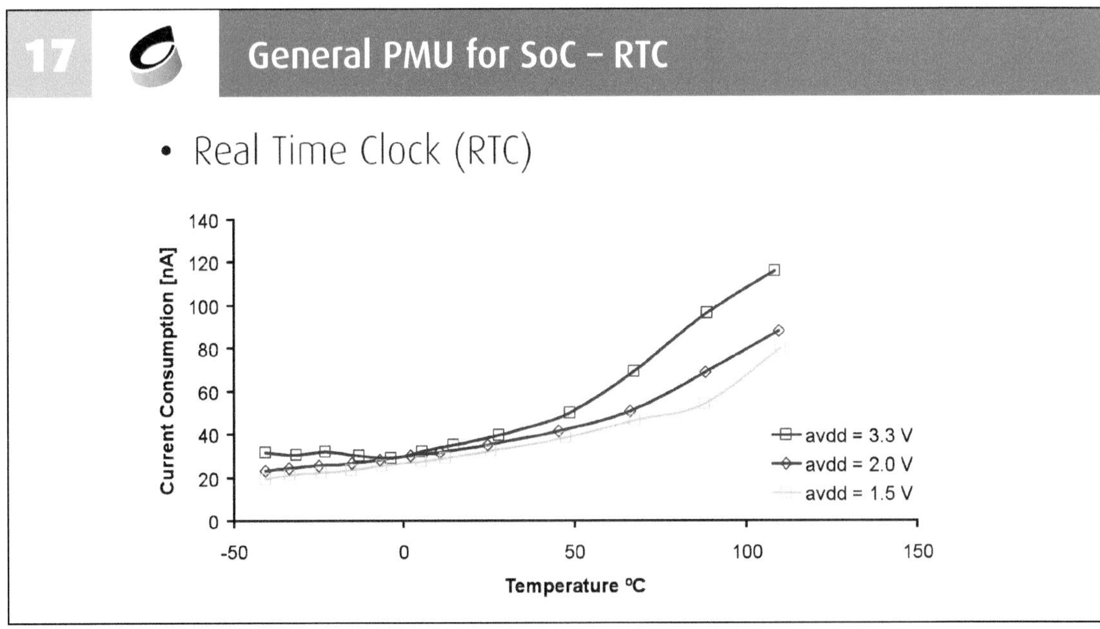

The plot shows the current consumption of the RTC as a function of the temperature. As the temperature increases so does the power consumption. Also notice that decreasing the supply voltage does not reduce significantly the power consumption, unless at high temperatures. To get even lower power consumption, the crystal can be used only from time to time. This greatly reduce the power consumption because when the crystal oscillator is powered down the parasitic capacitance losses of the crystal pads is zero.

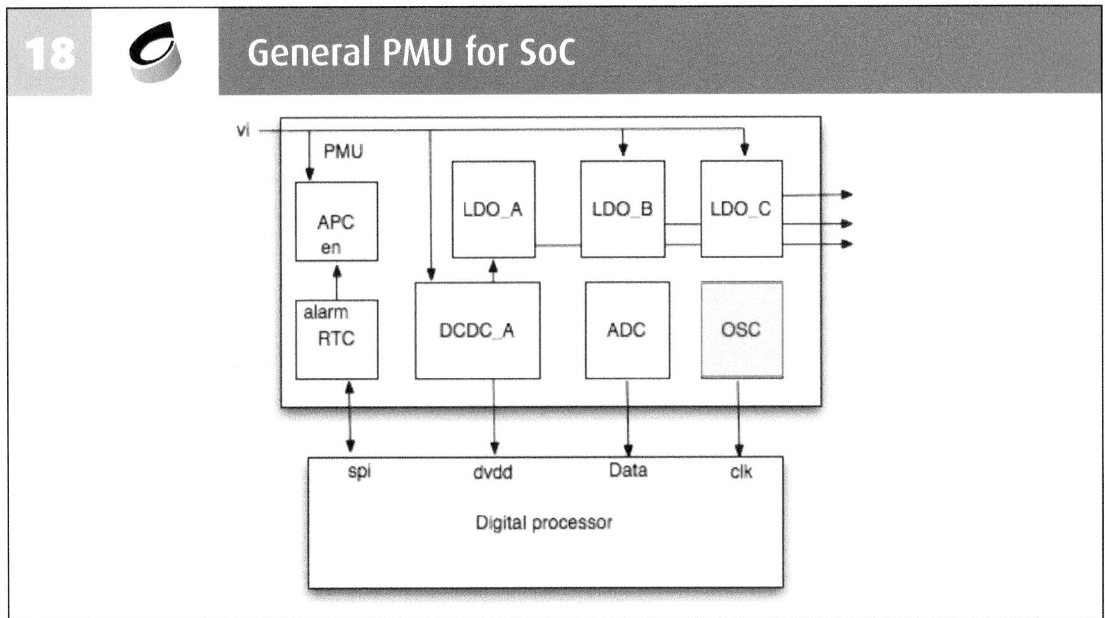

Lastly, other main block in a PMU is a programmable oscillator.

General PMU for SoC – OSC

- Programmable oscillator (2MHz to 160MHz):
  - Does not require one additional crystal
  - Frequency is dynamically programed with a digital word
  - Power scales with frequency
  - Lower power than PLL

This programmable oscillator is used to generate a clock signal that can be used for clocking DC-DC regulators or the digital logic. Programmable frequency and voltage can be used to implement dynamically voltage scaling (DVS).

General PMU for SoC – OSC

20

- Programmable oscillator (2MHz to 160MHz):
  - Uses the RTC 32768 Hz clock as reference
  - Multiplies the crystal frequency to obtain output
  - +/- 1% accuracy
  - Outputs lock="1" when freq within 1% range
  - Requires a biasing current from APC

The frequency can be quite accurate (within 1% error) because the crystal oscillator output is used to monitor the higher frequency oscillator, through a feedback loop. Hence high accuracy can be achieved.

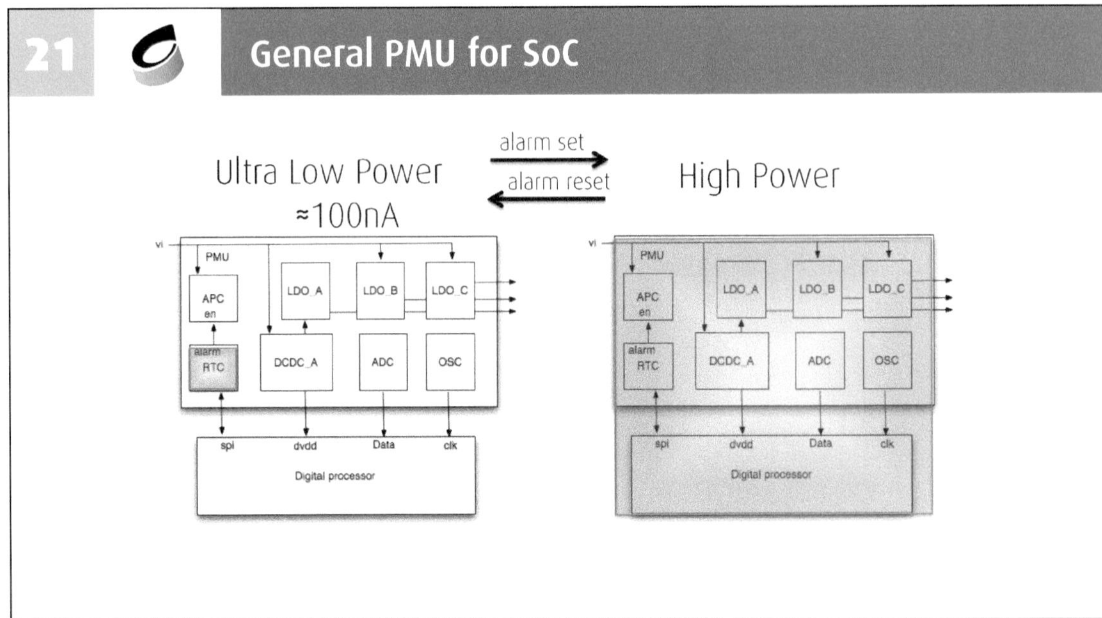

To ensure low power consumption the system must enter in a low power mode when everything can be turned OFF except the RTC. This is because the RTC keeps track of the time and thus it can trigger an alarm to wake up all system, forcing the high power mode after a predefined sleep time.

22 PMU for IoT - specs

- In IoT applications:
  - It is frequent to have a very tight energy budget
    - Systems powered with energy harvesting
    - Systems with limited volume
    - Systems powered by a non-replaceable battery
  - System needs to be compliant with multiple sensors
  - Multiple power modes and reset actions

Because of the IoT market constraints, more functionalities and requirements have been required for the PMU. Some examples are the capability of a PMU to collect/gather energy from the environment (energy harvesting) or to be able to operate during large periods of time (as some modules might have irreplaceable batteries). Also, most systems aim to be portable and, as result, they should have a limited volume. Since some IoT SoCs are designed to be generic, the PMU must be, as much as possible, transversal to the application, supporting multiple and configurable power modes, wake signals and reset actions.

PMU for IoT - specs

- Therefore, PMUs for SoC targeting IoT require:
  - Power configurability
    - In High Power - some voltage regulators can be kept disable
    - In Low Power - a configurable set of voltage regulators is kept active
  - Multiple *wake*, *power_on* and *reset* sources – with configurable polarity and masking

PMUs designed for the IoT market should present a power configurability feature, where voltage regulators are enabled/disabled in accordance to the required SoC operation mode. In other words, even at high power operation it might occur that a specific set of voltage regulators can be turned off (as they are not required) and at low power, just to ensure proper operation, some voltage regulators can be kept active. Moreover, the PMU should have multiple wake, power_on and reset signals, each compliant with a different sensor.

PMU for IoT – why the RTC?

- Power configurability
  - High Power – when an event wakes the system, a register identifies the regulators that must be excluded from power-on
  - Low Power – when the system goes to Low Power, a register identifies regulators that must be kept active

Registers that control power modes need to be placed in a domain always powered - RTC

Specific registers that define which regulators are enabled or disabled during high and low power modes are essential for this power configurability feature. However, these registers need to be always active and, as a result, they are placed in a domain that is always powered – the RTC. This way no information is lost when the system enters a low power mode.

25 

PMU for IoT – why the RTC?

- Multiple *wake*, *power_on* and *reset* sources with configurable polarity and masking
  - wake inputs can be active high or active low
  - wake inputs can be enabled or disabled
  - The system can be sesitive to power_on and/or reset signals depending on the present power state

Registers that control wake[i] functionality and the system power state need to be placed in a domain always powered - RTC

Since the RTC is always powered ON, the APC becomes a slave of the RTC because the latter knows the status of the system. Wake signals can be configured to be active high or low, and their functionality enabled or disabled. The system can also define the priority of each signal.

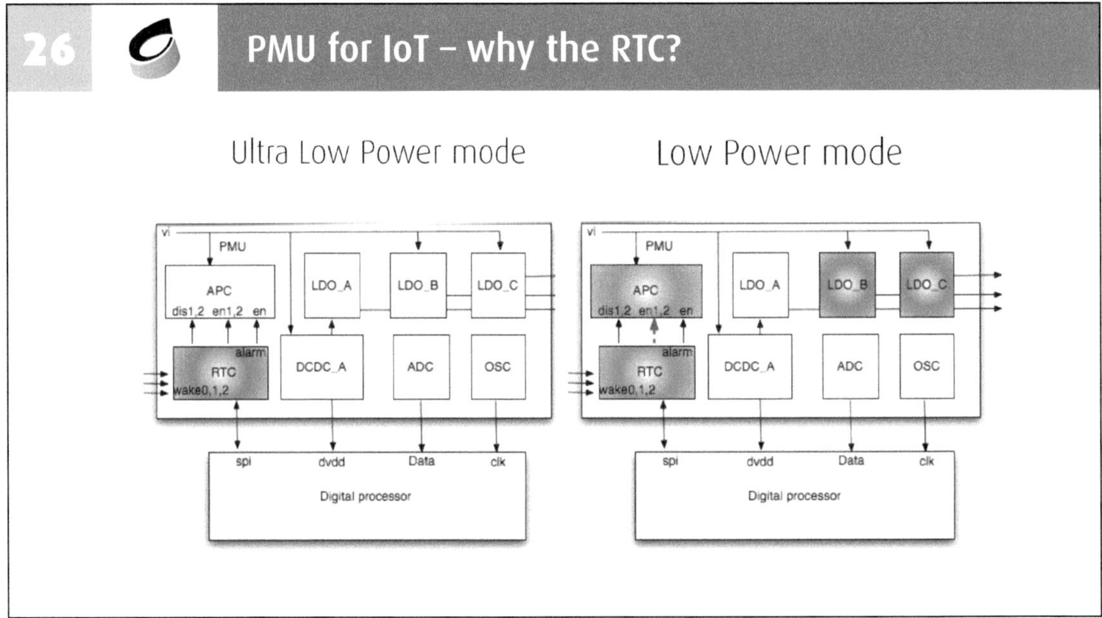

To further decrease power consumption of the PMU, an ultra low power mode can be created, apart from the low power mode. In this mode, only the RTC is powered ON and all the voltage regulators are turned OFF. This means that the SoC will be completely turned OFF. Even in low power mode, dynamic voltage scaling (DVS) can be used by decrease the supply voltage of the core itself, reducing the power consumption.

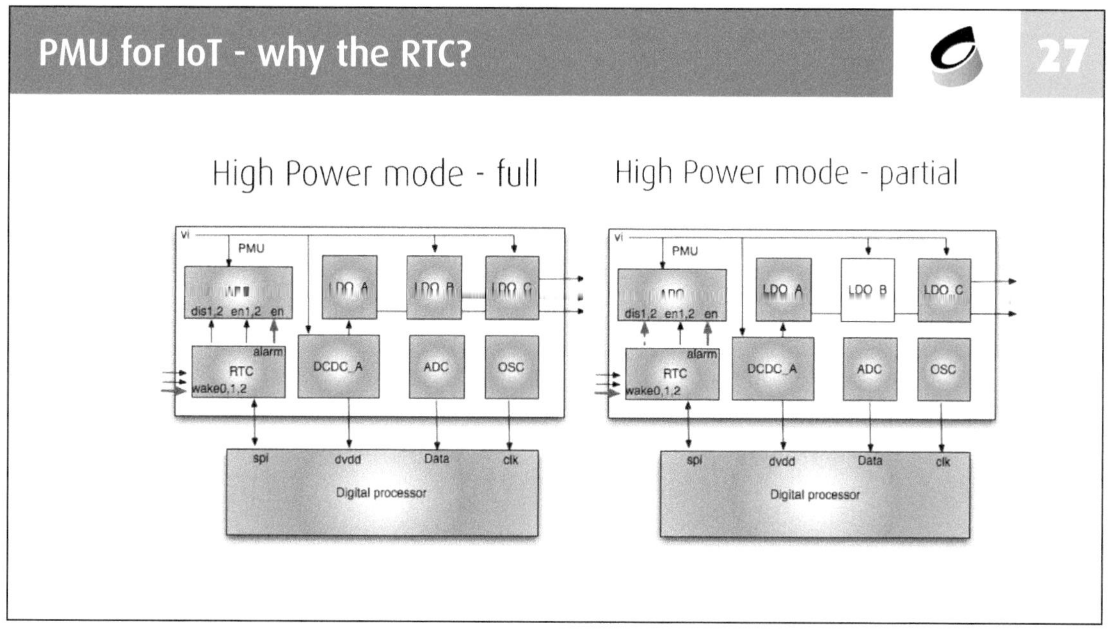

Since the RTC has the "disable" register, in the high power mode the APC can disable some of the regulators that are not being required from the SoC, ensuring power configurability and reducing current consumption. This flexibility feature is highly attractive to IoT applications.

PMU for IoT – RTC as pw controller

28

Cause register

Since the system can be wakened by different events (and not only the alarm), after a power-up, the processor needs to know the "cause" why it has just been powered.

- the cause register has bits which are individually set by the wake inputs or other wake events, like alarm or power-up

When the system wakes up it needs to know why, i.e., which event set it back on. Hence, a "cause" register is required. When the RTC receives the wake signal, it will set a bit in the cause register so that the processor can read this register and understand why the system is being turned on. One of the many causes possibly responsible for waking is the alarm, which triggers the system to be powered on.

29 

PMU for IoT – RTC as pw controller

Configurable Power at the RTC

Targeting leakage reduction in Ultra Low Power:

- The RTC SPI and output interface can be powered down when the system is in Ultra Low Power
- The RTC core logic can be supplied at a lower voltage without the need of level converters
- The internal blocks that are powered down, when the system is in Ultra Low Power, are configurable in a register

The RTC itself can reduce its power consumption in the ultra low power mode, through power gating. In other words, the RTC Serial Peripheral Interface (SPI) and output interface can be completely disabled. Not only that, but its core logic can be powered at a lower voltage (DVS), discarding the use of level converters. Again, this configuration can be saved in a register inside the RTC.

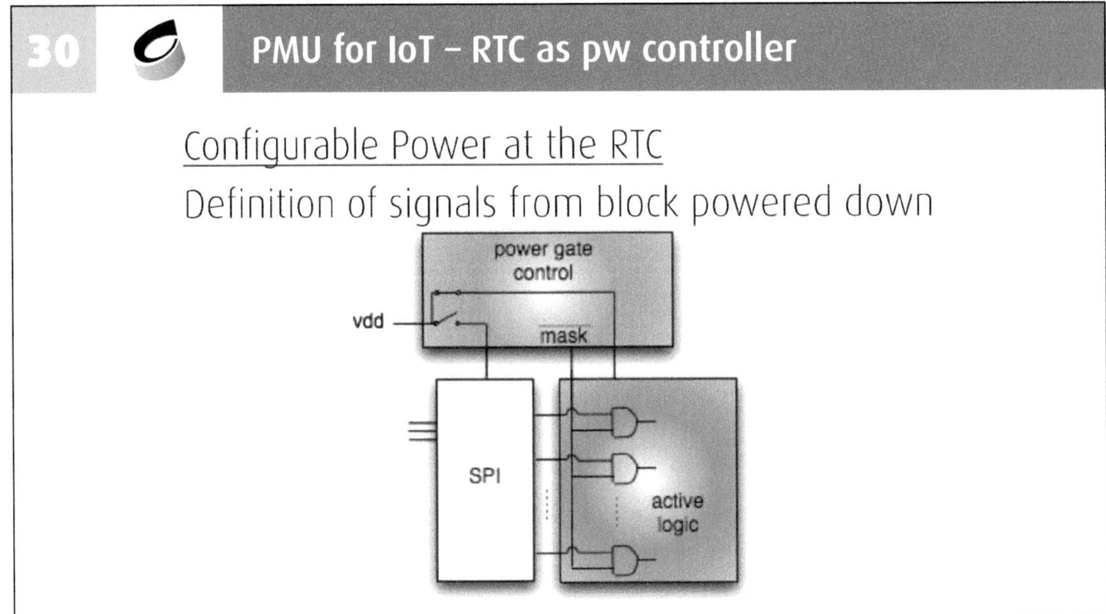

Power gate the SPI should not disable the logic that needs to remain active, for example counting the time and comparing with the alarm, which is always active despite being in low voltage. Hence, the digital signals of the active logic need to be masked. We need well defined logic levels all the time, which means that whenever a block is turned off, its output value should be well defined. To make sure of this, validation procedures are needed, which will be explained later.

PMU for IoT – RTC as pw controller

Configurable Power at the APC

- Inside the APC only the required blocks are powered for each operating mode
  - In ultra low power there is no voltage reference or bias current circuit enabled
  - In low power, one bit of the RTC (lcm) can induce a lower current mode on all active regulators, using lower bias currents from the APC

In ultra-low power mode, when disabling the voltage regulators that are not being used, the APC must also disable the auxiliary circuitry used for the proper function of the voltage regulators. For example, the DC-DC requires an oscillator to maintain the voltage regulated, hence when the DC-DC is not required the oscillator must also be turned OFF. Other examples are the disabling of bias currents, bandgap voltages, and so on. An alternative approach, also used, is to keep the voltage regulators working at very low current values, while ensuring that the output voltage is kept.

PMU for IoT – RTC as pw controller

Configurable Power at LDOs

- In low power, if lcm=1, all active regulators
  - use a lower reference current
  - reduce their static power scarifying speed

Not an issue since load regulation and PSRR are not relevant for this mode since the system is with minimum activity

This last approach can be easily programmed in the RTC by using a control bit (lcm=1). For example, in the LDO regulators, using a smaller value for the bias current can greatly reduce the power consumption at the cost of slowing down the regulator response to transient changes. However, this is not a problem because in this mode the system is in sleep mode, so there should not be any fast change.

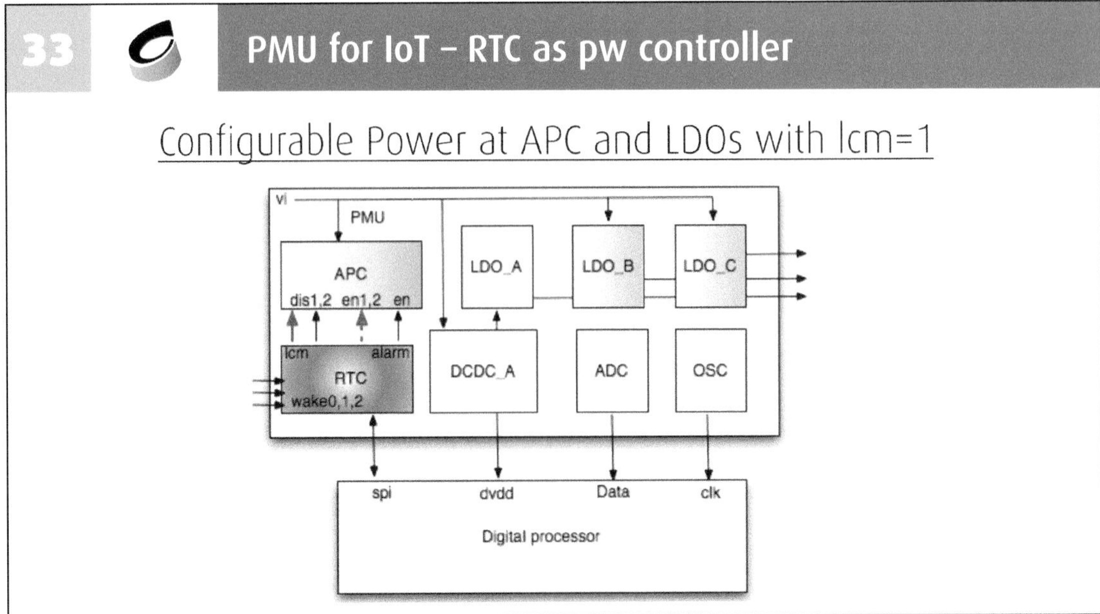

This ultra-low power mode also inside the APC, creates a new degree in reduction in power, where even if the alarm is OFF and the system is asleep, it is still possible to enable some regulators, and keep them in low current mode, thus the bias current is reduced.

PMU for IoT – RTC as pw controller

Configurable Reset

Targeting a flexible use in different scenarios, a reset event causes default values in a programmable subset of registers:

- time
- alarm1, alarm2
- general purpose
- cause
- config1, config2, ...
- trimming

The reset can also be configurable, i.e., different reset signals can be selectively used to force the default values of several registers (such as time, alarm, etc.), instead of resetting the whole system. The configuration of the action of each reset command is also kept in registers within the RTC.

PMU for IoT – Design for Ultra LP

- Logic functions need to be implemented with asynchronous digital circuits
- Sub-threshold is used in most analog design
- Large channel length transistors are used even for switching transistors with the purpose to limit leakage
- When in low power modes, circuits may be in non-continuous operation, being activated periodically by a dedicated clock from the RTC

One way of reaching ultra low-power operation is to use asynchronous digital circuits, preferred over synchronous ones, as the energy dissipated in toggling is greatly reduced in comparison. To reduce leakage, large channel length (> 500 um) transistors are used, sacrificing speed, even for the logic circuitry. Most of the devices operate in the sub-threshold region. Once again, these circuits can operate in a non-continuous fashion, monitored by the RTC.

PMU for IoT – Design for Ultra LP

Asynchronous digital

Synchronous digital circuits require a significant amount of energy just for the toggling of the clock tree.

- RTC logic needs to be an asynchronous design
- Asynchronous design is not part of the digital flow supported by well known EDA tools
- High validation effort is mandatory

Despite the inherent energy savings behind the use of an asynchronous design, most of the well-known EDA tools cannot validate it as well as synchronous designs are validated. Thus, most of the design relies heavily on the designer's experience and, as a result, validation procedures should be done with extreme care and precaution. System validation methods are detailed next.

37 

PMU for IoT – system validation

- Transistor level simulation
  - Only for individual blocks
  - Mandatory to detect leakage that results from improper crossing of power domains
- Digital verilog simulation
  - Excellent for digital blocks, value for analog blocks depend on the level of accuracy in the model
  - Digital verilog model is mandatory as frontend view – to use it for validation is possible with no design cost

At transistor level, only individual modules are extensively simulated, as the whole system takes large amounts of time and, therefore it is simulated at transistor level only at the final validation of project. At transistor level simulation, particular attention is given to the supply current at all possible operational modes. The aim of this validation is to detect leakage.

Digital Verilog simulation is carried out for the entire PMU module, benefiting from its highly accurate analysis of digital blocks. Analog blocks can also be evaluated, as long as their model representation is also accurate. Digital Verilog simulation can detect more than 90% of the errors in the design, in most cases.

38 

PMU for IoT – system validation

- Digital verilog simulation
  - Relevant analog signals are simulated with 64 bits

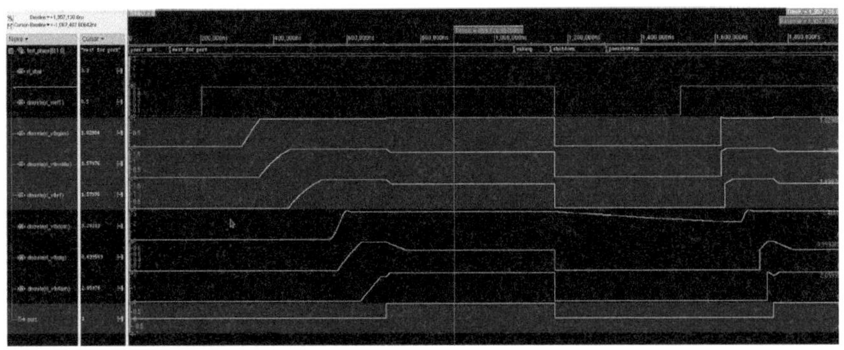

Other types of validation are also presented. In mixed level simulation, the validation of the whole system can take as long as transistor level simulations (with no clear advantage) and, as a result, slower than digital Verilog. Furthermore, as it combines both analog and digital simulations, proper interfaces must be configured. As an alternative, FPGA emulation is the fastest method and allows testing specific features for a particular module (such as the RTC, APC, etc.). Nevertheless, transistor level simulations are required afterwards in order to validate power domain crossings.

PMU for IoT – system validation

39

- Mixed level simulation
  - Much lower simulation speed than pure digital and with no clear advantage if individual blocks are simulate at transistor level
  - Requires additional effort for configuration of interfaces
- Emulation in FPGA
  - Fastest validation method
  - Enables test preparation
  - Does not validate the crossing of power domains

The graph on the slide shows a digital Verilog simulation result. In it the output of voltage regulators and enabling/disabling signals are plotted. The last trace (in green) is the enable signal from the digital interface. It can be seen that when it is triggered, the voltage regulators output is decreased because of the current being pulled from the digital. When the digital is disabled, some voltage regulators have a sinc source to pull down the voltage to zero. Others do not have a sinc source and their output is slowly discharged in time.

Summary (1)

40

- PMUs for IoT
  - can be awaked by different wake inputs - not a single alarm
  - can be configured into different power modes, minimizing the energy required for each action
  - use the RTC as power controller and permanent data storage

In sum, PMUs intended to meet certain specification imposed by IoT applications should be highly reconfigurable and programmable, i.e., they must have different alarm signals to trigger different functions and different power modes to minimize power consumption. The RTC works as the core block of these PMUs, as it can control the power of the unit and also work as a permanent storage unit, since it is the only block that is always powered on.

41 

Summary (2)

- PMUs for IoT
  - The RTC enables configurability for power modes in the RTC and APC and for reset actions
  - Ultra low power design is targeted
  - Asynchronous digital design, large channel length and subthreshold operation are generally used in the RTC and ultra low power modes of the APC
  - Periodic operation is an additional strategy to reduce power in all cores

The RTC should support different power modes (for itself and combined with the APC). Reset actions are also configurable. The PMU should be designed for ultra low power operation. This means using asynchronous digital design, large channel length and keep the transistors working in the sub-threshold region, especially in the RTC and ultra-low power modes of the APC. The disabling of internal blocks of both the APC and RTC when they are not required. All these modes of operation must be tested using all possible operational modes. To reduce the leakage current the crossing of power domains should be extensively verified.

42 

Summary (3)

- RTC and APC have different power modes, disabling internal blocks when not needed
  - Require extensive verification of possible power states vs. stimulus scenarios
  - The crossing of power domains needs to be extensively verified at transistor level in order to ensure no unnecessary leakage

CHAPTER 7

Connecting the IoT Node to the Cloud for Smart Adaptive Monitoring

João Pedro Oliveira,
Nuno Miguel Correia, Luis Bica Oliveira,
Rui Santos-Tavares
Universidade NOVA de Lisboa, NOVA School of Science and Technology and CTS-UNINOVA, Portugal

Complementing the theoretical material presented on all morning sessions, the proposed course includes 4 x 2 hours hands-on lab sessions, during two afternoons, where IoT demo projects will be implemented. The course demo project will consist on the design and implementation of a multi-sensor system using a programmable and commercial SoC platform. The same platform will have also to monitor the state of the battery (through voltage sensing) and use this information to feed the power management implemented algorithm.

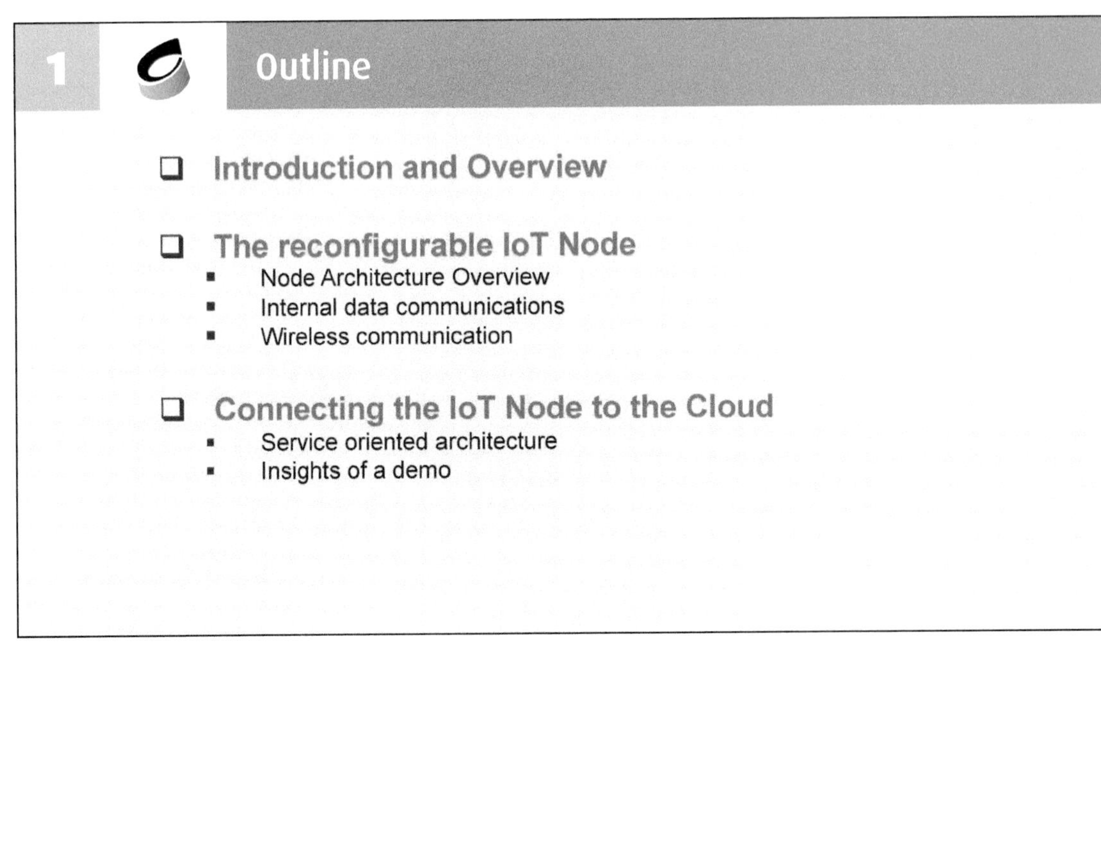

2

Outline

- Introduction and Overview
- The reconfigurable IoT Node
  - Node Architecture Overview
  - Internal data communications
  - Wireless communication
- Connecting the IoT Node to the Cloud
  - Service oriented architecture
  - Insights of a demo

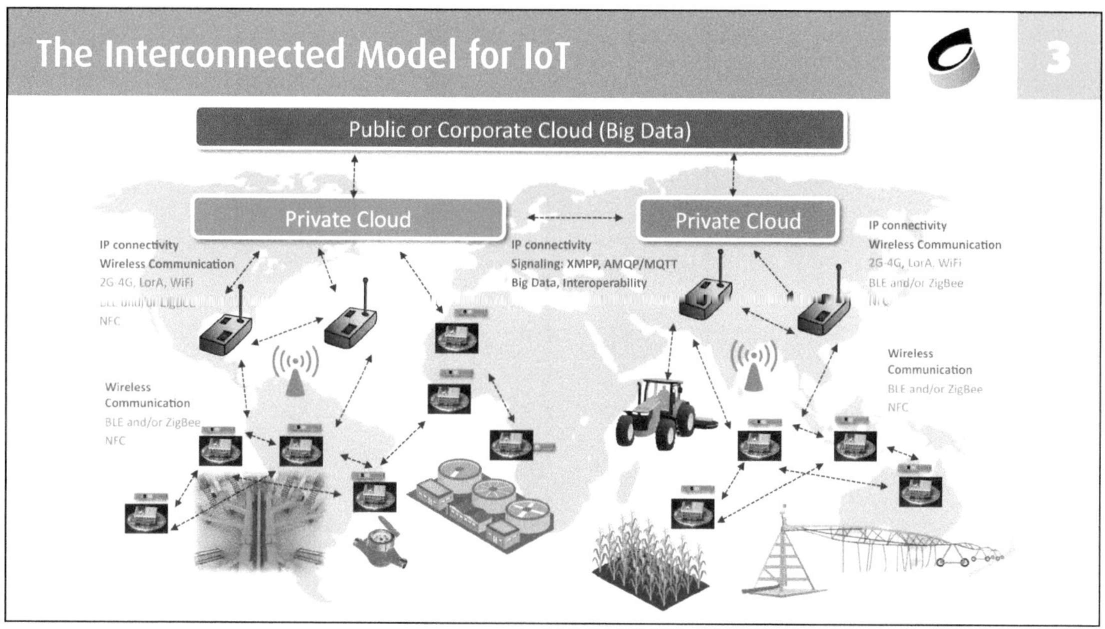

The technological development in the past couple of decades envisaged connecting people and computers, massificating the mobile networks and the internet. The next step is to bring the objects into this global ecosystem tracking an interconnected model for the Internet of Things (IoT).

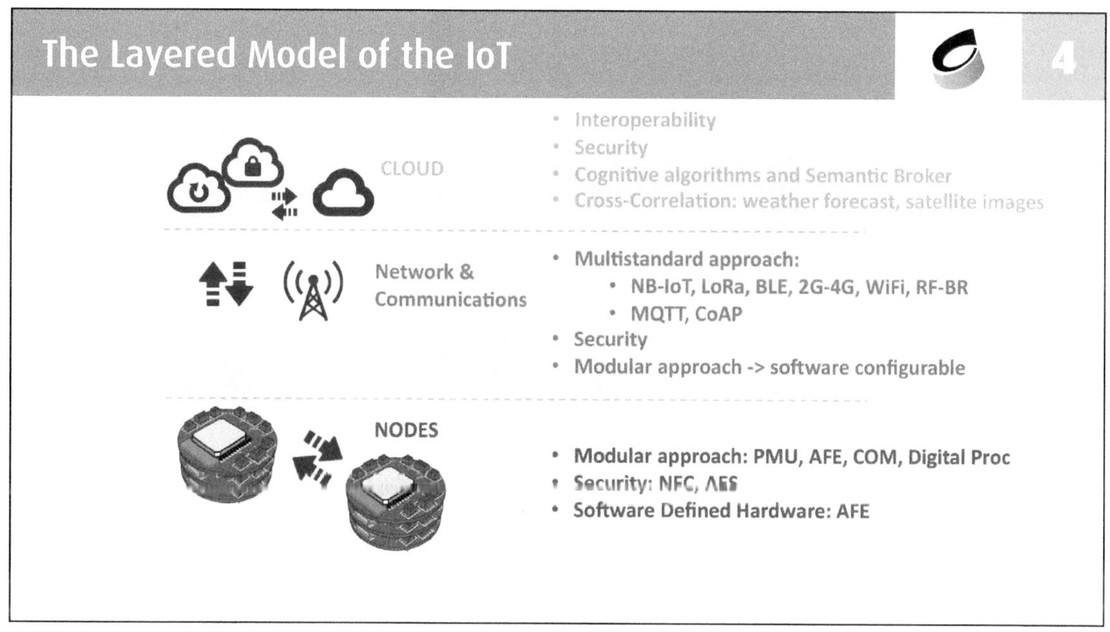

The deployment of the interconnected smart objects IoT ecosystem, is a multidisciplinary effort covering a tree layer model. The first one is formed by the objects themselves, which uses the services from the communications layer to connect to the high layer applications provided by the recent concept of Cloud services aggregation.

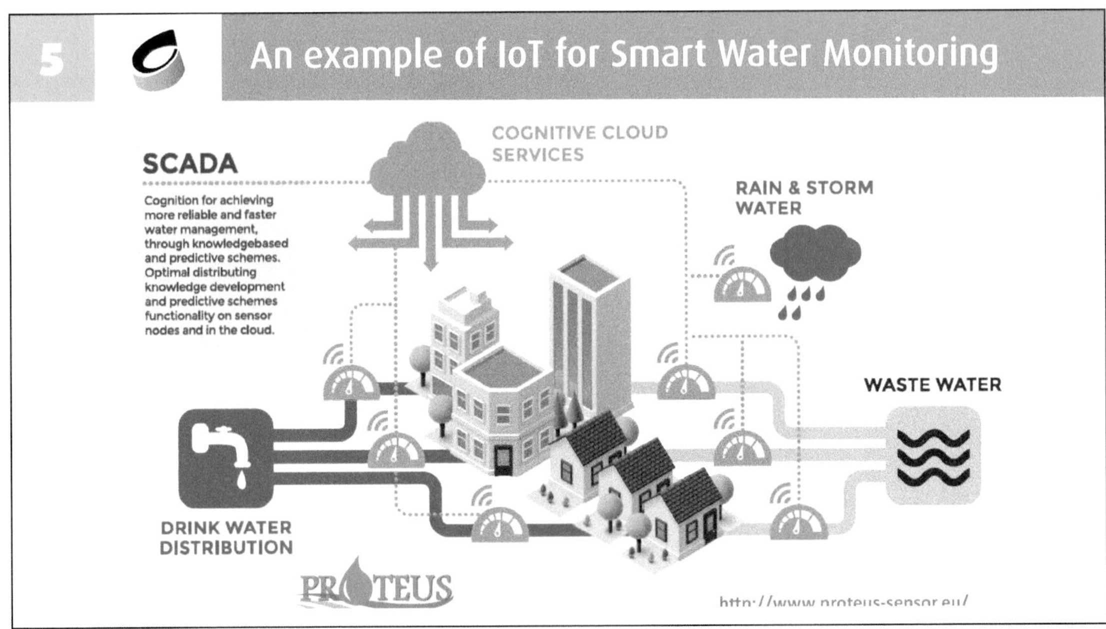

The identification and specifications of a "use case" is an important tool for drawing the path toward IoT systems. This slide describes one example addressed in the European H2020 project PROTEUS, which is related to the smart water monitoring. It demands a highly reconfigurable and adaptive multi-parameter IoT sensor that can be used either in the clean water segment or waste water part.

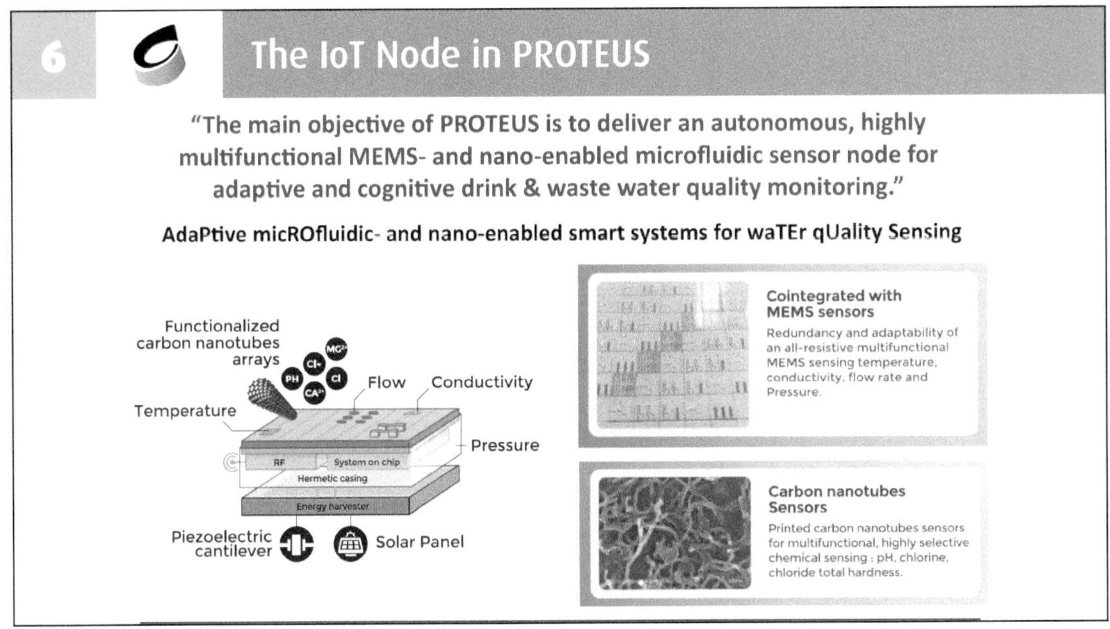

The previous requirements are answered by the PROTEUS project by developing a small size IoT multi-parameter sensor node that combines a set of heterogeneous technologies, including MEMS, CNT and CMOS. A successful IoT projects is intrinsically a multidisciplinary task.

Outline

7

- Introduction and Overview
- **The reconfigurable IoT Node**
  - Node Architecture Overview
  - Internal data communications
  - Wireless communication
- Connecting the IoT Node to the Cloud
  - Service oriented architecture
  - Insights of a Demo

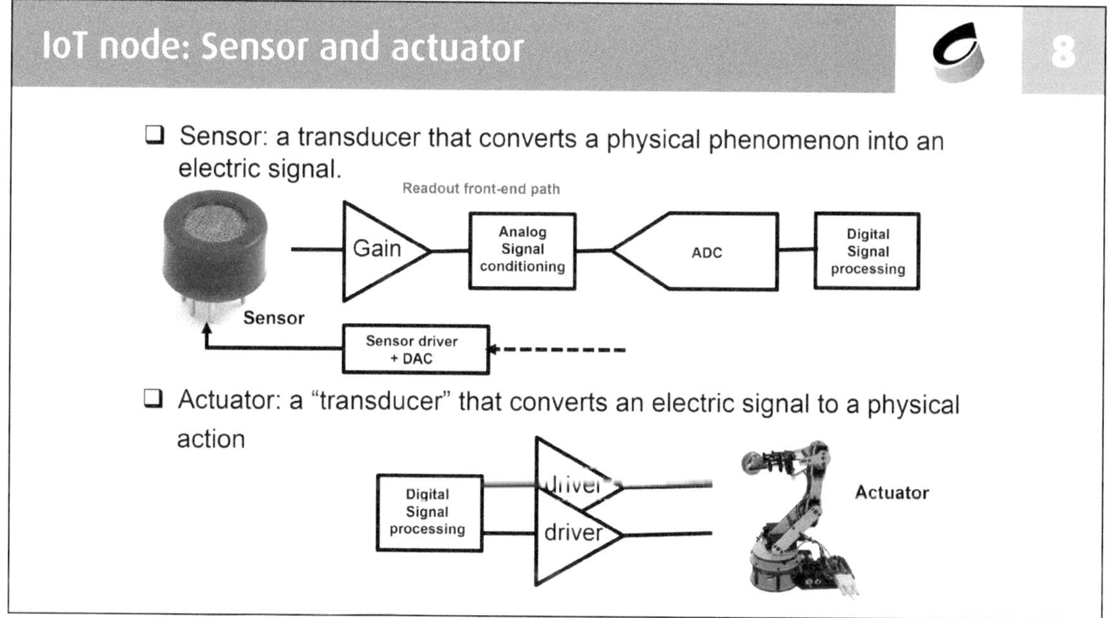

The IoT Node senses and actuates in the physical analog world through electrical transducers, and so it needs a certain form of signal processing. It usually involves both digital and analog processing meaning that the localization of the frontier between these two domains is an important aspect of the system to be considered.

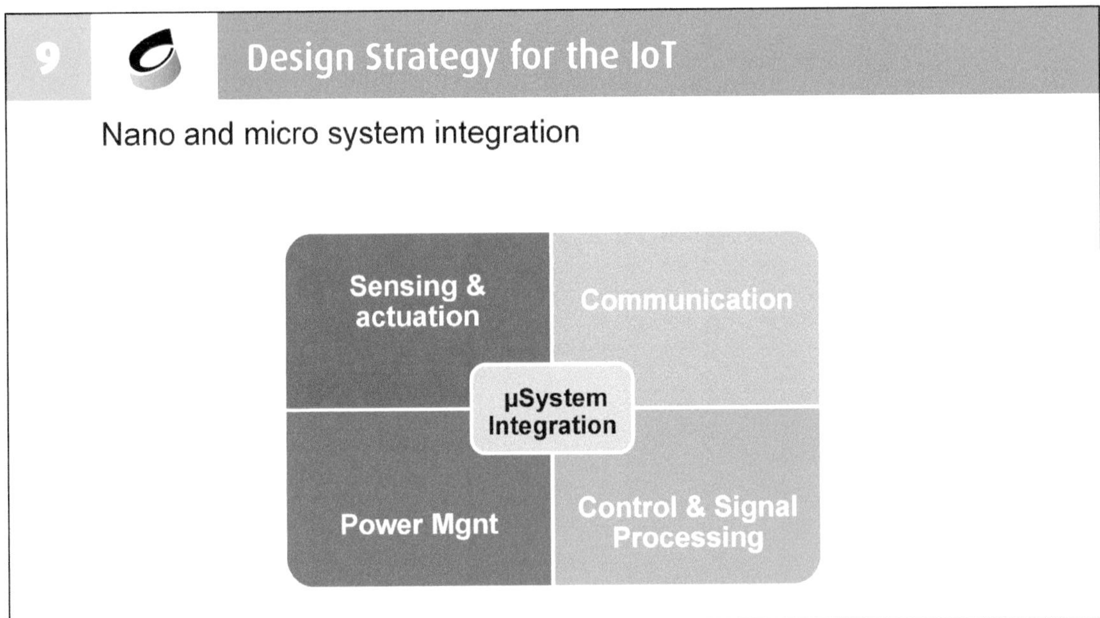

Considering the aspects previously mentioned, the co-design strategy for the IoT Node must address the sensing and actuation, communication, power and energy management, control and signal processing.

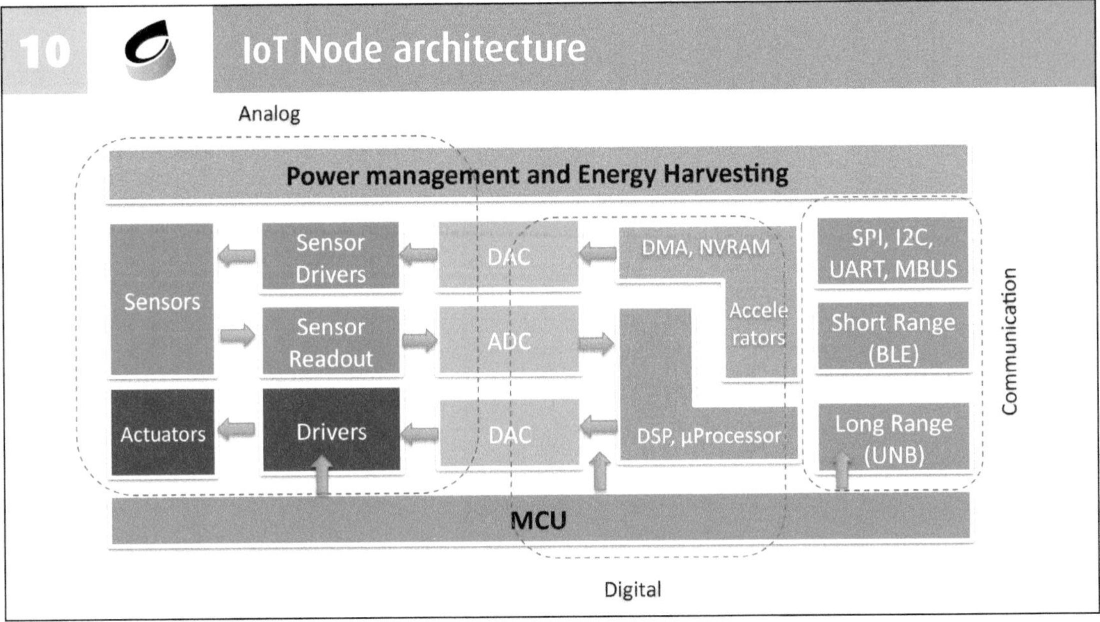

The IoT Node architecture includes sensors and actuators, actuators drivers, sensor readout, ADC and DAC blocks, digital processing and communications. Nowadays, both analog and digital parts can be implemented in low cost CMOS technologies while the sensors are usually implemented in diverse technologies, like MEMS.

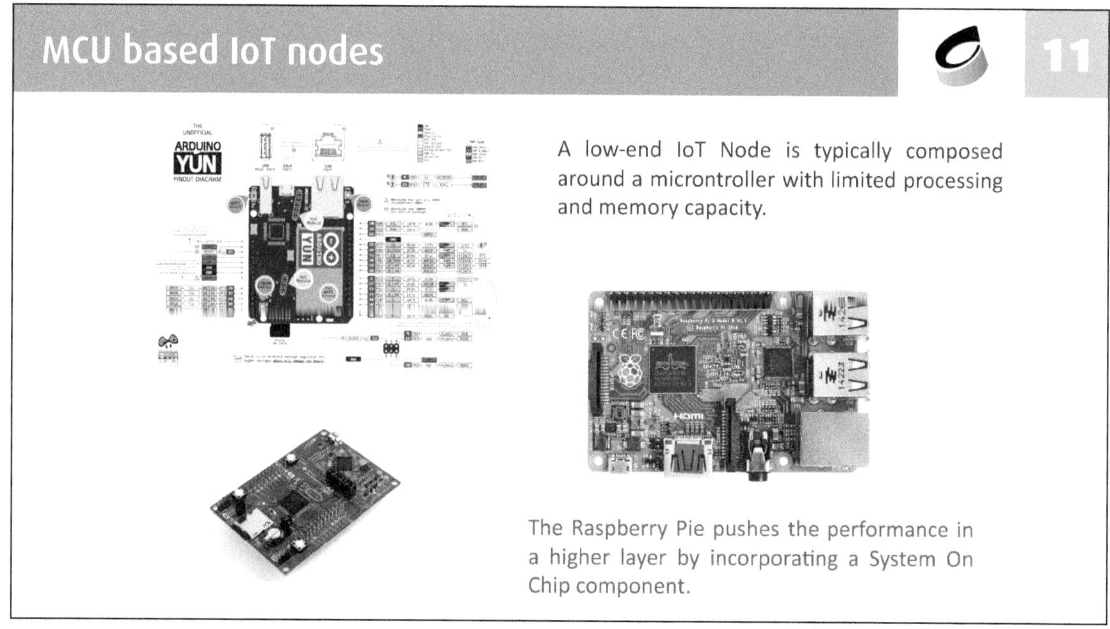

MCU based IoT nodes
11
ARDUINO
YUN
A low-end IoT Node is typically composed around a microntroller with limited processing and memory capacity.
The Raspberry Pie pushes the performance in a higher layer by incorporating a System On Chip component.

The Arduino platform facilitated the adoption of hardware handling, by incorporating a clever bootloader and providing an easy to use and free SDK kit. The Arduino strategy has been followed by several platforms, like the MSP430.

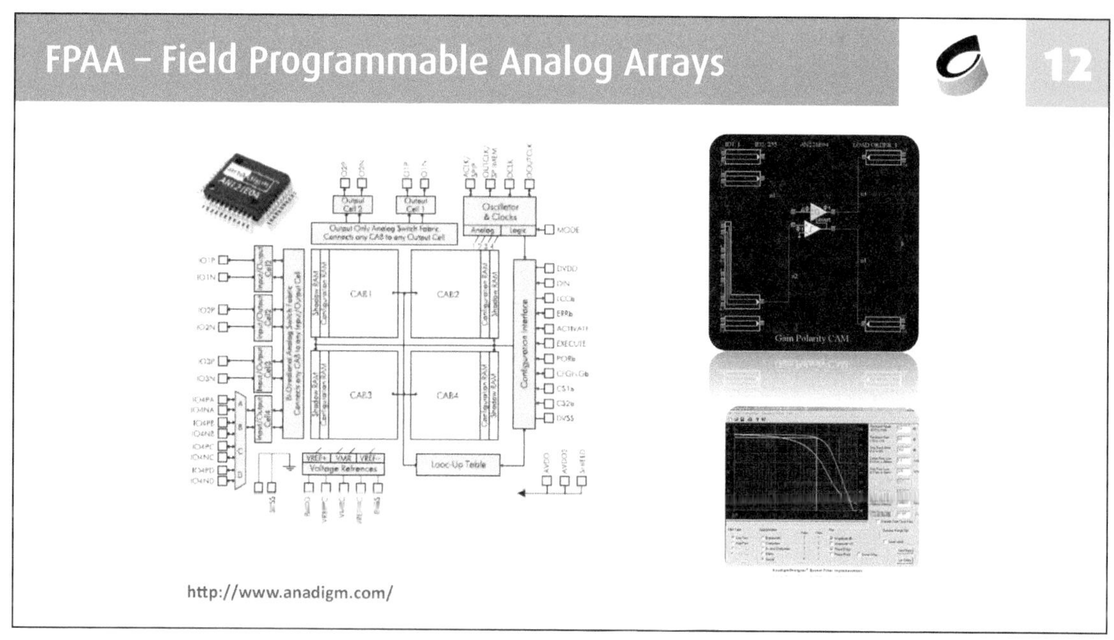

For solutions requiring high levels of reconfigurability at the analog part, the IoT Node benefits from the use of programmable analog front ends. One of the approaches for the latter is the FPAA from Anadigm.

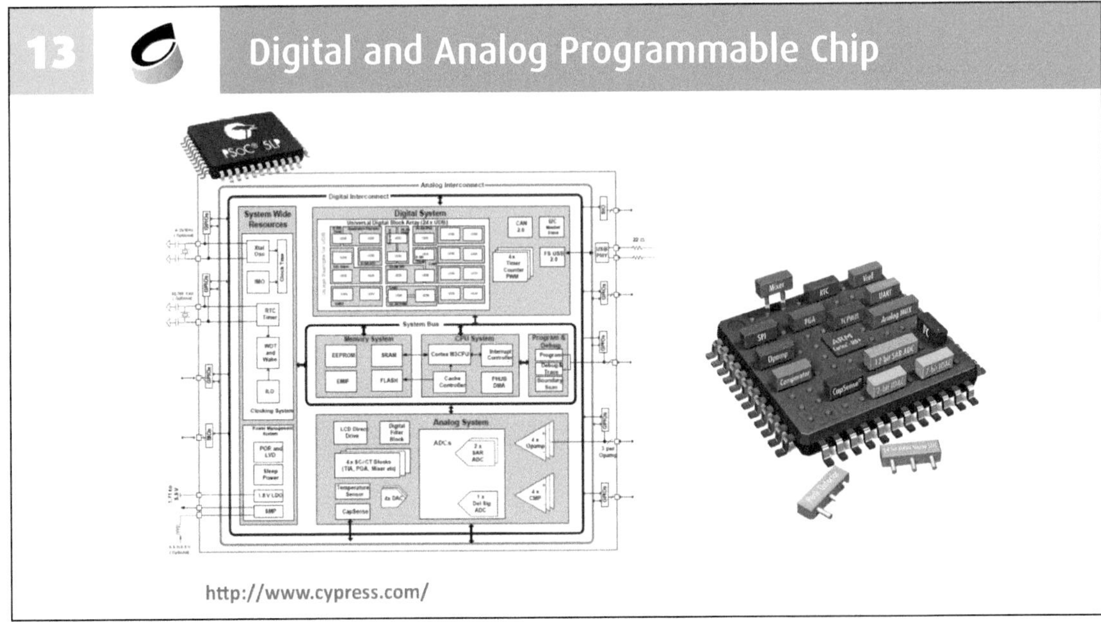

However, other solutions are available, depending on the architectures options selected by the manufacturer. The PSoC from Cypress, is one example of such approach. It integrates not only, reconfigurable analog building blocks for the analog front end, but also an ARM based digital processing unit as well as dedicated reconfigurable digital logic.

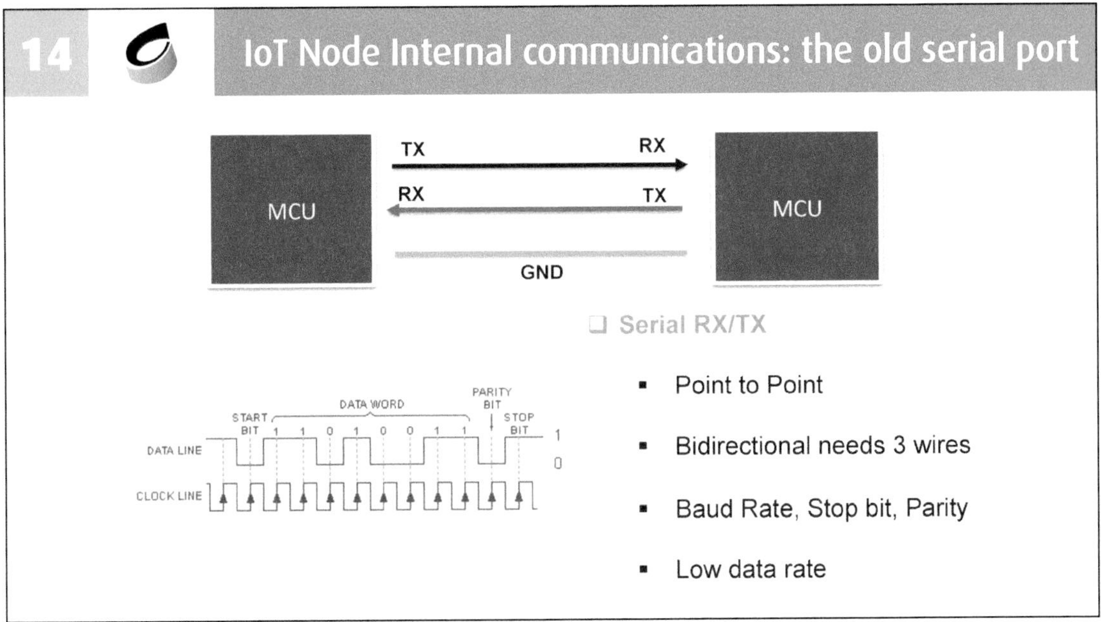

The serial port is still an effective way to connecting modules inside the IoT Node. It is an asynchronous interface in the sense no explicit clock line is needed. The clock signal is recovered using a start/stop bit training sequence. It is a low data rate communication that can easily reach 115 kbps. Not suitable for point to multipoint communications.

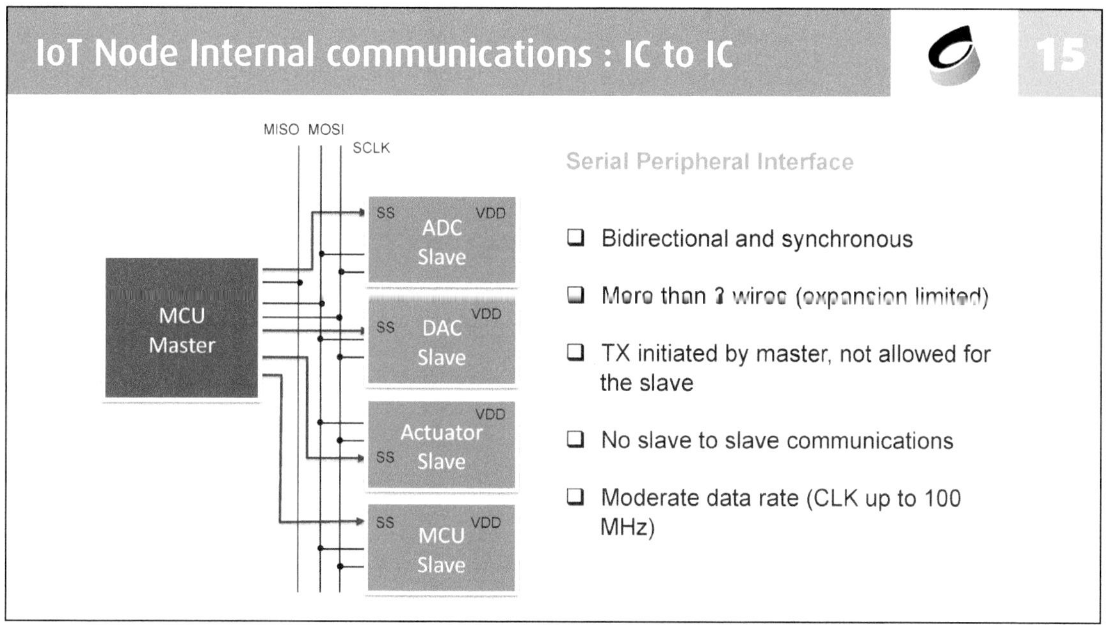
IoT Node Internal communications : IC to IC
15
MISO MOSI
SCLK
MCU
Master
SS
ADC
Slave
VDD
SS
DAC
Slave
VDD
Actuator
Slave
SS
VDD
SS
MCU
Slave
VDD
Serial Peripheral Interface
Bidirectional and synchronous
More than 3 wires (expansion limited)
TX initiated by master, not allowed for the slave
No slave to slave communications
Moderate data rate (CLK up to 100 MHz)

The serial peripheral interface SPI, is a serial interface which minimize the number of connecting lines, but when compared with the previous serial protocol it enables a common bus to be shared by several modules. It is a point to multipoint scheme where a master activates the select input of a slave module to establish a synchronous communication with it. It can reach higher data rates.

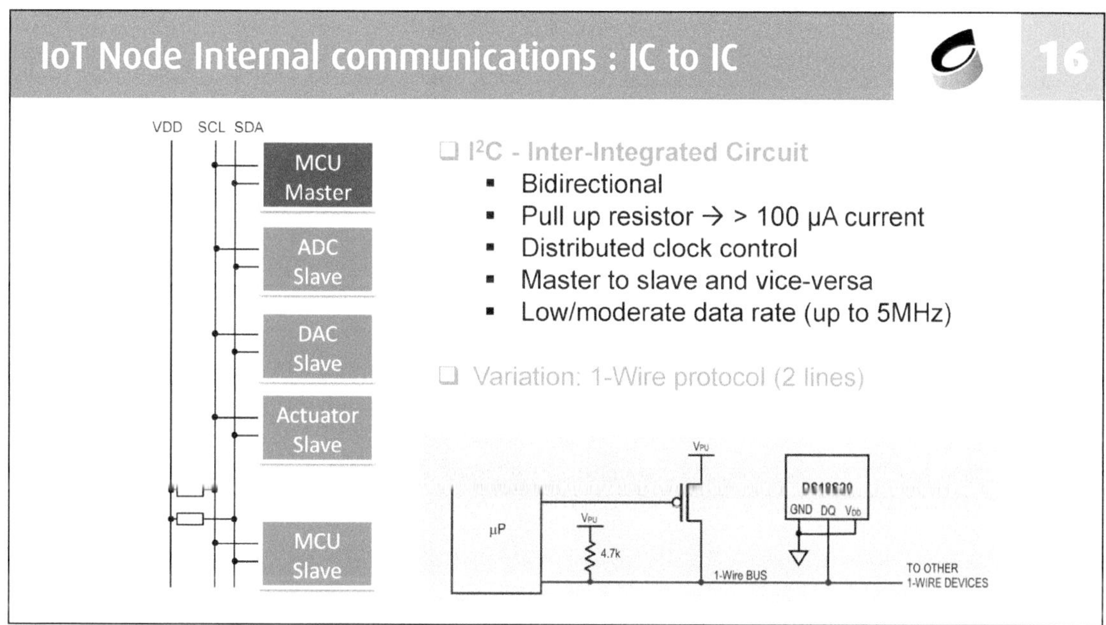
IoT Node Internal communications : IC to IC
16
VDD SCL SDA
MCU
Master
ADC
Slave
DAC
Slave
Actuator
Slave
MCU
Slave
I²C - Inter-Integrated Circuit
Bidirectional
Pull up resistor → > 100 μA current
Distributed clock control
Master to slave and vice-versa
Low/moderate data rate (up to 5MHz)
Variation: 1-Wire protocol (2 lines)
μP
VPU
VPU
4.7k
1-Wire BUS
GND DQ VDD
TO OTHER
1-WIRE DEVICES

Comparing to SPI, the I2C interface avoids a dedicated select line by using a protocol address scheme by which the master can establish a data link to each individual slave. It is a bidirectional communication protocol that can reach low to moderate data rates. Thus, is not suitable for high throughput applications. A similar addressing approach is used by the 1-Wire interface which eliminates the clock line and the supply power and data share the same line reducing the number of conductors to one (plus the ground).

17 

Wireless Communications

- Near Field Communications
  - Low Power
  - Contact range
- BLE (V4.0)
  - Low Power
  - Short Range
  - Moderate data rate
- IEEE 802.15.4, ZigBEE
  - Low power
  - Moderate range
  - Moderate Data rate
- UltraNarrow Band (UNB)
  - Long range
  - Low Data Rate

In a common IoT communication approach, the IoT Node establish a connection to a gateway which handles the transfer of data to the cloud. The typical wireless technologies used to reach this IoT Gateway Node can get different radio distances. The Bluetooth Low Energy can reach 10 to 20 meters while the IEEE 802.15.4/ ZigBEE can go up to several hundred of meters. Also, by using a multi-hop mechanism, the ZigBee can reach the gateway located at several kilometres of distance. The emerging Ultra Narrow Band radio technology can reach long distances, up to 20 to 30 kilometres, due to high sensitivity receiver which benefits from extreme narrow band radio channels.

18 

Wireless Connectivity

Model	Protocol	Frequency	txPower	Sensitivity	Range *
XBee-802.15.4-Pro	802.15.4	2.4GHz	100mW	-100dBm	7000m
XBee-ZB-Pro	ZigBee-Pro	2.4GHz	50mW	-102dBm	7000m
XBee-868	RF	868MHz	315mW	-112dBm	12km
XBee-900	RF	900MHz	50mW	-100dBm	10Km
LoRa	RF	868 and 900MHz	14dBm	-137dBm	22Km
WiFi	802.11b/g	2.4GHz	0dBm - 12dBm	-83dBm	50m-500m
GPRS Pro and GPRS+GPS	-	850MHz/900MHz/ 1800MHz/1900MHz	2W(Class4) 850MHz/ 900MHz, 1W(Class1) 1800MHz/1900MHz	-109dBm	- Km - Typical carrier range
3G/GPRS	-	Tri-Band UMTS 2100/1900/900MHz Quad-Band GSM/ EDGE, 850/900/1800/1900 MHz	UMTS 900/1900/2100 0,25W GSM 850MHz/ 900MHz 2W DCS1800MHz/ PCS1900MHz 1W	-106dBm	- Km - Typical carrier range
Bluetooth Low Energy	Bluetooth v.4.0 / Bluetooth Smart	2.4GHz	3dBm	-103dBm	100m

Source: "Gases Pro, Technical guide", Libelium

Complementing ISM license free digital radio standards, the mobile network play an important role for connecting the gateway to the Cloud:.2.5G, 3G and 4G/LTE are the most common solutions that can be used.

Outline

19

- Introduction and Overview
- The reconfigurable IoT Node
  - Node Architecture Overview
  - Internal data communications
  - Wireless communication
- **Connecting the IoT Node to the Cloud**
  - Service oriented architecture
  - Insights of a demo

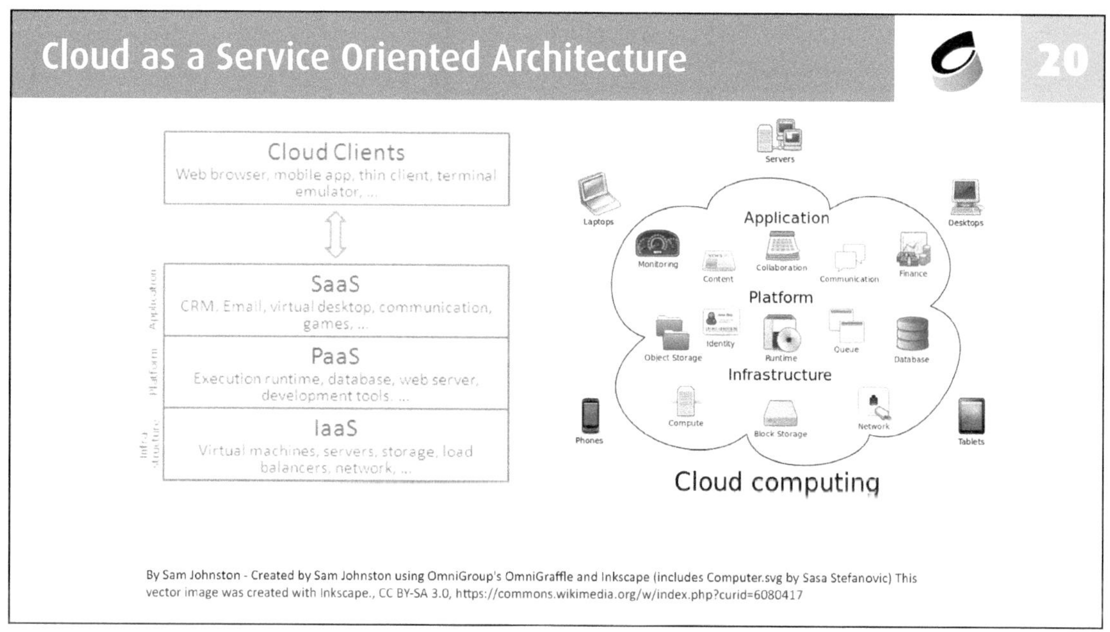

The interconnected internet servers evolved to a distributed architecture where services' access location has been decoupled from the need to know the exact "location" of the server. This Cloud concept was then emerged where three level of service access has been available, depending on the abstraction level the Cloud client wants: SaaS, PaaS and IaaS.

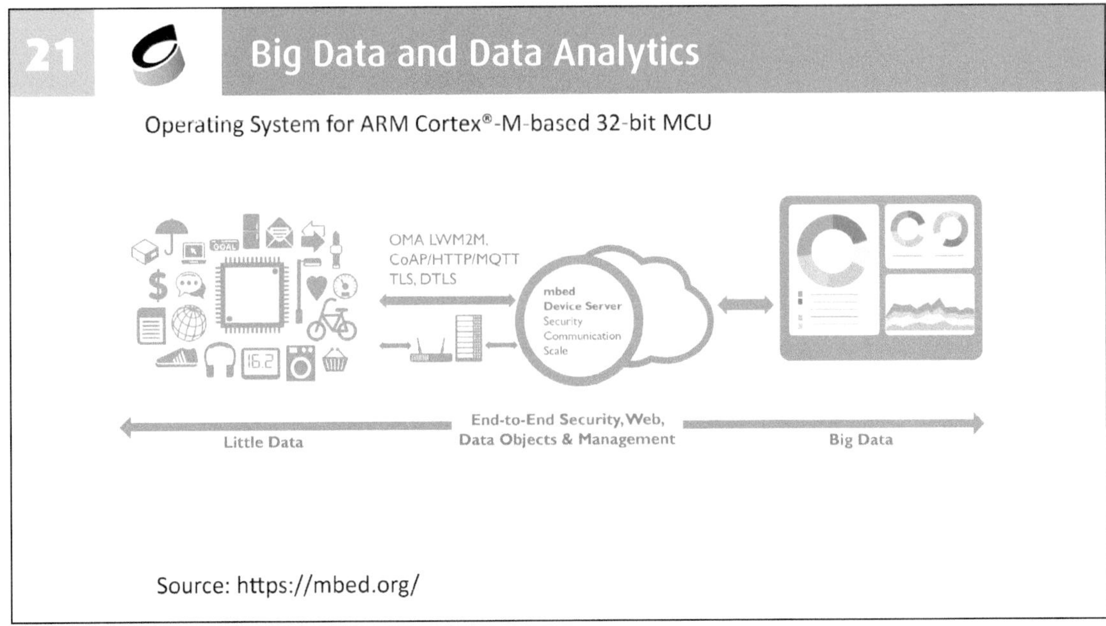

As shown in the diagram, the multiple IoT nodes generates a significant amount of data which must be exchanged to the cloud using a secure and reliable channel. This "Big Data" concept pushes the requirements for database technologies, emerging new paradigms as non-relational databases, but also implies the use of advanced "Data Analytics".

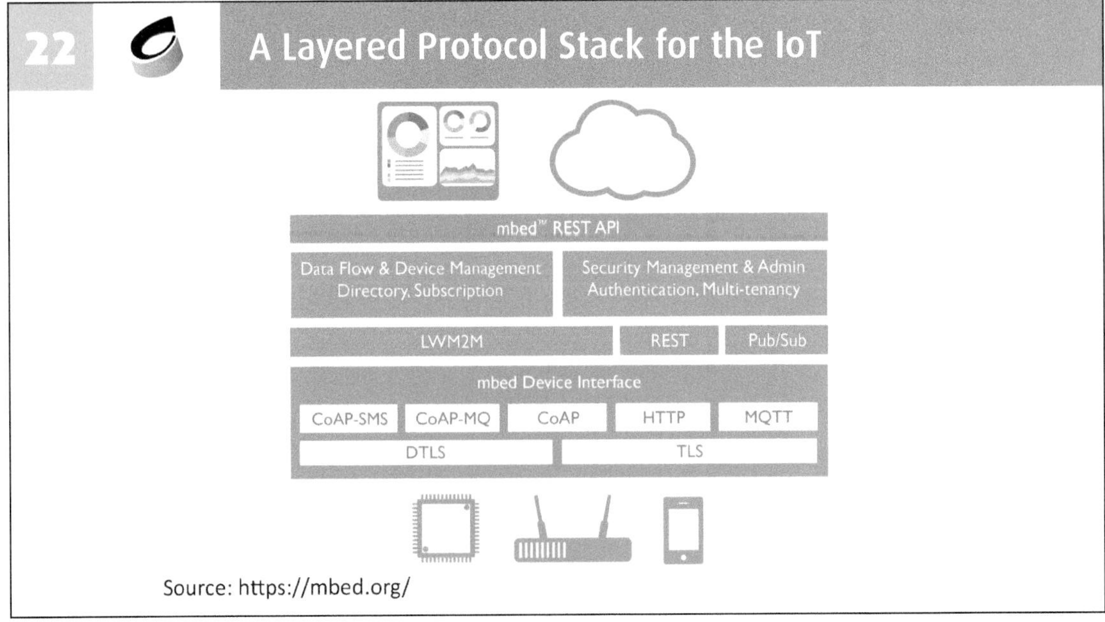

In the IoT protocol stack model presented in the diagram, the top layers uses common technologies already in use in the Internet, like RESTful Web services. However, for the lower layers, and due to the limited resources of the IoT nodes, new simplified protocols have emerged like the Constrained Application Protocol (CoAP) or the Message Queue Telemetry Transport, MQTT.

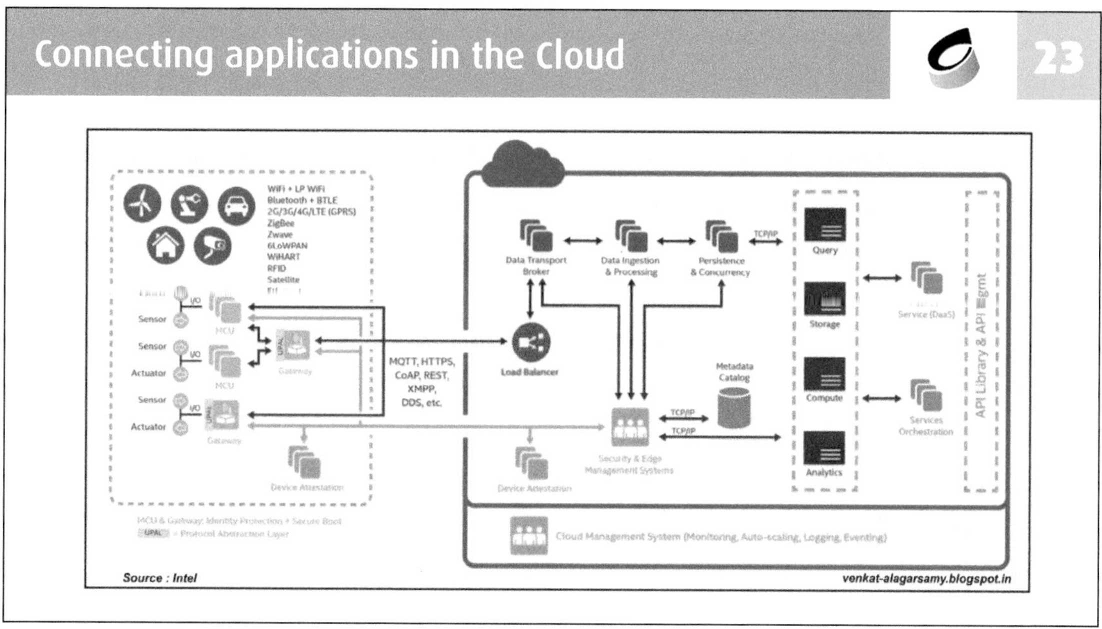

CoAP is designed to facilitate the interface with HTTP and at the same time supporting multicast, very low overhead, and simplicity, which is a key factor for IoT. CoAP can run on most devices that support UDP.

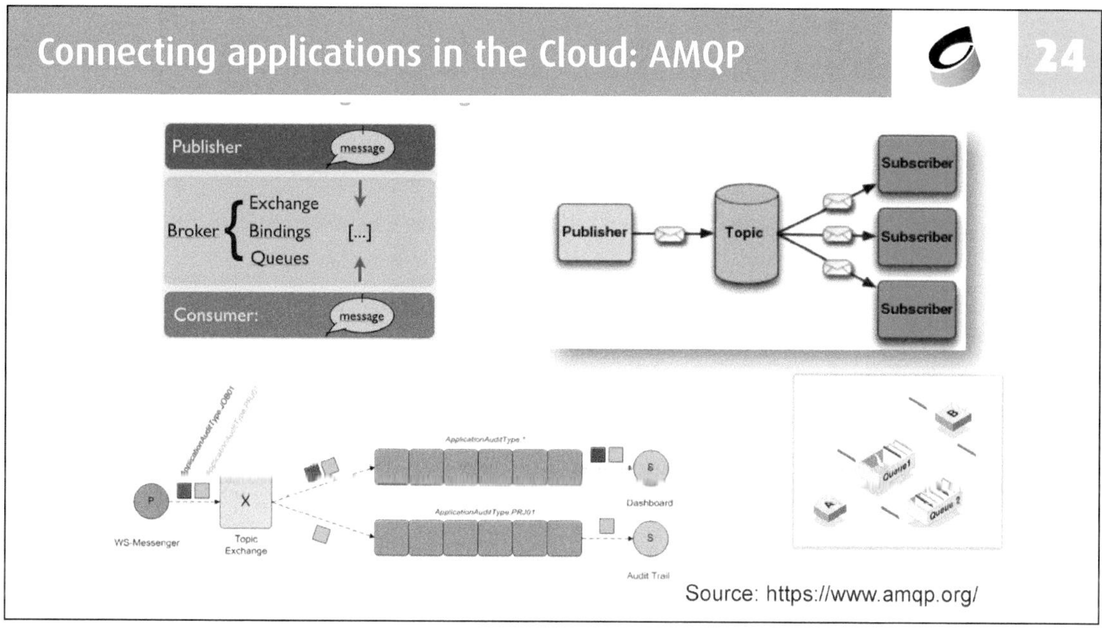

The Advanced Message Queuing Protocol (AMQP) is an application and message oriented protocol. Messages are exchanged within the system using a publish-subscribe mechanism supported on queues and routers. The subscription is made by Topics which are also important for the message publish phase. This technology is widely used for point to multipoint communication.

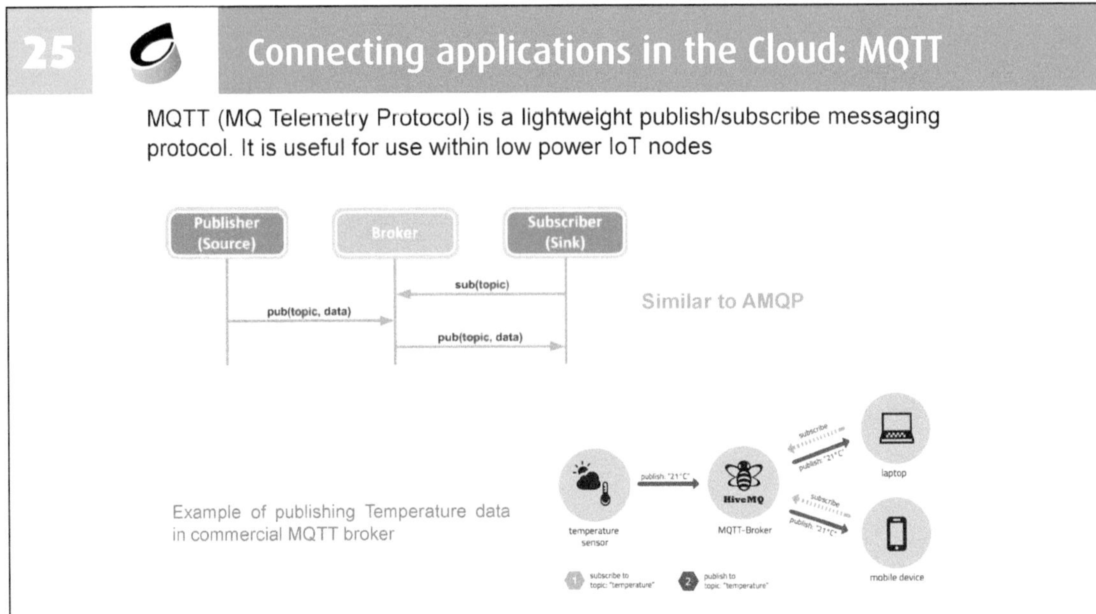

Connecting applications in the Cloud: MQTT

Similar to AMQP
Example of publishing Temperature data in commercial MQTT broker

MQTT is an open standard application layer protocol for message-oriented connectivity based on publish-subscribe paradigm. Publish-subscribe messaging pattern requires a message broker, as shown in the diagram. It is a simplified message based protocol for use on top of the TCP/IP protocol. It is optimized for connections with remote IoT nodes where network bandwidth is limited. The broker distributes messages to clients considering the topic attribute.

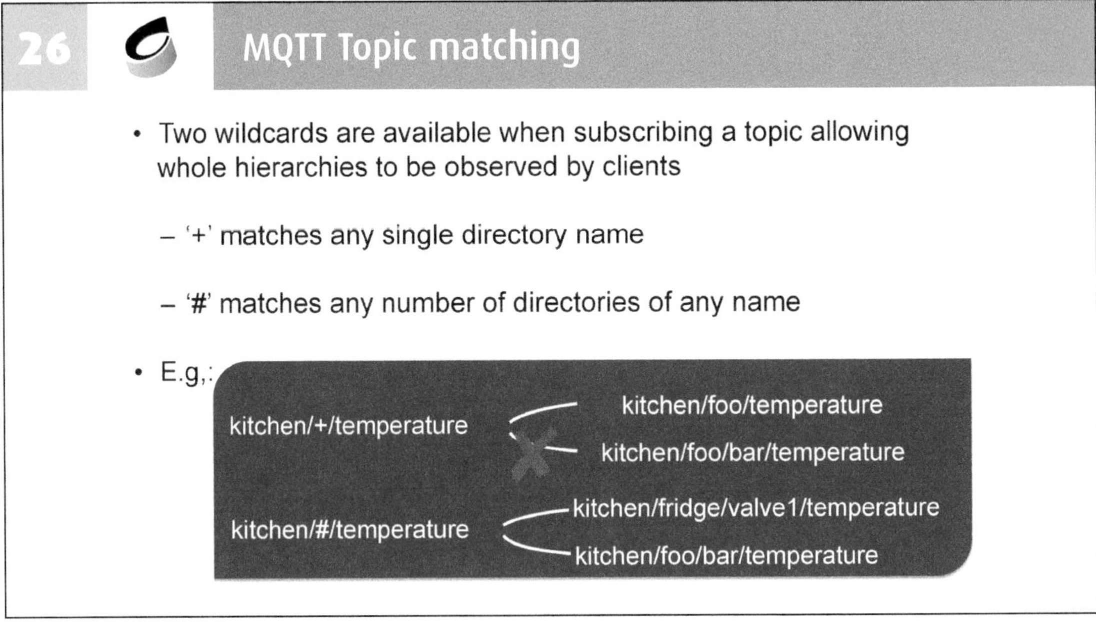

26
MQTT Topic matching
Two wildcards are available when subscribing a topic allowing whole hierarchies to be observed by clients
– '+' matches any single directory name
– '#' matches any number of directories of any name
E.g,:
kitchen/+/temperature
kitchen/foo/temperature
kitchen/foo/bar/temperature
kitchen/#/temperature
kitchen/fridge/valve1/temperature
kitchen/foo/bar/temperature

The subject of the information of a publish-subscribe message is designated by Topic. In MQTT, Topics are organized hierarchically into topic trees where the '/' character distinguishes subtopics. In the tree structure, subtopics can be leaf-nodes (with no further subtopics), or intermediate ones (with subtopics below in the tree).

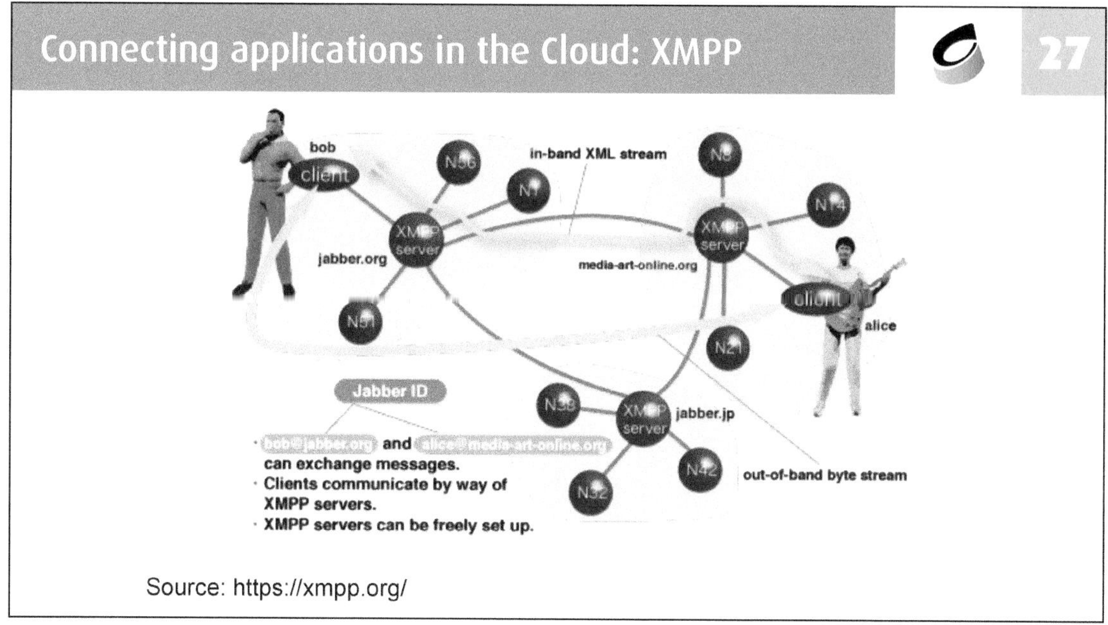

The eXtensible Messaging and Presence Protocol (XMPP) is suitable for applications requiring a point to point application connectivity and the availability of store and forward mechanism. The XMPP protocol offers streaming XML elements to exchange messages and presence information, seemly to real time. It was designed for use with basic/simple electronic devices. The XMPP layer plane is composed by XMPP servers responsible to route messages between XMPP clients.

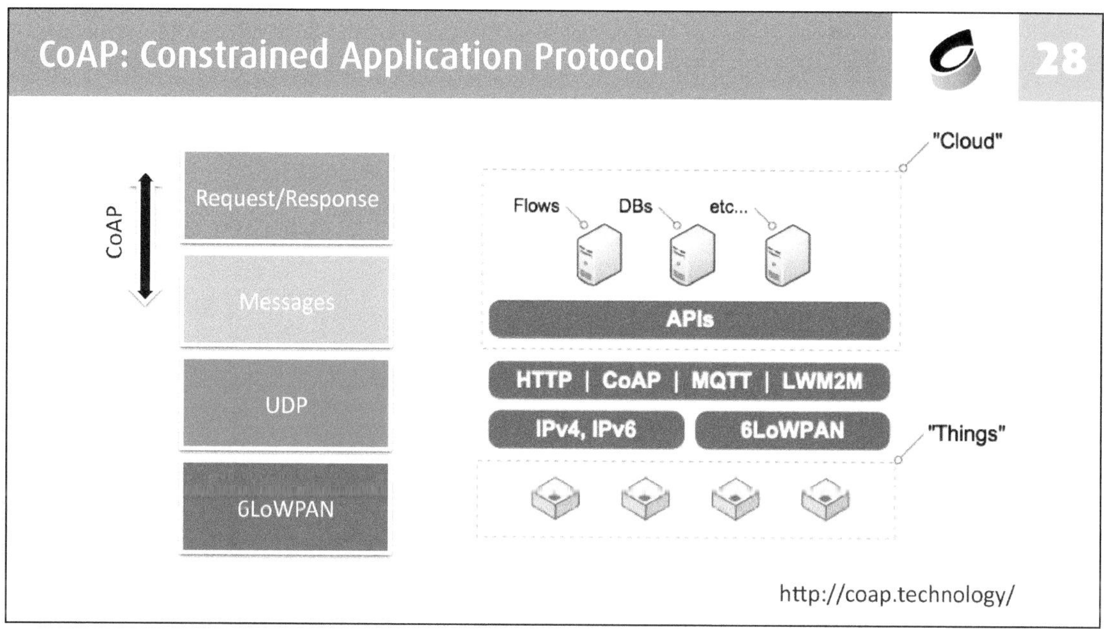

CoAP is a document transfer protocol but it is not a replacement for HTTP. It is ideal for constrained devices and networks, i.e. it is specialized for M2M applications by using a very efficient RESTful protocol. CoAP packets are much smaller than HTTP TCP flows since it runs over UDP, not TCP. Clients and servers communicate through connectionless datagrams. Because CoAP is datagram based, it may be used on top of SMS or other packet based communications protocols.

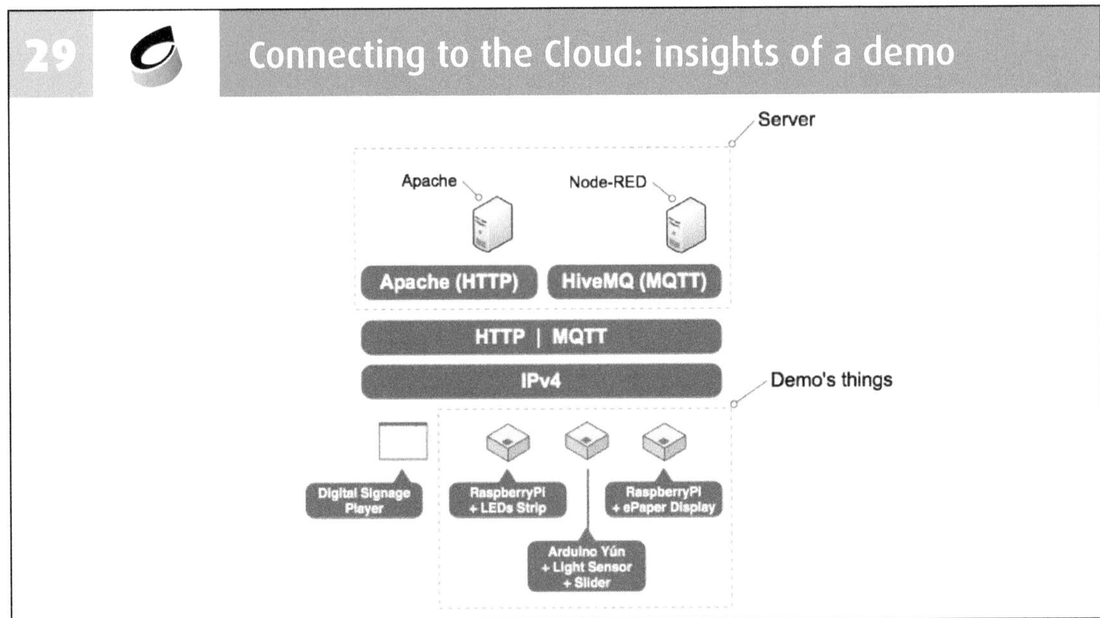

The demo setup is composed by 4 different terminal IoT nodes, a layer of transport mechanism and cloud servers.

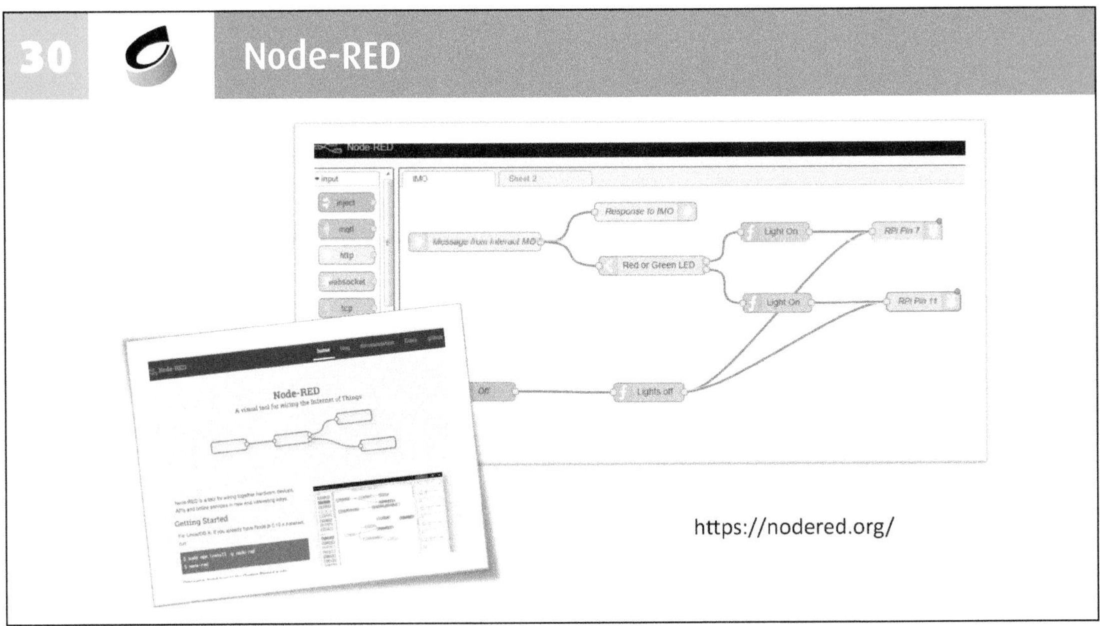

Node-RED is a programming tool for wiring together hardware devices, APIs and online services in a Drag and Drop approach. It gives a browser-based editor to facilitate the setup of flows using a range of nodes that can be deployed to its runtime in a single-click.

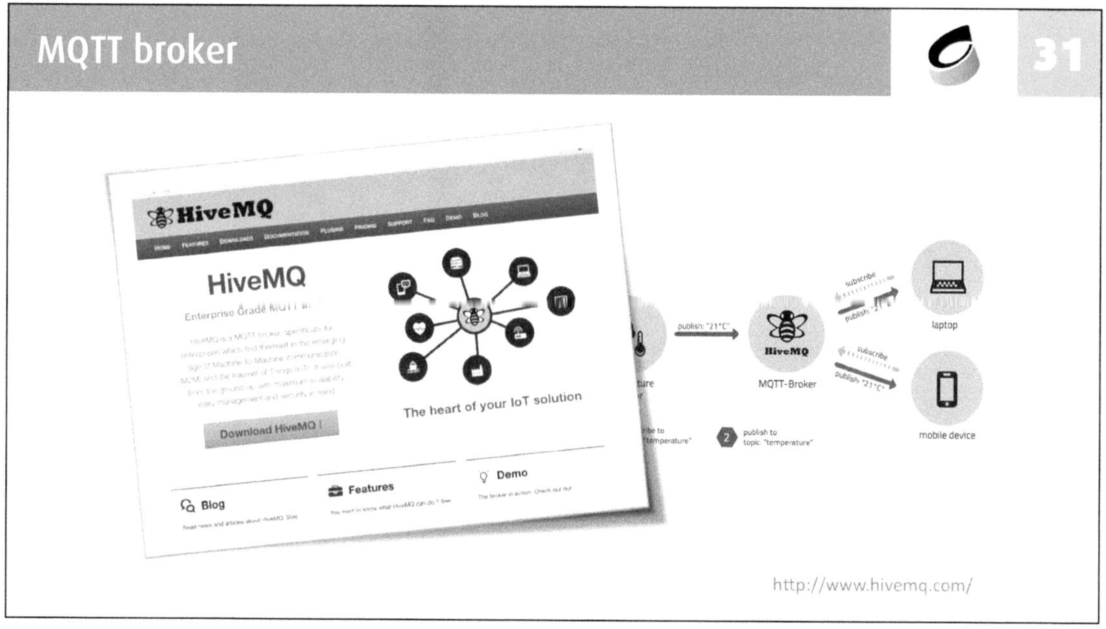

To provide a MQTT service, a broker must be deployed. Several MQTT brokers online services are available nowadays. One of the key aspects to consider is the level of security supported by the chosen broker technology, namely SSLv3, TLS 1.0, TLS 1.1 or TLS 1.2.

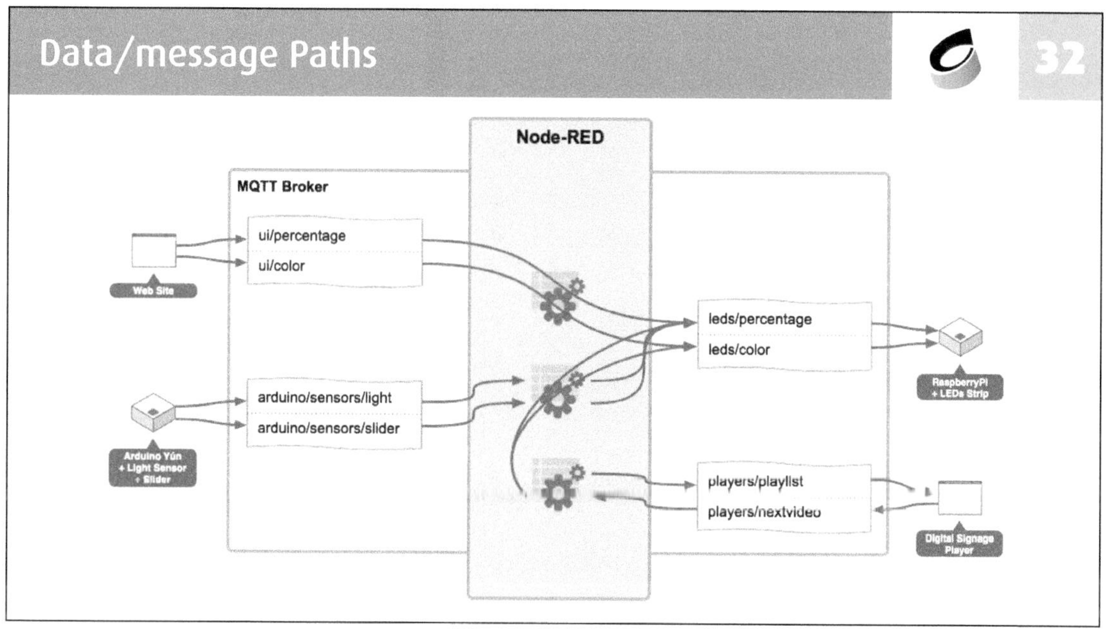

The MQTT broker is used as an entry point for the remote IoT Node while Node-RED defines the interaction between objects.

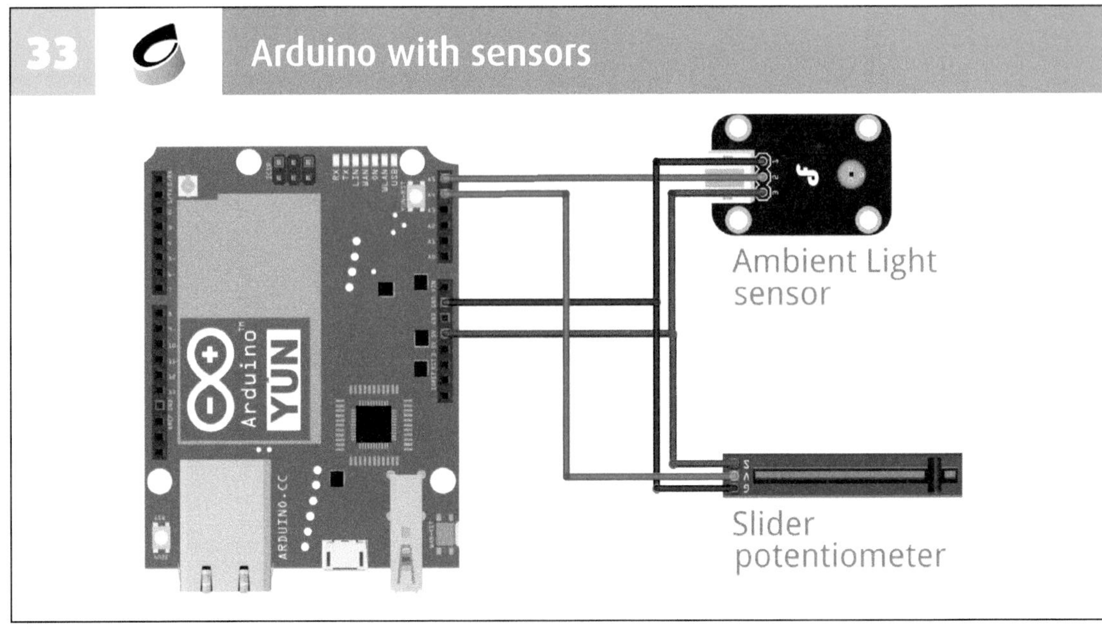

The terminal IoT Node is composed with an Arduino Yún platform with integrated WiFi module. Analog outputs from the slider and the light sensor are connected directly to an Arduino analog input port. These input signals are digitalized by a 10-bit SAR ADC. The WiFi module integrates a TCP/IP protocol stack which is used to establish a MQTT session to the Cloud server.

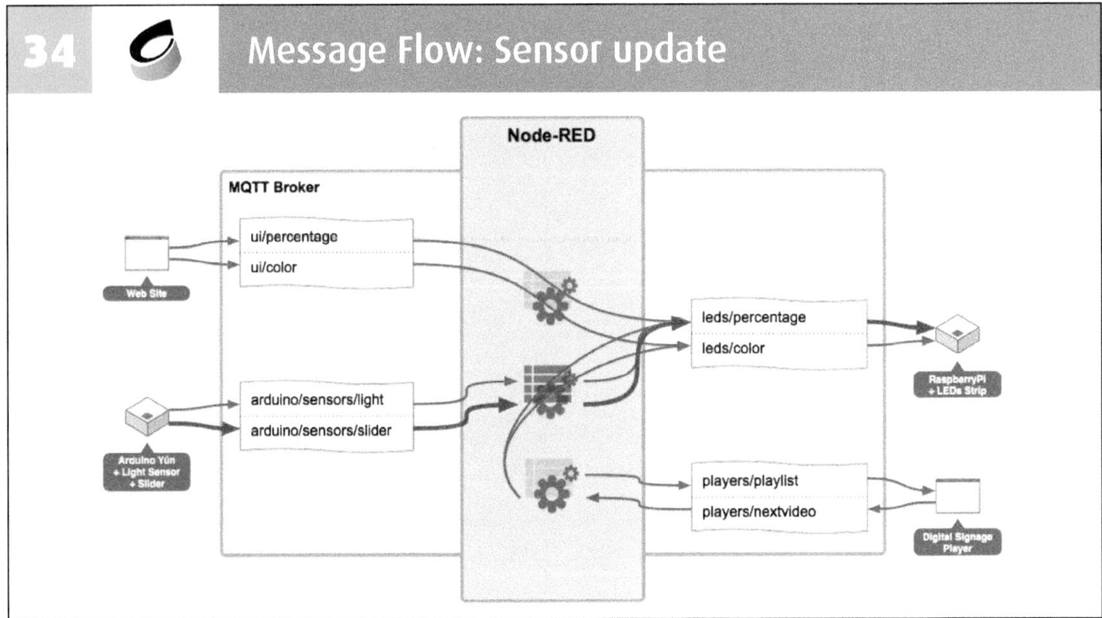

In this sequence flow, the slider connected to the Arduino Yún platform generates a message for the MQTT topic "arduino/sensors/slider". The objects that have subscribed this topic can consume this information, specifically, the light intensity of a LED strip.

Arduino YUN w/ sensors

35

Code for the interface with the sensors

```
// sensor pins configuration
const int LIGHT_SENSOR_PIN = 5;
const int SLIDER_SENSOR_PIN = 5;

// current sensor values
int sliderValue = 0;
int lightValue = 0;
int tempSensorValue = -1;


// only runs once
void setup()
{
    // use pins as inputs
    pinMode(LIGHT_SENSOR_PIN, INPUT);
    pinMode(SLIDER_SENSOR_PIN, INPUT);
}
```

```
// runs forever
void loop()
{
    // read current sensor value
    tempSensorValue = analogRead(SLIDER_SENSOR_PIN);
    if (tempSensorValue != sliderValue) {
        sliderValue = tempSensorValue;

        // send value update to MQTT topic
        mqttClient.publish("sensors/arduino/slider", sliderValue);
    }

    tempSensorValue = analogRead(LIGHT_SENSOR_PIN);
    if (tempSensorValue != lightValue) {
        lightValue = curValue;

        // send value update to MQTT topic
        mqttClient.publish("sensors/arduino/light", lightValue);
    }

    delay(50);
}
```

The code needed by the Arduino Yún platform for publishing the slider data is straightforward. Having the MQTT object created, the available "publish" method is enough to send data to the MQTT topic "sensors/arduino/slider".

CHAPTER 8

BLE Radio Architectures and Design for the IoT Market

Augusto Marques
Sandeep Perdoor
Aura Semiconductor
Bangalore, India

Chapter contents: Short range wireless markets; Overview of the BLE standard, BLE Radio Specifications: 1) RX, TX, Synth specifications; 2) Current market specifications; BLE Architecture and Design of Functional Blocks; Specifications of radio building blocks; Architecture and Design of radio building blocks; Performance of state of the art BLE Radios.

A. Short range wireless markets

B. Overview of the BLE standard

C. BLE Radio Specifications:
  a. TX, RX, Synth specifications
  b. Current market specifications

D. BLE Architecture and Design of Functional Blocks:
  a. Specifications of radio building blocks
  b. Architecture and Design of radio building blocks

E. Performance of state of the art BLE Radios

1 

Outline

- Introduction
- BLE Radio Specifications
- BLE Radio Architecture
- BLE Radio Design
- BLE Radio Performance

This chapter will focus on the Architecture and Design of state of the art Bluetooth Low Energy Transceivers for the IoT Market.

2 

Outline

- Introduction
- BLE Radio Specifications
- BLE Radio Architecture
- BLE Radio Design
- BLE Radio Performance

We will present the standard specifications, and typical market requirements, describe Architectures commonly used, and the Design of the building blocks.

Finally, achieved performance in implementations done in 55nm and 40nm CMOS technologies will be presented.

Short Range Wireless : Markets

3

1. PC & Peripherals
2. Mobile Phones & Smart Phones
3. Automotive
4. Consumer Electronics
5. Medical & Health, Sports & Fitness
6. Home Automation
7. Retail & Location-Based Services
8. Wearables

Currently, the short range wireless market is very fragmented:

- in customers : customers with completely different volume requirements and requests
- in applications : very different end applications, with different cost basis

From a big picture point of view, the majority of the short range products have the target markets presented in the slide

Short Range Wireless : Technologies

4

	Voice	Data	Audio	Video	State
Bluetooth ACL / HS	x	Y	Y	x	x
Bluetooth SCO/eSCO	Y	x	x	x	x
Bluetooth low energy	x	x	x	x	Y
Wi-Fi	(VoIP)	Y	Y	Y	x
Wi-Fi Direct	Y	Y	Y	x	x
ZigBee	x	x	x	x	Y
ANT	x	x	x	x	Y

State – low bandwidth, low latency data

Low Power

There are multiple technologies being used at the core of the short range wireless products. The predominant technologies being used are Bluetooth, WiFi, ZigBee or ANT.

The short range wireless transmission is normally used to transmit voice, data, audio, video or basic state information. When state information is transmitted, small number of bits need to be transmitted, requiring low bandwidth and low latency.

Bluetooth Low Energy is often used for these applications, and it is the focus of this presentation.

5 BLE

- BLE is a wireless personal area network (WPAN) technology designed and marketed by Bluetooth Special Interest Group (SIG)
- BLE is marketed as Bluetooth SMART

- BLE was introduced in 2010 as part of Bluetooth 4.0 release
- BLE operates in the same ISM band as Bluetooth : 2400-2480MHz

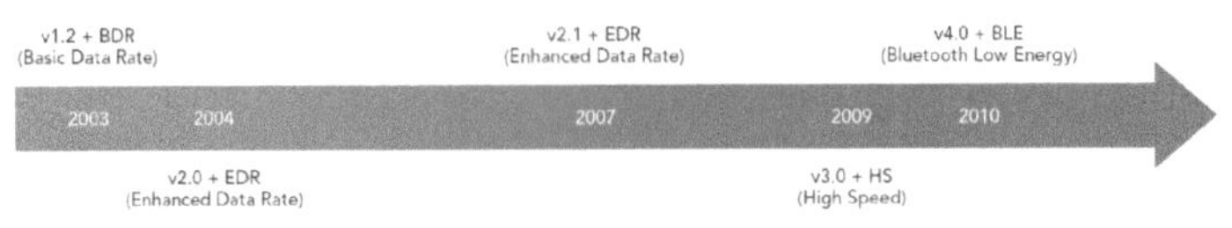

Bluetooth Low Energy, simply abbreviated as BLE, is a wireless personal network marketed as Bluetooth SMART.

Devices using BLE operate in the ISM band from 2400-2480 MHz.

The most recent version of Bluetooth standard is 5.0. This was released in Dec 2016.

From a BLE point of view, Release 5.0 standardizes two new modes of operation:

1. 2 Mbps mode of operation using GFSK modulated data
2. Long Range mode of operation, intended to provide 125 kbps and 500 kbps, supporting a lower sensitivity (equivalent to longer range).

BLE Market Adoption

- BLE is seeing the highest adoption among BLE, ZigBee and ANT technologies. Because :
  - Classic Bluetooth has very high attach rate to the mobile and PC market
  - BLE "sensors" can talk to Classic Bluetooth devices
  - By adopting BLE on the sensor side, companies ensure the immediate interoperability with a huge number mobile and PC devices
- BLE devices are expected to cross 1.2B units by 2020

Over the last few years, BLE is having a very wide market adoption – higher than any other short range communication protocol.

Fundamental reason is that BLE devices, used in peripherals, can communicate with Bluetooth devices (in dual mode of operation) already existent in personal computers, tablets, cell phones.

Dual mode of operation means that new Bluetooth devices (used in the computers and phones) will support the main Bluetooth mode of operation, as well as the simpler BLE mode of operation.

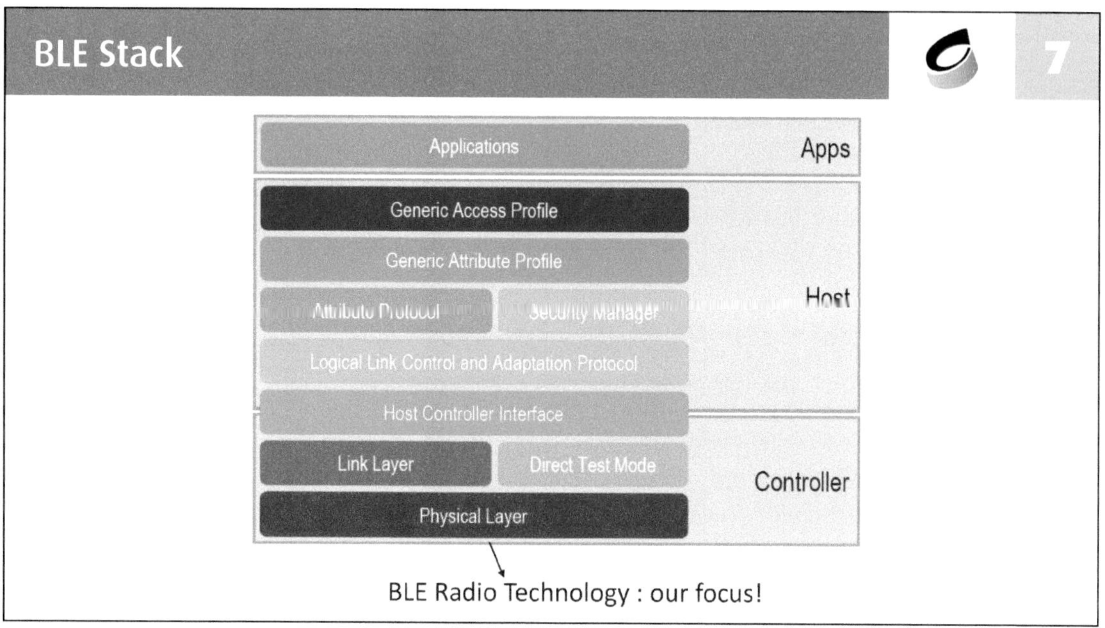

As in other communications systems, BLE is supported by multiple layers of a protocol stack.

At the bottom of the stack we have the BLE radio technology that is responsible for the transmission and reception of bits over the air.

In the remaining of this presentation we will focus precisely on this radio technology.

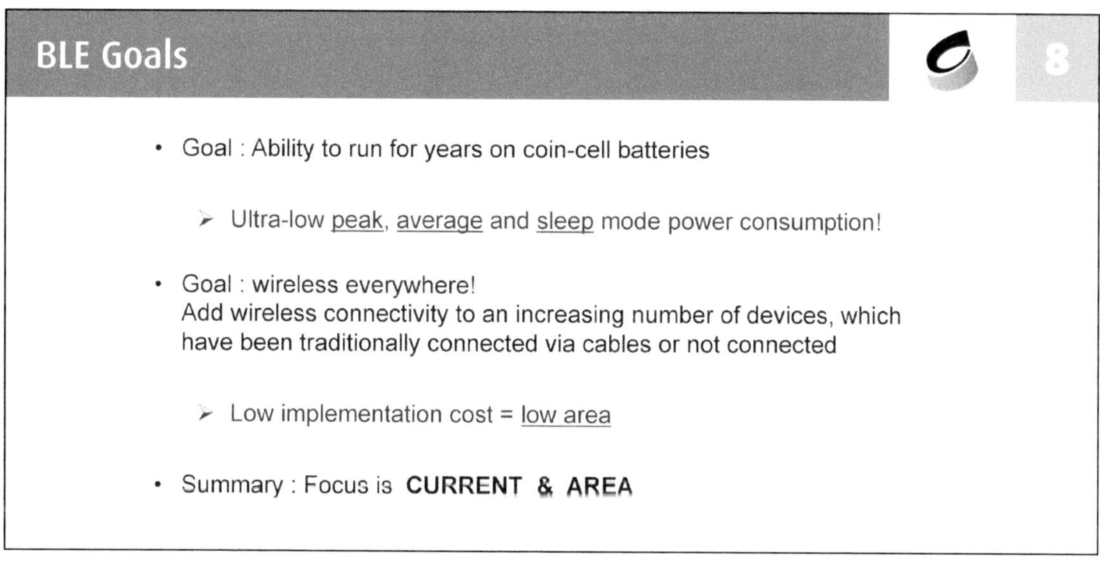

What are the critical goals while designing a BLE radio?

First objective is to be able to run for a long period of time on even a coin cell battery.

This means that not only average current needs to be reduced, but also that peak and sleep mode current should be reduced as much as possible!

Second objective is to make BLE available everywhere! Therefore, incremental cost of adding BLE should be made very small.

Consequently the area of the radio much be made as small as possible!

9 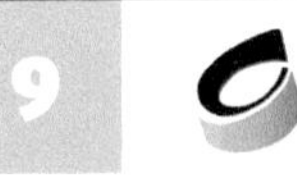

BLE Overview (1)

- Data rate and modes :
  - 1 MBPS
  - 2 MBPS
  - LR = Long Range
- Packet :

LSB			MSB
Preamble (1 octet)	Access Address (4 octets)	PDU (2 to 257 octets)	CRC (3 octets)

- Modulation : Gaussian Frequency Shift Keying (GFSK)
  - Modulation index : h = 0.5
  - Gaussian filtering : BT = 0.5

The data is GFSK modulated (with modulation index = 0.5), and Gaussian filtering (with BT = 0.5) is being used to reduce the spectral bandwidth used by each channel.

The package structure is relatively simple, as shown in the slide.

One point is worth mentioning: the preamble is quite short (8 ms), making the AGC, preamble detection, carrier frequency offset estimation and correction somewhat challenging.

Note that in release 5.0 two additional modes were introduced – 2Mbps and Long Range.

In Long Range, data encoding is used to trade off effective data rate for range of operation.

In this mode, the effective data rate is either 500kbps or 125kbps depending upon the encoding used.

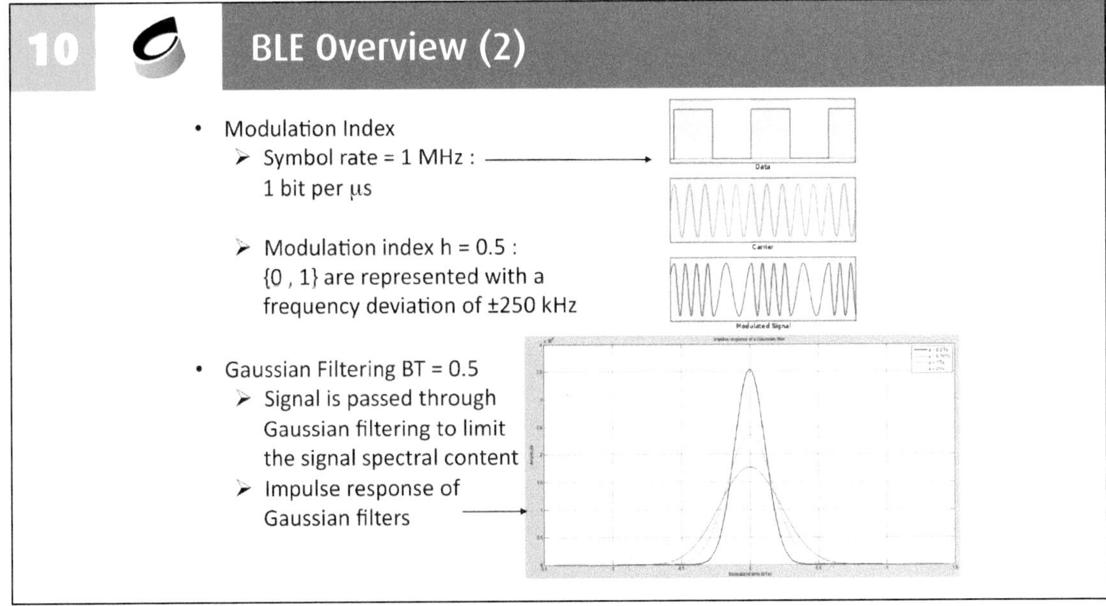

10
BLE Overview (2)
Modulation Index
Symbol rate = 1 MHz :
1 bit per µs
Modulation index h = 0.5 :
{0 , 1} are represented with a
frequency deviation of ±250 kHz
Gaussian Filtering BT = 0.5
Signal is passed through
Gaussian filtering to limit
the signal spectral content
Impulse response of
Gaussian filters
Data
Carrier
Modulated Signal

Just to clarify the concept of modulation index.

Assume first that no filtering was used.

In that case, the 1 Mbps bit stream would be used directly to modulate the frequency by ±250kHz around the carrier frequency, as illustrated in the slide.

Such abrupt frequency deviation would have a broad spectrum.

Therefore, in order to reduce the spectral content, a Gaussian filter with BT=0.5 is first applied, which reduces considerably the spectral leakage.

BLE Overview (3)

11

- Communications channels :
  - 40 channels, on a 2 MHz spacing
  - carrier frequency range from 2402 MHz to 2480 MHz
- Bluetooth frequency hopping capability is retained in BLE

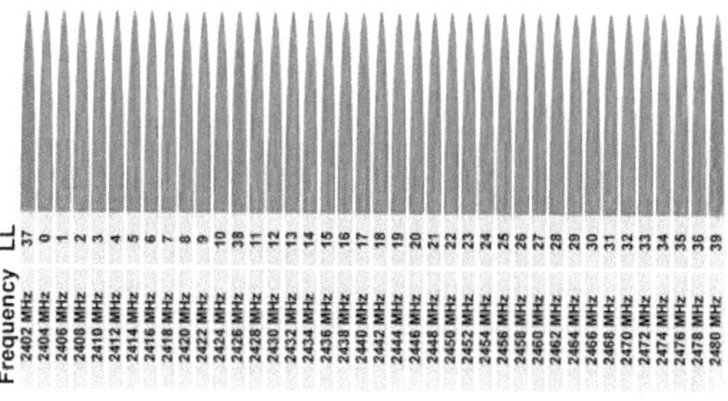

Communication is done over 40 channels on a grid of 2MHz, covering a range from 2402 MHz to 2480 MHz. Note that 3 channels are reserved for advertising. The remaining 37 channels are used for data transfer. Advertising is used to establish a link between two BLE compatible devices.

Once the link is established, one of the 37 data channels is used for data transfer as per the frequency hopping scheme specified by the standard.

Outline

12

- Introduction
- BLE Radio Specifications
- BLE Radio Architecture
- BLE Radio Design
- BLE Radio Performance

We will now focus on the BLE Radio Specifications.

13 

BLE Tx Specifications (1)

- In-band spurious emissions
  - Standard specs :
  - Market : ACPR ≈ -50 dBc

Frequency offset	Spurious Power
2 MHz (|M-N| = 2)	-20 dBm
3 MHz or greater (|M-N| ≥ 3)	-30 dBm

- Out-of-band spurious emissions
  - No BLE specification
  - FCC emissions specification of -42 dBm
  - Market performance : HD2 & HD3 < -45 dBm
- Frequency tolerance
  - Center frequency deviation
    - BLE specification : ±150 kHz
    - Market performance : ±250 kHz
    - This spec will impact mostly the receiver DEMOD and CFO
  - Drift during packet transmission : < 50kHz
    - Open loop VCO direct modulation architecture might struggle with this spec
  - Drift rate : < 400 Hz/μs

Radios in the market should definitely comply with the specification set by the standard.

Often due to competition, the acceptable market performance needs to be considerably higher.

The slide shows the standard specifications, and the typical market specs for current radios.

Points worth noting:

1. ACPR is typically at -50 dBc
2. HD2 & HD3 of the transmitted signal should be better than -45 dBm to -48 dBm

14 

BLE Tx Specifications (2)

- Output power :
  - min ≈ -30 dBm, typ = 0 dBm, max ≈ 3 dBm
  - ramp up / down time ≈ 2 μs
- Modulation accuracy
  - 225 kHz ≤ Δf1 ≤ 275 kHz — Pattern 1 = 00001111
  - Δf2 ≥ 185 kHz — Pattern 2 = 01010101
  - Δf2 / Δf1 ≥ 0.8
- Extra
  - PN@3MHz ≈ -130 dBc/Hz
    - if tx signal is a blocker, then PN@3MHz should be small contributor to rx noise
  - FM SNR ≈ 30 dB
    - transmitter signal should have good purity, to not impact rx chain
- Supply
  - **Voltage : 1V**
  - **Current : 6 – 10 mA** (depending on power level)

The TX output power for short range IoT applications is about 0 dBm (max around 4 dBm).

For not so short range applications, the output power often can be 10 dBm to 15 dBm.

From a TX design point of view, particular attention should be paid to the phase noise at 3 MHz offset.

Reason is that the phase noise skirt of the transmitters should have a negligible impact on the receiver chain of the transceiver used to receive the signal.

This often puts the TX phase noise requirement at 3MHz offset at about -130 dBc/Hz.

To reduce the radio power dissipation, the radio supply voltage should be around 1.0V to 1.2V, with a current consumption of around 6 mA for state of the art performance BLE radios.

BLE Rx Specifications (1)

- Sensitivity level
  - Receiver input power at which BER is 0.1%
  - Corresponding PER : 0.999^376 ≈ 70% ⇒ PER = 30%
    Note : measured with a packet of 376 bits
  - BLE sensitivity spec : -70 dBm
    Current market spec : -96 dBm
- Receiver NF
  - SNR ≈ 11 dB : demodulator SNR for BER=0.1%
  - SNR = Psens - (-174 dBm/Hz + dB10(BW) + NF)
  - NF = -96 + 174 + dB10(1.2M) - 11 = 6.2 dB

State of the art BLE radios are achieving a sensitivity of about -95 to -96 dBm

This corresponds to a NF of approximately 6 dB, which is not so difficult to achieve.

Note that it is not so difficult to achieve a lower NF, thus a better sensitivity.

However, that will increase somewhat the current consumption and/or die area.

So, going for another 1-2 dB in sensitivity improvement is not so appealing due to the current/area penalty.

Sensitivity is defined as the receiver input power level at which BER is 0.1%

However, for conformance testing what is measured is the equivalent PER level of 30%

BLE Rx Specifications (2)

16

- In-band Interference performance

Interferer Type (with Pwanted = -67 dBm)	BLE spec	Market
Co-channel Interference C/I	+21 dB	+10 dB
1 MHz Interference C/I	15 dB	-4 dB
2 MHz Interference C/I	-17 dB	-35 dB
3 MHz Interference C/I	-27 dB	-49 dB
Image Interference C/I	-9 dB	-27 dB

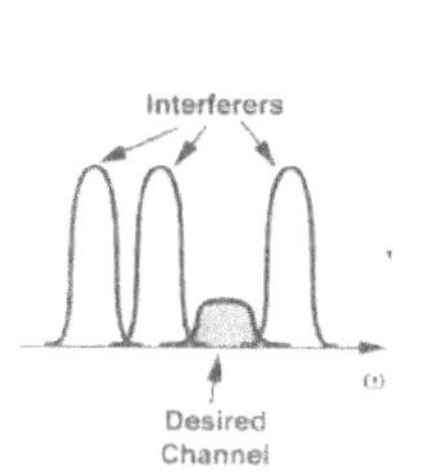

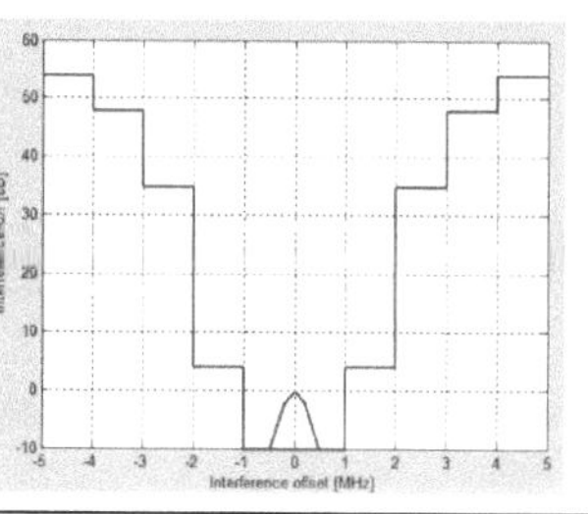

It also presents typical performance of current radios in the market.

Standard states these tests should be done with the wanted signal at -67dBm which is 3dB higher than BLE specified sensitivity.

To be truly compliant with the spirit of the spec, similar performance should also be achieved with the wanted signal at a power level 3 dB higher than the radio achieved sensitivity (not the target sensitivity).

The slide shows the BLE specs for the in-band interferer at different frequency offsets.

17 

BLE Rx Specifications (3)

- Intermodulation characteristics

3M/6M Test (with Pwanted = -64 dBm)	BLE spec	Market
Modulated & CW signal power	-50 dBm	-32 dBm

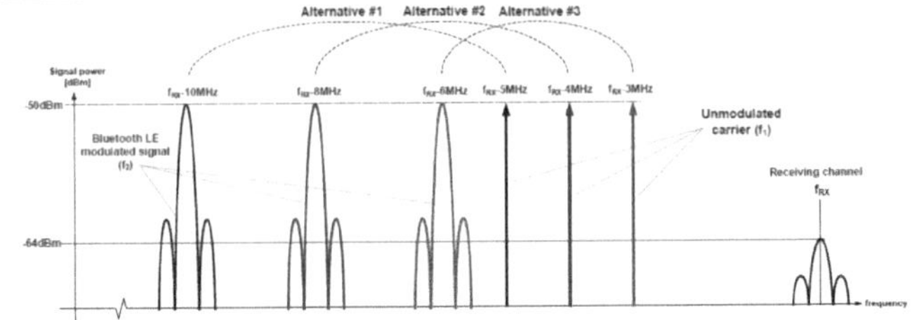

- IIP3 requirement
  - IM3 = 3*Pin - 2*(C/I) - 2*IIP3
  - Pin-IM3 ≈ SNR (11 dB) and Pin = -64 dBm, then IM3 = -75 dBm
  - IIP3 = 0.5*(-3*64+2*32+75) = -26 dBm

Current slide illustrates the intermodulation scenarios used to test the radio.

From this, we can derive the IIP3 spec that can be used in the radio design.

Note that in this test the wanted signal is at -64 dBm, i.e. 6 dB above the BLE specified sensitivity.

To be compliant with the spirit of the spec, similar performance should be achieved with the wanted signal at a power level 6 dB higher than the radio achieved sensitivity.

18 

BLE Rx Specifications (4)

- Out-of-band Interference performance (wanted signal at -67 dBm)

Interfering Signal Frequency	Interfering Signal Power Level	Measurement resolution
30 MHz – 2000 MHz	-30 dBm	10 MHz
2003 – 2399 MHz	-35 dBm	3 MHz
2484 – 2997 MHz	-35 dBm	3 MHz
3000 MHz – 12.75 GHz	-30 dBm	25 MHz

- Max usable signal level
  - BLE spec : -10dBm
  - Market :
    - ~5 dBm with PER < 30%
    - ~10 dBm maximum input signal
- Supply
  - **Voltage : 1 V**
  - **Current : 6 – 10 mA**

The maximum usable signal is specified at -10 dBm.

Often radios need to do much better than this. At minimum they should achieve PER<30% until approximately 5 dBm.

They should also tolerate an input signal at 10 dBm without damaging the part. This is critical, in particular, in deep submicron designs, where the transistors have stringent voltage regions for safe operation.

To reduce the radio power dissipation, the radio supply voltage should be around 1.0 V to 1.2 V, with a current consumption of around 6 mA for state of the art performance BLE radios.

Outline

19

- Introduction
- BLE Radio Specifications
- BLE Radio Architecture
- BLE Radio Design
- BLE Radio Performance

We will now focus on the BLE radio architecture.

Radio Block Diagram

20

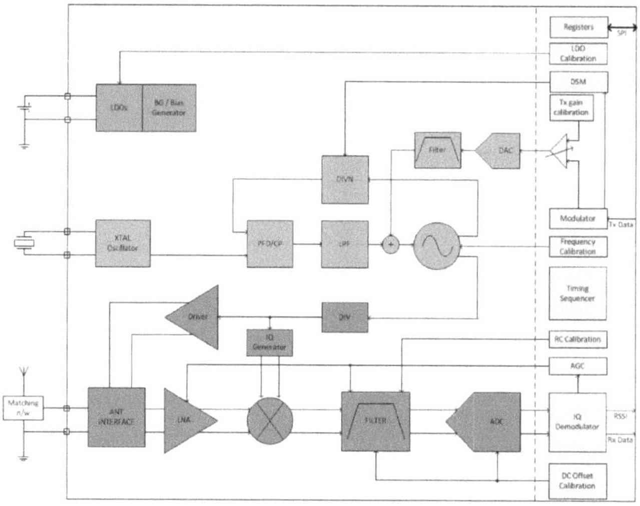

Transmitter is composed of:

- Antenna interface, where the receive and transmit chains are merged
- TX driver providing the TX power/gain programmability
- The phase modulated signal in generated by the PLL.
- Single PLL used for both RX and TX modes
- XTAL oscillator subsystem that generates the reference clock from an external crystal
- Several miscellaneous blocks: LDOs, bandgap voltage, RC calibration, etc.

A typical BLE radio architecture is presented in the slide.

Receiver is composed of:

- Antenna interface, which might have a single ended or differential interface
- LNA followed by a passive mixer
- Baseband filter, which includes a TIA and a complex bandpass filter
- ADC, typically delta-sigma or SAR with about 60 dB dynamic range

Digital subsystem implementing two types of functions:

- Calibration engines to reduce process/temp variations and simplify analog design
- Signal processing functions needed to perform the modulation / demodulation of the stream of raw bits

21 

TX Architecture

- Considerations :
  - Total dynamic range from ~-30 dBm to 0 dBm
    - Need digitally controlled TX driver
  - Modulation mask accuracy
    - PLL bandwidth ≈ 100's kHz
    - Signal bandwidth has contents till ≈ 500 kHz
  - Two alternatives often used
    - Two point modulation (DIVN & VCO)
      Requires calibration of the relative LP and HP gains
    - Open loop VCO modulation
      Requires careful handling of the frequency drift
      More difficult for LR support

The total dynamic range is about 30 dB. Typically the power is ramped up from about -30 dBm to about 0 dBm in approximately 2 μs.

The power ramp up is often implemented with a programmable TX driver, implemented by slicing the PA in unit cells.

Modulation accuracy considerations, and phase noise optimization, often dictate a PLL with a bandwidth of approximately 100 kHz – 200kHz.

However, the signal bandwidth is considerably larger. Hence to faithfully follow the signal, typically a dual point modulation strategy is employed.

A low pass path can easily be implemented by modulating the frequency with a delta-sigma modulator driving the divide by N.

A high pass path can be implemented by driving the VCO directly, through an independent varactor.

In order to achieve a good modulation accuracy, the relative gain of these two paths needs to be well controlled, which often requires a relative gain calibration.

Open loop modulation techniques are not often used.

In particular, when support of the BLE Long Range mode is needed, since the packages are very long, it is quite difficult to meet the frequency drift specifications.

RX Architecture

22

- Considerations :
  - Total dynamic range from ~-100 dBm to ~10 dBm
    - AGC algorithm to control analog gain depending on the signal and blocker conditions
    - DCOC to null chain offsets and to use full range of the ADC
  - Huge blockers in the presence of weak signals
    - Filter blockers around the IF frequency selected
    - Implement complex filter of I/Q signals to have equal tolerance to blockers on both sides of the signal
    - Plan filtering in analog and digital to reduce complexity, current consumption and die area
  - Avoid flicker noise impact on the NF of the receiver
    - Down convert signal to a low IF : around 0.75 – 2 MHz

In receive mode the dynamic range required is in excess of 100 dB.

This high dynamic range can be supported with relatively simple analog blocks, provided the analog gain is properly controlled by an AGC algorithm sensitive to both signal and blocker conditions.

The possible existence of huge blockers near a weak wanted signal, requires a significant amount of filtering before the signal reaches the ADC.

Typically this filter is done at the IF frequency. Furthermore, to have a symmetric blocker performance, the implemented filter is often designed as a complex band pass filter centered at IF.

Due to the large analog gain, any significant DC at the input can get gained up considerably and reduce the total ADC range.

In order to alleviate this problem, a DC offset calibration can be done prior to the burst that will reduce systematic DC components.

The IF frequency should be low enough to reduce the current consumption of the analog blocks, but high enough to not get impacted by flicker noise.

Normally the IF frequency is selected to be around 0.75 MHz – 1.5 MHz for the 1 Mbps case, and around 1.5 MHz – 2.0 MHz for the 2 Mbps case.

23 

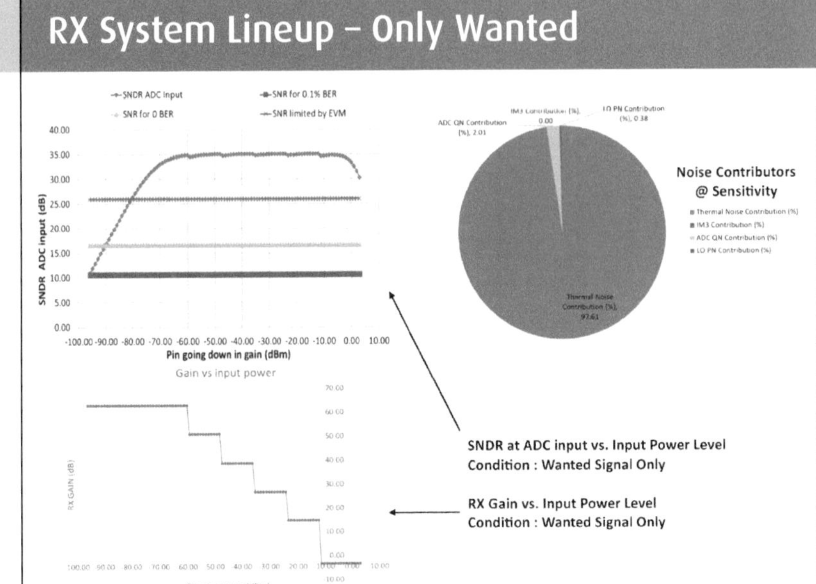

Slide presents RX lineup when only the wanted signal is presented at the antenna.

Starting from sensitivity, the achieved SNR increases progressively, crossing the value required for error free reception at about -90 dBm.

The SNR is maintained quite high until about 5 dBm, by changing the lineup gain at appropriate power levels.

It can also be seen that at sensitivity, the total noise is almost completely limited by thermal noise.

24 

RX System Lineup – Wanted + 3MHz Blocker

SNDR at ADC input vs. Input Power Level
Condition : wanted signal and +45dBc blocker at 3MHz offset

Noise Contributors @ -67 dBm

Slide presents RX lineup when the wanted signal and a 3 MHz blocker are simultaneously presented at the antenna – for an I/C of +45 dB.

Identically, for this case, the SNR increases progressively, crossing the value required for error free reception at about -90 dBm.

The SNR is maintained quite high until about -45 dBm for the wanted signal, corresponding to 0 dBm of the blocker signal.

This is achieved again by appropriately changing the lineup gain. Note that the points where the gain is changed are different from the previous case presented.

In order to achieve very good performance in all field conditions, the AGC has to be carefully implemented to account for multiple possible scenarios.

The noise contributors for a wanted power at -67 dBm, with a 3 MHz blocker, are also shown in the slide.

Here we can see that multiple contributors limit the achieved performance, hinting that it is important to do a detailed accounting of all major noise contributors in multiple blocking scenarios and as a function of the wanted signal power level.

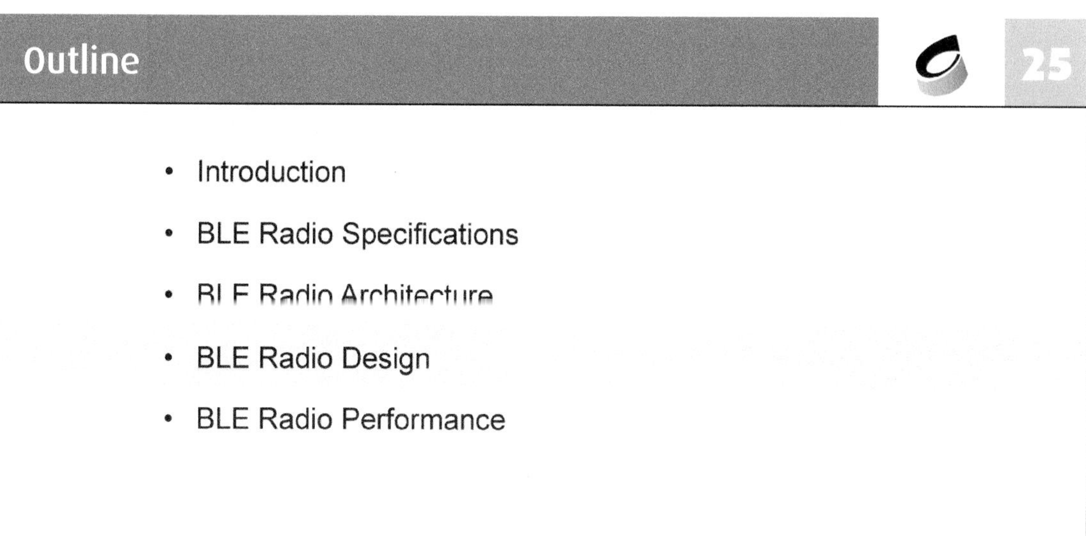

We will now focus on the design of the building blocks required for a BLE radio.

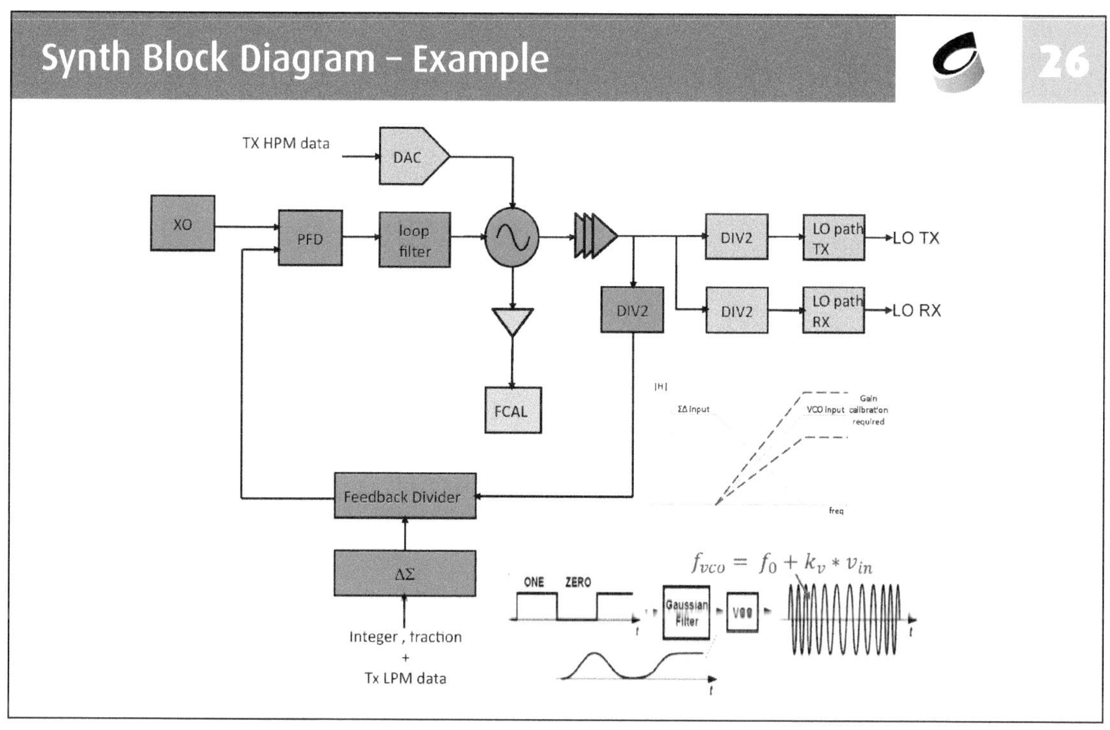

This slide presents a typical architecture of a PLL including LO dividers for RX and TX.

The modulation ports needed to implement dual point modulation for TX are also shown:

a) HPM is implemented by driving directly the VCO

b) LPM is implemented using a delta sigma to control the feedback divider

To achieve good modulation accuracy, the relative gain of the two paths needs to be calibrated before the burst starts.

27 

XO Specifications

Specification	
Voltage supply	1.2-3.6 V
Current Consumption	300 μA
Settling time - 20 ppm	0.5 ms
XTAL frequency	26 – 32 MHz
XTAL Load Capacitance	6 – 10 pF
Duty Cycle	47.5 – 52.5%
Phase Noise 1 kHz 10 kHz 100 kHz 1 MHz	 -127 dBc/Hz -140 dBc/Hz -144 dBc/Hz -144 dBc/Hz

This slide presents the typical specifications of the crystal oscillator block.

Two specifications are worth mentioning:

- Current consumption of the oscillator should be as low as possible, and typically for 24-32 MHz oscillator frequencies it is possible to achieve about 200μA - 300μA of current consumption over process and corners.
- There is a strong focus on reducing the settling time of the crystal oscillator. Currently crystal oscillators used in BLE radios can achieve <0.5ms of settling time, by using to a variety of tricks to reduce settling time. These tricks are often implemented by resorting to dynamic control of the oscillator current and/ or dynamic control of the oscillator capacitance.

28 

XO Design

- Challenges :
  - How to achieve Low Phase Noise with reduced current consumption?
  - How to achieve Fast Settling Time with XTALs with small motional cap?
  - How to minimize the reference spurs?

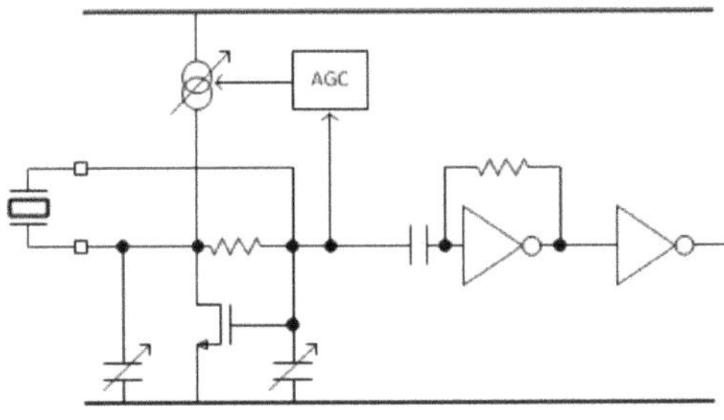

The design of a crystal oscillator is relatively straight forward, and does not present significant challenges.

Challenges / difficulties come from need to achieve low phase noise, without increasing current consumption, and without increasing area (i.e. large filters can not be used).

XO Design

28

For current IoT solutions the crystals being used are physically quite small (small packages are being used), implying that the motional capacitances are very small (< 2 fF).

Note that settling time is inversely proportional to motional capacitance ...

Still the settling time needs to be <0.5ms, presenting some design challenges.

Hence the core oscillators structures being used are very simple, similar to the one presented in the slide, but are highly optimized exploiting several analog tricks.

For example, the AGC can be very noisy and large, hence it is often eliminated completely.

The current sources and the capacitors are often controlled dynamically during start up.

Similarly, several analog tricks are often employed to reduce the magnitude of the reference spurs caused by the oscillator.

PD & CP Specifications

29

Specification	
Voltage supply	1.0 V
Current Consumption	≈ 50 µA
Reference frequency	26 – 32 MHz
PSRR @ 1MHz – 2MHz	≈ 30 dB
Gain programmability	≈ 20 dB
Pre charge settling time	< 5 µs
Non linearity @ operating point	≈ 0.5%
Phase Noise	
1 kHz	-133 dBc/Hz
10 kHz	-146 dBc/Hz
100 kHz	-150 dBc/Hz
1 MHz	-150 dBc/Hz

Slide presents typical specifications of the phase detector / charge pump structure.

Conventional designs are often used for this block, and the only significant effort is to make sure it is as small as possible.

30 

PD & CP Specifications

- PD and CP performance for BLE is quite relaxed (easy to achieve)
- Challenges :
  - Achieve required phase noise performance in a small area i.e. with very limited filtering
  - Programmability to support different reference frequencies often needed Typically increases the noise performance of the PD/CP structure
  - Achieve a relaxed / moderate linearity, but without any increase in current consumption or area
- Summary :
  - Relatively easy design, but more challenging than it seems when you want low area and low current

The design is relatively simple and does not present significant challenges.
The challenge is to implement the block in a low area, if possible consuming no current.

31

PD & CP Design

- XOR
  - Type I operation
  - Input fref duty cycle has to be good
  - Comparison at 2x fref lower DSM noise
  - Simple, low area & low current
  - Difficult to implement programmability
- SR
  - Type I operation
  - Comparison at 1x fref
  - Can be changed to operate at 2x fref
  - Simple, low area & low current
  - Difficult to implement programmability
- PFD
  - Type II operation
  - Comparison at 1x fref
  - More complex, higher current & area
  - Simple to implement programmability
  - Avoid operating at origin

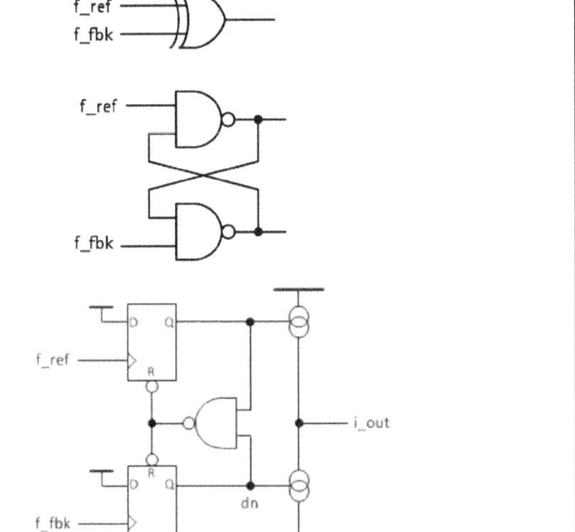

The linearity requirement of the PD / CP for a BLE system are quite relaxed. That is, a significant amount of delta sigma noise fold over can be tolerated before the overall noise gets affected.
Therefore very simple structures like a basic XOR or SR phase detector can be used.
However, their gain is fixed and it is not easy to adapt the structure to be able to program the gain.
For this reason, the conventional phase-frequency detector is probably the most used structure.

LPF Specifications

32

Specification	
Voltage supply	1.0V
Current Consumption	0 μA
Reference frequency attenuation	≈ 80 dB
Gain programmability	≈ 20 dB
Phase Noise contribution relative to total noise	≈ 0 dB

Slide presents the filter specifications.

Two aspects are worth mentioning:

- the filter is normally a passive structure, consuming no current
- it is important to achieve a very good attenuation of the reference harmonics to make sure they do not deteriorate the system performance

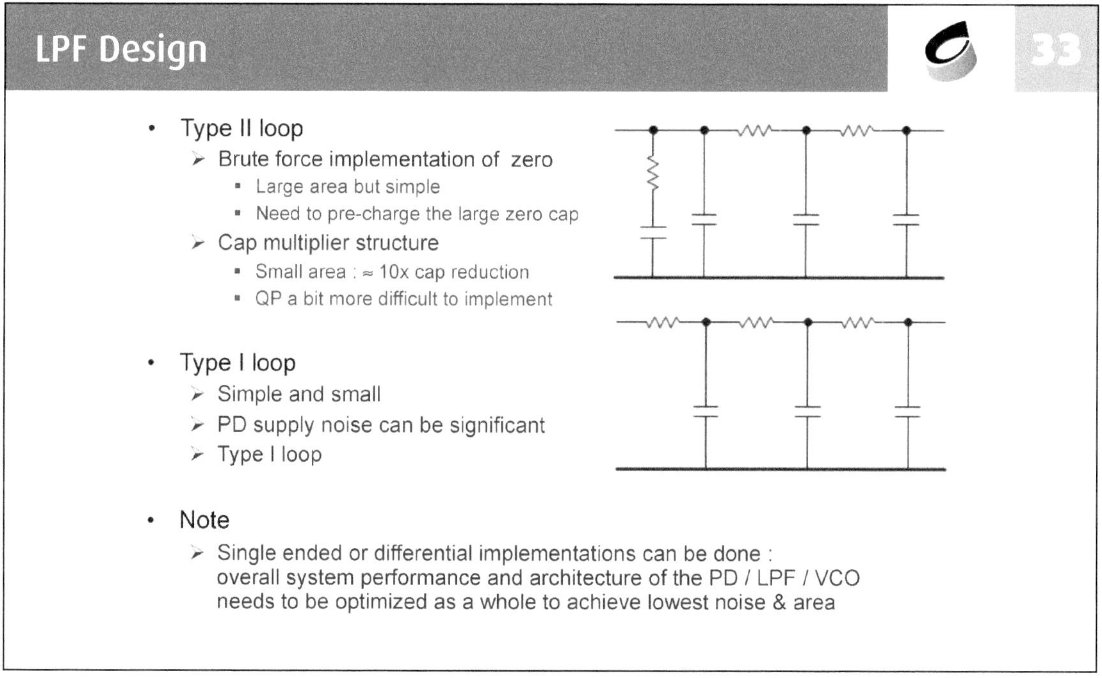

LPF Design
33
Type II loop
Brute force implementation of zero
Large area but simple
Need to pre-charge the large zero cap
Cap multiplier structure
Small area : ≈ 10x cap reduction
QP a bit more difficult to implement
Type I loop
Simple and small
PD supply noise can be significant
Type I loop
Note
Single ended or differential implementations can be done : overall system performance and architecture of the PD / LPF / VCO needs to be optimized as a whole to achieve lowest noise & area

For type II loops, often a brute force approach to implement the zero is used, as shown in the slide.

However, the area required for the zero capacitor is huge.

Often capacitor multiplier techniques are used to reduce the implementation area.

Such techniques come however with some extra implementation complexity and area.

For BLE it is possible to use a type I loop, which is much simpler and smaller.

In both cases, it is possible to implement the filters with single ended or differential implementations, and there are relative advantages and disadvantages.

The best solution requires a careful optimization of the overall architecture of the phase detector, loop filter and VCO, by taking into consideration target noise, current and area.

34 

VCO Specifications

Specification	
Voltage supply	1.0V
Current Consumption	< 1 mA
VCO frequency range	4800 – 4960 MHz
Phase Noise	
10 kHz	-65 dBc/Hz
100 kHz	-90 dBc/Hz
3 MHz	-130 dBc/Hz

- PN < ≈ 700 kHz determined FM SNR contribution to RX noise
- PN 3 MHz determined by required 3 MHz blocker specification

Slide presents typical VCO specifications, for a frequency planning where the VCO is at twice the band frequency. Key specification of the VCO is normally the 3 MHz phase noise that should be around -130 dBc/Hz (referred to the LO frequency).

35 

VCO Design

- Independent LDO
  - At low frequencies ≈100kHz, LDO noise up conversion is critical:
    - Low noice / low area LDO design
    - Minimize VCO supply pushing
  - DCDC ripple can impact degradation of blocking performance
    - Achieve reasonable PSRR at 1-2MHz
- Two analog tuning ports
  - vtune used to close the PLL loop
  - vmod used as HP port for dual modulations architecture
- Low loss digitally controlled cap bank
  - Frequency calibration to improve settling time and reduce varactors Kv
  - Relatively easy to design since VCO tuning range is small

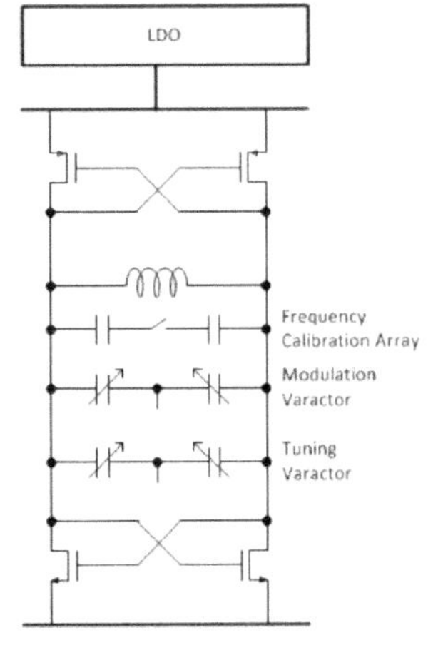

VCO Design

36

- Dominant noise sources :
  - LDO noise is upconverted by VCO pushing
    Reduce cause (noise) and minimize transfer function:
    - Design a low-noise LDO !
    - Find and optimize each individual contributor to supply pushing
    - Groszkowski effect : reduce harmonic generation in the amplifier
  - Current source for biasing : can be eliminated by correct sizing of the amplifier to overcome loss with just enough margin over PVT and freq.
  - Amplifier determines noise, and its size is determined by total loss.
    Then all loss components need to be minimized.
    Organize the VCO in independent parallel branches, and for each branch
    - Calculate effective series conductance and series capacitance/inductance
    - Reduce each loss contribution independently, no matter how small it is!
- Miscellaneous
  - Optimization of the inductor Q is be far the most important aspect to reduce the total loss – if no other loss path is missed !
  - Poly, diffusion and metal filles inside the VCO core are mandatory : minimize the size of the dummy fills to limit Eddy current losses
  - Don't forget to model NQS nature of the amplifier transistors

DIVN Specifications

37

Specification	
Voltage supply	1.0V
Current Consumption	≈ 300 μA
Phase Noise 1 kHz 10 kHz 100 kHz 1 MHz	 -133 dBc/Hz -146 dBc/Hz -150 dBc/Hz -150 dBc/Hz

- Divider needs to be linear over range of all N values:
  delta-sigma shaped quantization noise cannot be down converted
- Often possible to avoid zero phase start with small impact on the PLL settling time
- Frequency calibration can be combined with DIVN circuit to reduce area, or alternatively an independent FCAL block can be implemented

Slide shows typical feedback divider specifications

Since the divider is running at approximately 5 GHz, it is very easy to consume a significant amount of current.

Therefore, it is important to focus on a compact, optimized design that consumes a few hundred uA's.

The overall noise specs are quite relaxed. Hence, a significant amount of delta-sigma noise fold over can be accepted.

Nevertheless, it is still important to make sure that the feedback divider is linear enough over the range of all codes to not impact the noise performance.

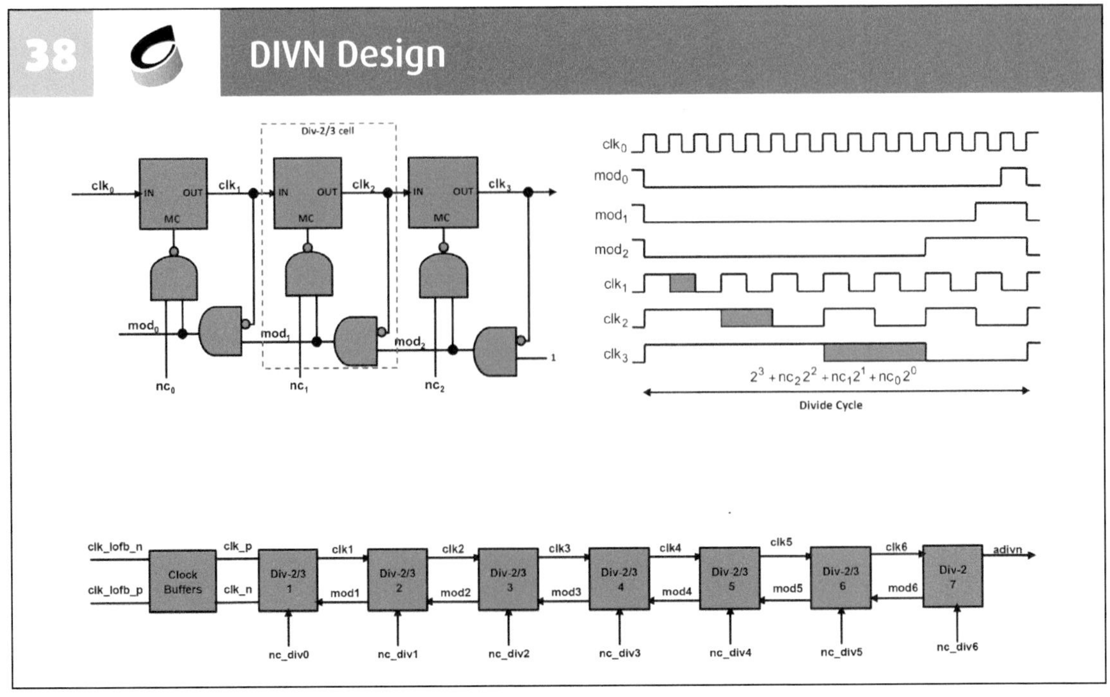

39 RX Specifications

Specification	
Voltage supply	1.0V
Current Consumption	6 mA
Image Rejection	32 dB
Analog gain (5-6 settings)	0 – 65 dB
Gain variation over PVT	±3 dB
Gain transient settling	≈ 500 ns
NF @ maximum gain (total chain)	≈ 6 dB

Slide shows the typical RX chain specifications.

The following specs are worth mentioning:

- Image rejection of even 30 dB to 32 dB is sufficient, which is possible to achieve with careful layout and without any calibration scheme
- Analog gain control needs to be higher than ~60 dB in order to support the total dynamic range, with a relatively inexpensive 10 bit ADC.
- The amplitude settling (when gain is changed) needs to be relatively fast so that the AGC can be done in a short time at the beginning of the preamble.
- The relaxed NF of about 6 dB should be used to reduce the current consumption and area!

LNA Specifications

40

Specification	
Voltage supply	1.0V
Current Consumption	1 mA
NF @ maximum gain (for total chain)	≈ 5.5 dB
IIP3 @ intermodulation power levels	≈ -26 dBm
Programmable gain range	≈ 60 dB
Gain settings	4-5

Slide presents the typical LNA specifications. Worth mentioning are the need to operate at 1 V, with a current consumption of approximately 1 mA.

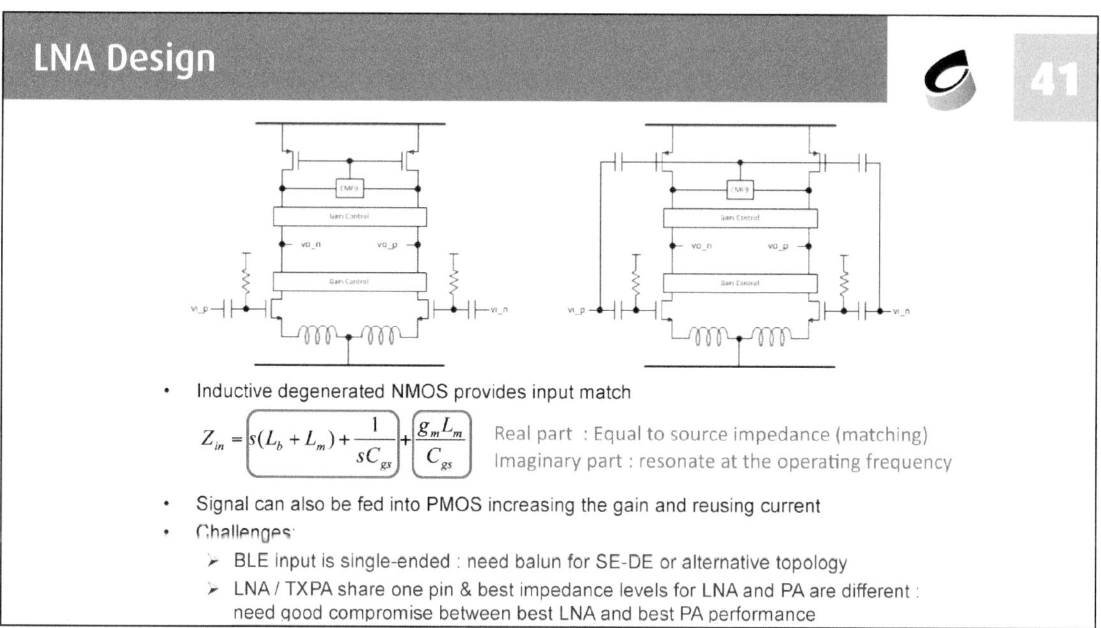

LNA Design
41
Gain Control
vo_n
vo_p
Gain Control
vi_p
vi_n
Gain Control
vo_n
vo_p
Gain Control
vi_p
vi_n
Inductive degenerated NMOS provides input match
$Z_{in} = s(L_b + L_m) + \frac{1}{sC_{gs}} + \frac{g_m L_m}{C_{gs}}$
Real part : Equal to source impedance (matching)
Imaginary part : resonate at the operating frequency
Signal can also be fed into PMOS increasing the gain and reusing current
Challenges:
BLE input is single-ended : need balun for SE-DE or alternative topology
LNA / TXPA share one pin & best impedance levels for LNA and PA are different : need good compromise between best LNA and best PA performance

Slide presents typical LNA topologies using inductive degeneration to match the antenna to the input.

Often the input signal is fed to both the NMOS and PMOS devices to increase the achieved gm for a given current.

The LNA implementation challenge results from the fact that the LNA and the TX PA need to drive the same pin (single ended). The optimum impedance levels for the LNA and the PA are not the same, and some compromise needs to be reached with respect to the impedance level at the LNA/PA port.

42 TIA / CBPF Specifications

Specification	
Voltage supply	1.0V
Current consumption	≈ 1 mA
Gain/Phase matching	45 dB
Output referred offset – after calibration	< 10 mV
Bandwidth (depends on supported modes)	700 kHz – 2 MHz
Filter order (Butterworth)	2 / 3
IF frequency	≈ 750 kHz – 2MHz

- Depending on the architecture, often 6 dB to 20 dB of gain programmability is required
- Bandwidth programmability is often needed to support multiple datarates
- Corner frequency correction / calibrations over process corners is normally needed (so that the performance is consistent and stable over corners)
- Design itself is not difficult. Often the complexity comes from the significant level of programmability that is required

43 TIA / CBPF Design

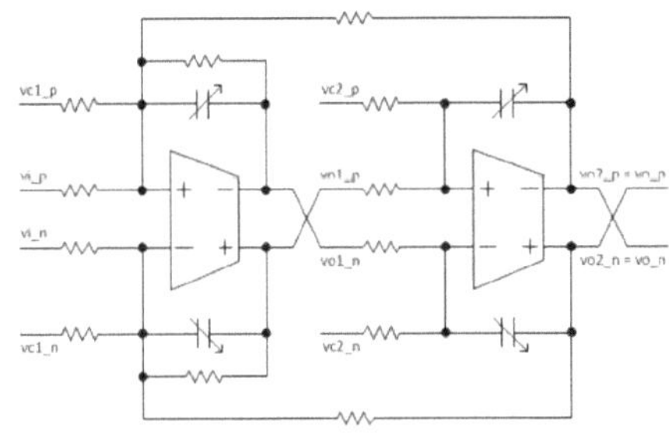

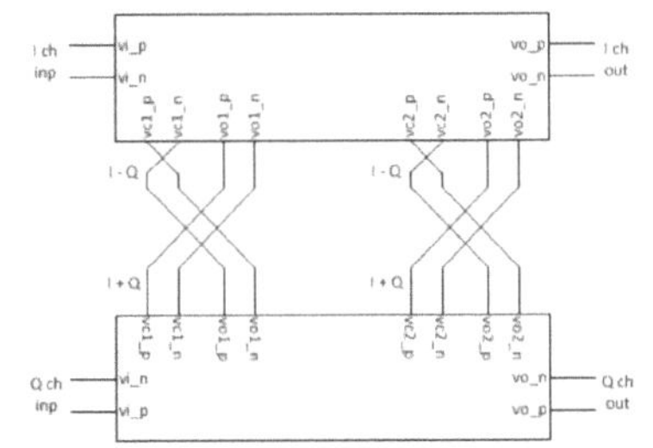

- Mixer interfaces with TIA or directly with CBPF, i.e. TIA built in CBPF
- R & C are programmable to support
  - ➢ different BWs,
  - ➢ different IFs (1 Mbps & 2 Mbps)
  - ➢ different gains
  - ➢ RC calibration
- CBPF provides symmetric filtering at blocker frequency offsets without saturating ADC
  - ➢ Internal variables should not saturate!
- Current DAC required for DC offset calibration

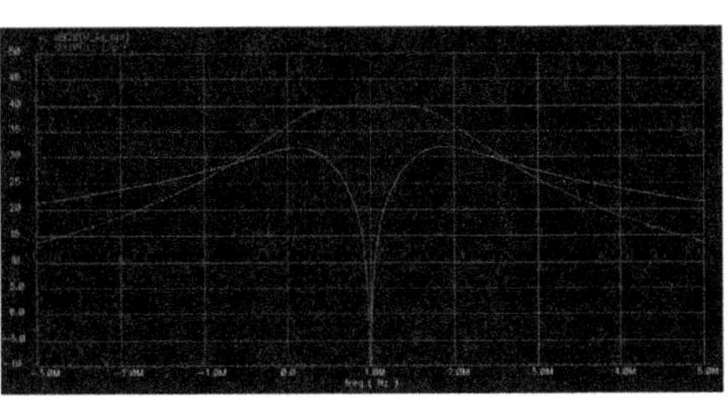

ADC Specifications

44

- Typical Alternatives:
  - SAR ADC : 10 bit running at ≈ 48-64 MHz
  - CT Delta-Sigma ADC : 3rd order running at ≈ 48-64 MHz

Specification	
Voltage supply	1.0V
Current Consumption (two channels + reference)	≈ 500 μA
Dynamic range	≈ 60 dB
Effective number of bits	9 – 10 bit
IQ mismatch	≈ 40 dB
Input differential voltage	≈ 1.6 V

Both SAR and Delta-Sigma ADC's are often used in BLE designs. In the next slides we will present a few design considerations for both alternatives.

ADC SAR Design

45

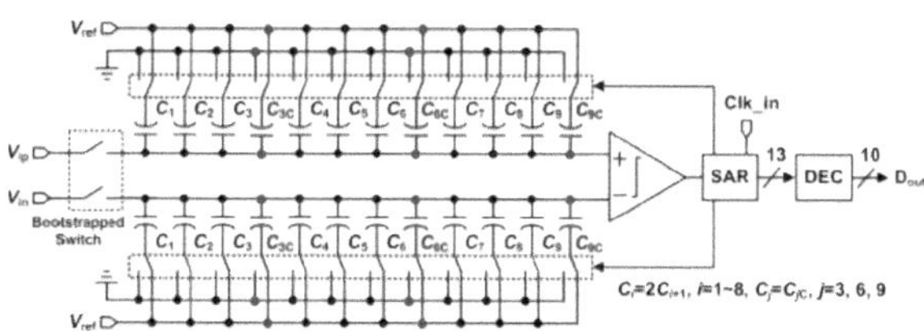

- Often limited by the reference: either high current, or large area in bypass capacitors
  - Exploiting redundancy to improve the performance of the ADC under incomplete settling of the reference & cap network is very powerful
- Often supporting a rail-to-rail differential input signal is difficult
  - Input sampling devices often need bootstrapping
  - Sampling cap non-linearity can be critical in SARs using top plate sampling
- Asynchronous SARs use efficiently all available time, and eliminate the need to generate a high speed clock
- Monotonic switching scheme (towards GND) can be used to reduce the current consumption of the switch cap network
- Need to deal with blockers across the complete out of band region
  - Filtering before the ADC needs to be is sufficient to deal with all blocker cases

46 

ADC CT Delta Sigma Design

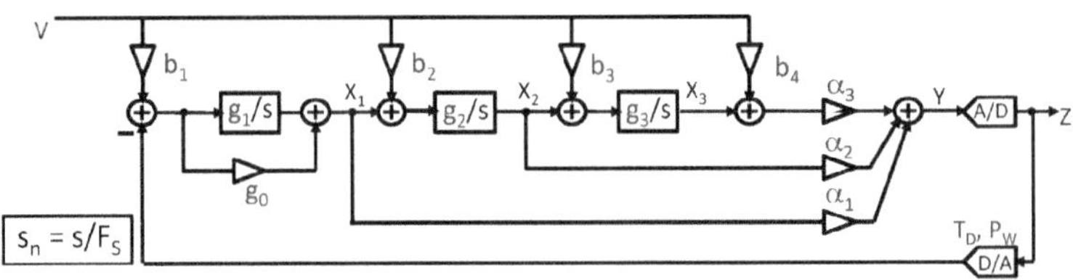

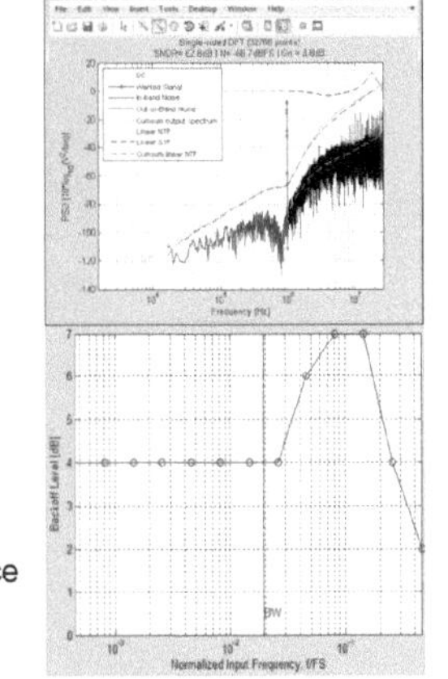

- Back-off level typically increases for blockers (e.g. 3MHz)
  - Design ΔΣ parameters for these signal conditions
- Need to compensate excess loop delay of the structure
  - Typical delay is TD = T/2 and pulse width PW = T (NRZ)
  - Add extra DAC, feeding back to Y
  - Design ΔΣ parameters for intrinsic compensation
- Typical implementation
  - All integrators implemented with RC-amplifier
  - A/D is simple comparator
  - Summation node done with passive resistive network
  - Feedback D/A is current steering or switched R
- NTF/STF models should match exactly the simulated performance
  - Need to understand well how to model quantizer

47 

TX DAC Specifications

Specification	
Voltage supply	1.0V
Current Consumption	≈ 100 μA
Number of bits	8 bit
Clock frequency	24 – 32 MHz
Images	< ≈60 dB
Bandwidth	few MHz

- Current steering DAC : e.g. 6 bit thermometer + 2 bit binary
- Output stage is amplifier driving directly the VCO varactor
  - Programmable CM output voltage
  - Programmable gain to support Kv calibration and normal TX operation
- Conventional design with no major challenges

TX Driver Specifications

48

Specification	
Voltage supply	1.0V
Current Consumption	3 – 4 mA
Power delivered to antenna	0 – 3 dBm
Min power level	-30 dBm
Power ramp up time – digitally controlled	≈ 2μs
Harmonics HD2 HD3	 -45 dBm -45 dBm

TX Driver Design

49

- GFSK modulation (constant envelope) allows use of efficient, non-linear driver with :
  - Significant reduction of current consumption
  - Significant generation of 2HD and 3HD : still need to comply with emission specs!
- The impedance requirements of the TX driver and of the LNA require a careful compromise between the best RX and TX performances
- For open loop TX architectures, frequency accuracy specs impose strict control of :
  - coupling between VCO and TX driver
  - changes in impedances after the VCO is settled

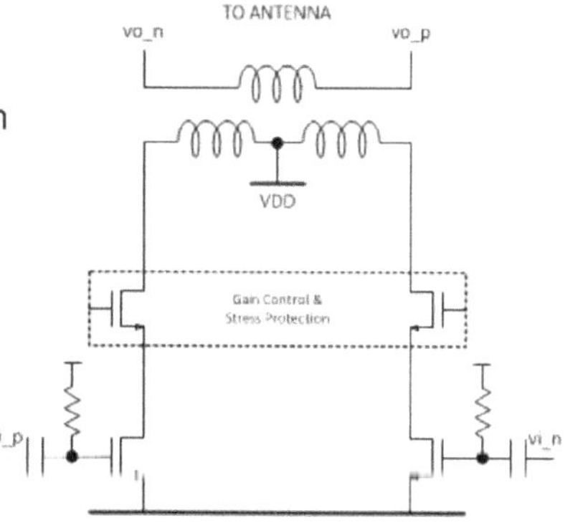

50  Verification

- Relentless market & customer pressure
  - Design time needs to be minimized
  - Time to market is critical : new features added continuously
  - Significant design errors on revA are not acceptable !
- Solid Mixed Signal Verification strategy is very important
  - Signal representation such that there is no switching at RF enables complete radio simulation from antenna to bits
  - Digital design : models in RTL
  - Analog design : models in Verilog using wreal
- Include as much effects as possible in the MSV strategy
  - All calibration loops between analog and digital included
  - Noise effects should be included
- Final goal
  - Complete chain from antenna to bits can be simulated
  - RX: All signal and blocker conditions with PER can be simulated
  - TX : All modulation accuracy metrics can be simulated

51  Radio Layout

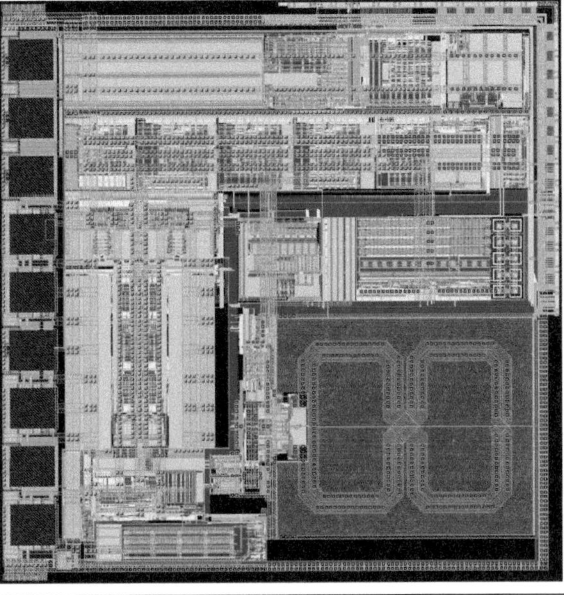

Slide presents the layout of a BLE transceiver implemented in TSMC 55nm process.

Similar design is also implemented in TSMC 40nm and SMIC 55nm, for the same performance targets, including area and current consumption.

The total area is less than 0.5 mm2, including all pads and the interface with digital.

Outline

52

- Introduction
- BLE Radio Specifications
- BLE Radio Architecture
- BLE Radio Design
- BLE Radio Performance

In the next slides we will present the measured radio performance of a state of the art BLE radio implementation.

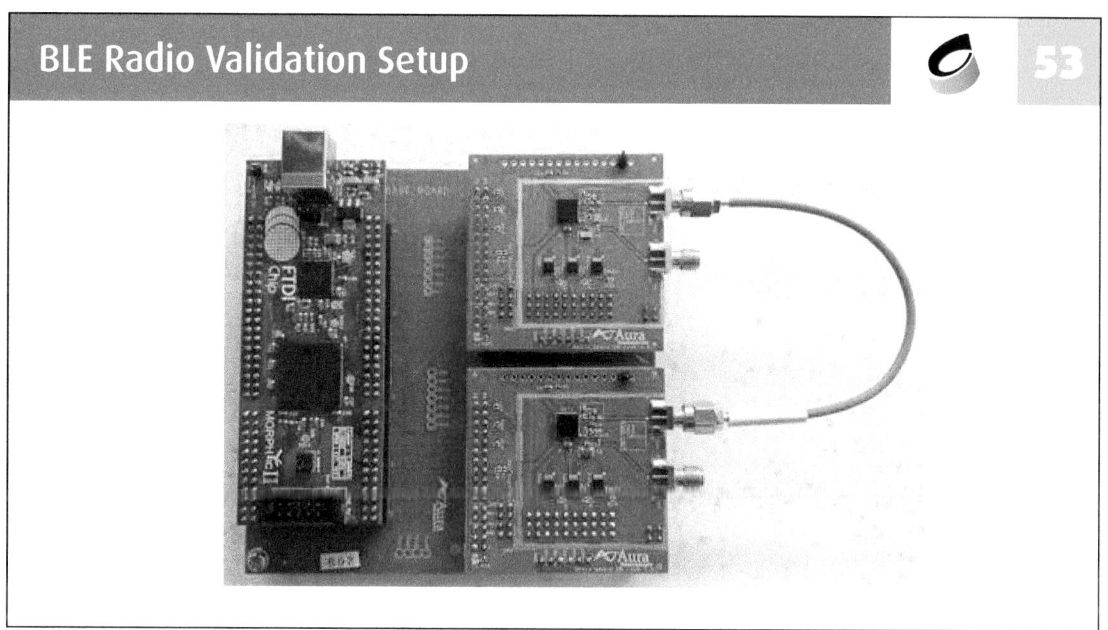

Two identical radio boards are assembled next to each other, and controlled by a base board.

One of the boards can be configured for TX, and the other for RX, and PER can be calculated for 1500 packets of data as specified by the standard.

The same setup can be used to characterize performance of individual radio blocks.

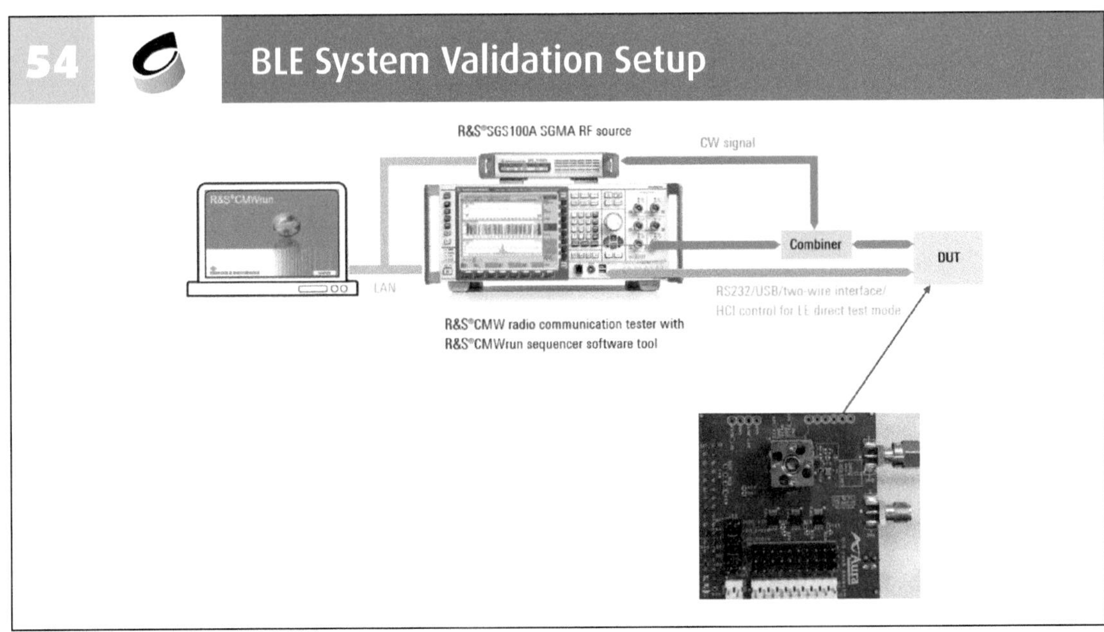

Slide shows the setup used to characterize the performance of the complete BLE system using a R&S CMW system.

55

BLE Radio Performance

TX Specification	Measured
Output Power (typical setting)	0 dBm
Output Power (max setting)	2.0 dBm @ 1.00V core 6.5 dBm @ 1.50V core
TX DAC Images	< -85 dBc
DF2 10101010 pattern	217 KHz
DF1 11110000 pattern	250 KHz
In Band Emissions (2 / 3 MHz offsets)	-61 / -67 dBc
TX Harmonics (3rd, 2nd) at 1.0dBm output power	-50 / -51 dBc
TX Harmonics (3rd, 2nd) at 6.5dBm output power	-44 / -40 dBc
Phase Noise @ 1MHz	-119 dBc / Hz
Phase Noise @ 3MHz	-130 dBc / Hz
Integrated Phase Noise (1kHz - 700kHz)	0.9 deg rms
DC Current Consumption (from line @ 1.2 V - analog)	6.3 mA
DC Current Consumption (from 1.0 V - digital)	0.4 mA

Slide shows the measured radio performance in transmit mode.

BLE Radio Performance

56

RX Specification	Measured
Sensitivity	-96 dBm
Image Interference (C/I) wanted signal @ -67dBm ; BLE spec < -9 dB	-34 dB
Co Channel Interference (C/I) wanted signal @ -67dBm ; BLE spec < 21 dB	9 dB
1 MHz Interference (C/I) wanted signal @ -67dBm ; BLE spec < 15 dB	-5 dB
2 MHz Interference (C/I) wanted signal @ -67dBm ; BLE spec < -17 dB	-42 dB
3 MHz Interference (C/I) wanted signal @ -67dBm ; BLE spec < -27 dB	-49 dB
6 MHz Interference (fref/4) (C/I) wanted signal @ -67dBm ; BLE spec < -27 dB	-53 dB
12 MHz Interference (fref/2) (C/I) wanted signal @ -67dBm ; BLE spec < -27 dB	-50 dB
3M / 6M Blocking Test (C/I) wanted signal @ -64dBm ; BLE spec > -50dBm	-29 dBm
DC Current Consumption (from vline supply of 1.2V)	6.2 mA
DC Current Consumption (from vdig supply of 1.0V)	1.0 mA

Slide shows the measured radio performance in receive mode.

References

57

1. Jansz Groszkowski, "The interdependence of Frequency Variation and the Harmonic Content, and the problem of constant-frequency oscillators," Proceedings of the Institute of Radio Engineers, Vol. 21, No 7, pp. 953-981, July 1933
2. S. Levantino, C. Samori, A. Bonfanti, S. Gierkink, and A. L. Lacaita, "Frequency dependence on bias current in 5-GHz CMOS VCO's: Impact of tuning range and flicker noise up-conversion," *IEEE J. Solid State Circuits*, vol. 37, no. 8, pp. 1003–1011, Aug. 2002.
3. A. Ismail and A. A. Abidi, "CMOS differential LC oscillator with suppression up-converted flicker noise," in *Dig. Tech. Papers Int. Solid State Circuits Conf.*, 2003, vol. 1, pp. 98–99.
4. A. Jerng and C. G. Sodini, "The impact of device type and sizing on phase noise mechanisms," *IEEE J. Solid-State Circuits*, vol. 40, no. 2, pp. 360–369, Feb. 2005.
5. S.-J. Yun, C.-Y. Cha, H.-C. Choi, and S.-G. Lee, "RF CMOS LC-oscillator with source damping resistors," *IEEE Microw. Wireless Compon. Lett.*, vol. 16, no. 9, pp. 511–513, Sep. 2006.

57  References

6. A. Bonfanti, S. Levantino, C. Samori, and A. L. Lacaita, "A varactor configuration minimizing the amplitude-to-phase noise conversion in VCOs," *IEEE Trans. Circuits Syst. I*, vol. 53, no. 3, pp. 481–488, Mar. 2006.

7. N. N. Tchamov and N. T. Tchamov, "Technique for flicker noise up-conversion suppression in differential LC oscillators," *IEEE Trans. Circuits Syst. II*, Exp. Briefs, vol. 54, no. 11, pp. 481–488, Nov. 2007.

8. A. Bevilacqua and P. Andreani, "On the bias noise to phase noise conversion in harmonic oscillators using Groszkowski theory," in *Proc. Int. Symp. Circuits Syst.*, 2011, pp. 217–220.

9. Andrea Bonfanti, Federico Pepe, Carlo Samori, and Andrea L. Lacaita, "Flicker Noise Up-Conversion due to Harmonic Distortion in Van der Pol CMOS Oscillators," *IEEE Trans. Circuits Syst I*, vol. 59, no. 7, pp. 1418-1430, July 2012.

10. E. A. Vittoz,M.G. R. Degrawe, and S. Bitz, "High-performance crystal oscillator circuits: Theory and application," *IEEE J. Solid State Circuits*, vol. 23, no. 3, pp. 774–783, Jun. 1988.

11. T. D. Gavra and I. A. Ermolenko, "An ultrashort-wave quartz oscillator with automatic amplitude control," *Telecommun. Radio Eng.*, vol. 30, pp. 133–134, 1975.

12. Kevin J. Wang, and Ian Galton, "A Discrete-Time Model for the Design of Type-II PLLs With Passive Sampled Loop Filters," *IEEE Trans. Circuits Syst. I*, vol. 58, no. 2, pp. 264–275, Feb. 2011.

13. Derek Shaeffer, and Thomas Lee, "A 1.5-V, 1.5-GHz CMOS Low Noise Amplifier," *IEEE J. Solid-State Circuits*, vol. 32, no. 5, pp. 745–759, May 1999.

14. A. Mirzaei, H. Darabi, J. Leete, X. Chen, K. Juan, and A. Yazdi, IEEE"Analysis and Optimization of Current-Driven Passive Mixers in Narrowband Direct-Conversion Receivers," *IEEE J. Solid-State Circuits*, vol. 44, no. 10, pp. 2688, Oct. 2009.

15. A. Mirzaei, H. Darabi, J. Leete, Y. Chang, "Analysis and Optimization of Direct-Conversion Receivers With 25% Duty-Cycle Current-Driven Passive Mixers," *IEEE Trans. Circuits Syst. I*, vol. 57, no. 9, pp. 2353–2366, Sep. 2010.

16. M. V. Krishna et al., "Design and analysis of ultra lowpower true single phase clock CMOS 2/3 prescaler," *IEEE Trans. Circuits Syst. I*, Reg. Papers, vol. 57, no. 1, pp. 72–82, Jan. 2010.

17. Zhiming Deng, and Ali M. Niknejad, "The Speed–Power Trade-Off in the Design of CMOS True-Single-Phase-Clock Dividers," *IEEE J. Solid State Circuits*, vol. 45, no. 11, pp. 2457–2465, Nov. 2010

18. J. Yuan et al., "A true single-phase-clock dynamic CMOS circuit technique," *IEEE J. Solid-State Circuits*, vol. SC-22, no. 5, pp. 899–901, Oct. 1987

References

19. F. Kuttner, "A 1.2-V 10-b 20-Msample/s nonbinary successive approximation ADC in 0.13-μm CMOS," *ISSCC Dig. Tech. Papers*, pp. 176-177, Feb., 2002

20. J. Craninckx and G. Van der Plas, "A 65fJ/Conversion-Step 0-to-50MS/s 0-to-0.7mW 9b Charge-sharing SAR ADC in 90nm Digital CMOS," *ISSCC Dig. Tech. Papers*, pp. 246-247, Feb., 2007.

21. V. Giannini, P. Nuzzo, V. Chironi, A. Baschirotto, G. Van der Plas, J. Craninckx, "An 820μW 9b 40MS/s Noise-Tolerant Dynamic-SAR ADC in 90nm Digital CMOS," *ISSCC Dig. Tech. Papers*, pp. 238-239, Feb. 2008.

22. C. C. Liu, S. J. Chang, G. Y. Huang, and Y. Z. Lin, "A 10-bit 50-MS/s SAR ADC with a monotonic capacitor switching procedure," *IEEE J. Solid-State Circuits*, vol. 45, no. 4, pp. 731–740, Apr. 2010.

23. Gerhard Mitteregger, Member, IEEE, Christian Ebner, Member, *IEEE, Stephan Mechnig*, Member, IEEE,

24. Thomas Blon, Christophe Holuigue, Ernesto Romani, "A 20-mW 640-MHz CMOS Continuous-Time ADC With 20-MHz Signal Bandwidth, 80-dB Dynamic Range and 12-bit ENOB," *IEEE J. Solid State Circuits*, vol. 41, no. 12, pp. 2641–2649, Dec. 2006

CHAPTER 9

Nanosensors, from fundamentals to Internet of Things deployments

Bérengère Lebental
Research Scientist at IFSTTAR
Paris, France

The chapter discusses the new sensing opportunities provided by nanotechnologies.

After a brief overview of the current status of nanotechnologies, it presents a review of existing types of nanosensors. The different options for fabrication are summarized before moving on to detailing some of their electrical and sensing features. Two of the challenges of nanosensors, reproducibility and reliability, are explained in depth. Finally, we discuss how to integrate nanosensors into IoT devices to carry out their deployments in the field.

1 Outline

- Nanosensors for improved IoT
- Fundamentals on nanomaterials
- Fundamentals of nanosensors
- A case study on carbon nanotube sensors
- Deployments of nanosensors
- Reliability of nanosensors
- Summary

In the previous chapter the sensing aspect was treated as a black box. In this presentation this black box will be examinated. The questions discussed here are: 1. "What is this black box?"; 2. "What is inside it?" and 3. How to use this black box?

The focus of this talk is in sensors, more particularly those based on nanotechnologies and nanomaterials.

Two sections compose this presentation. The first section presents a general discussion about the topic and the second one presents examples of nanotechnology-based sensors that might come out on the market in the next few years. The goal of this talk is to make the reader more aware to the parameters he or she may encounter in research sensors and commercial ones.

2 Outline

- Nanosensors for improved IoT
- Fundamentals on nanomaterials
- Fundamentals of nanosensors
- A case study on carbon nanotube sensors
- Deployments of nanosensors
- Reliability of nanosensors
- Summary

Let us start with how these types of sensors fits into the framework of Smart Cities.

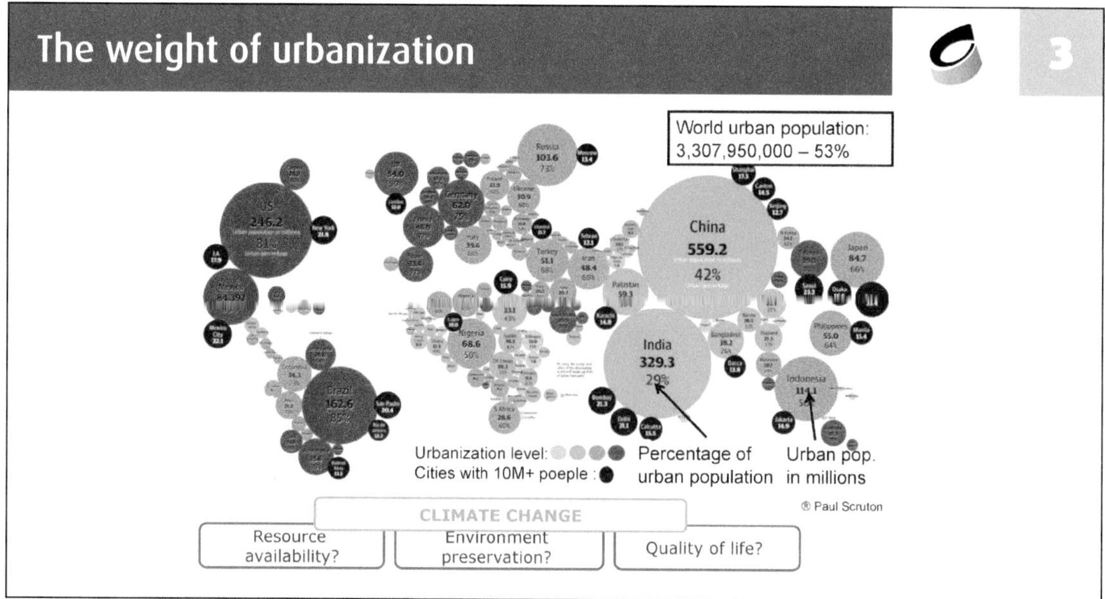

This Figure presents the weight of urbanization in the world. For example, in the left we have US with low urban population, but with high urbanization rate. And in the top right we have China with very high urban population, but with low urbanization. Despite the diversity of urbanization scenarios, cities are all associated with the same wealth of challenges: resource availability, environment preservation, and citizens' quality of life. These issues are expected to worsen with climate changes.

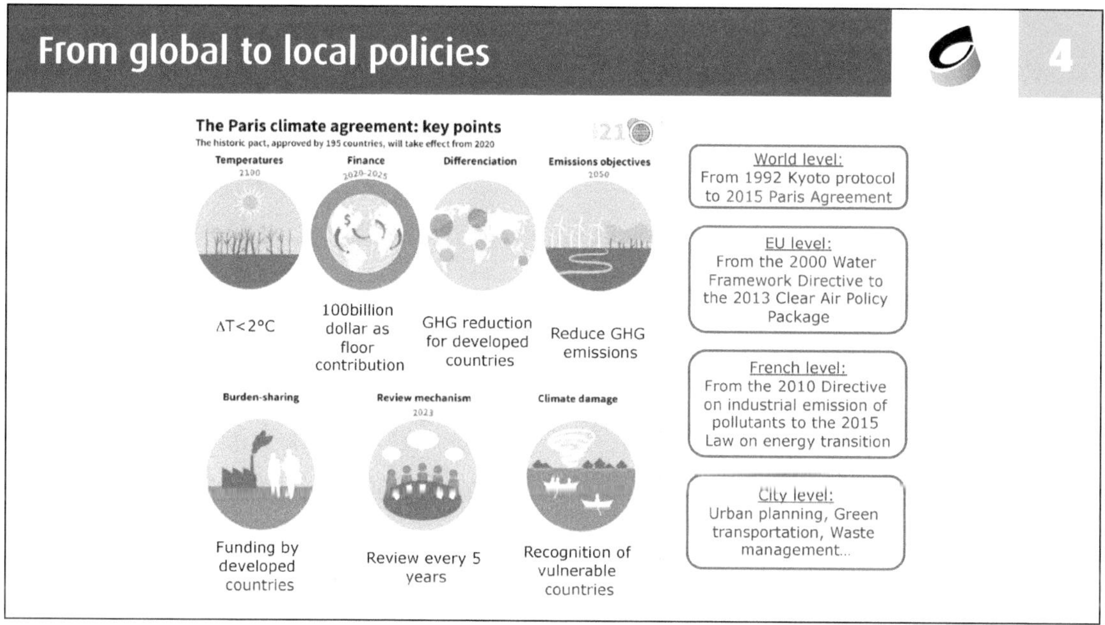

A set of policies have been and are being developed at all levels to reduce these urban impacts, notably climate changes. The latest one is the 2015 Paris Agreement, setting guidelines for actions toward climate change mitigation. The Paris Agreement is recent and global, but it has been preceded by a wide range of policies at all institutional level, from worldwide to European to national to city level. For instance, at city level, Paris is closing downtown highways as well as prohibiting access to more polluting vehicles.

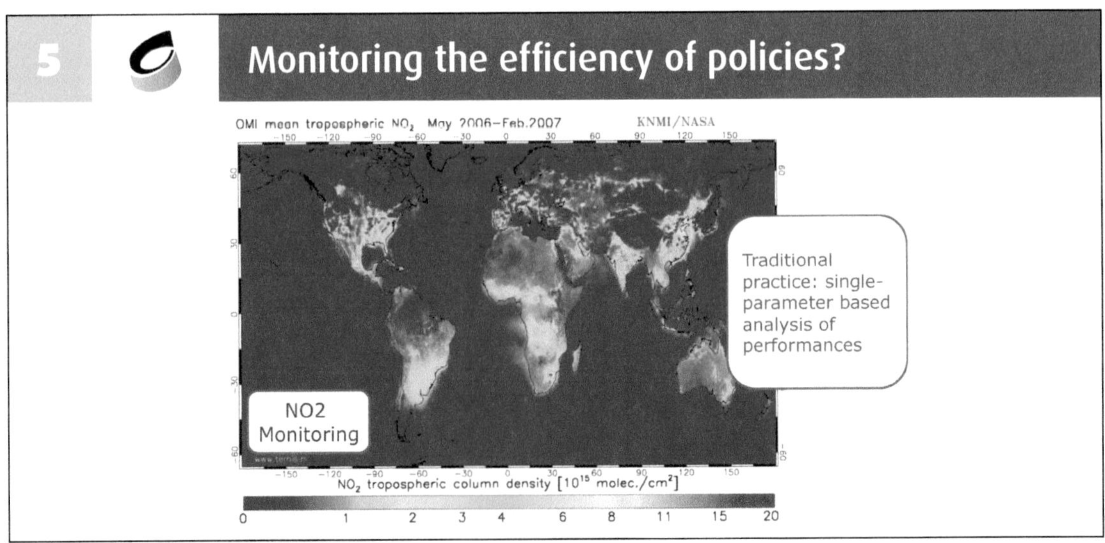

And so it becomes necessary to monitor the efficiency of those policies. When you apply a new policy you want to verify how well it works. For example if you apply a worldwide policy to reduce NO2 emissions, you want a system that is able to check that those levels are actually decreasing following application of the policy. The most classical approach is to monitor the main parameter of interest (NO2 in this example) across the area of interest (here worldwide by satellite). This approach is relatively well mastered today. However, it does not account well for externalities of a policy, meaning the unexpected, indirect impacts of the policy. For instance, reducing greenhouse emissions in an industry may lower its rentability and thus have a detrimental impact on job market, although the target of NO2 reduction is achieved satisfactorily and benefits both people health and the environment.

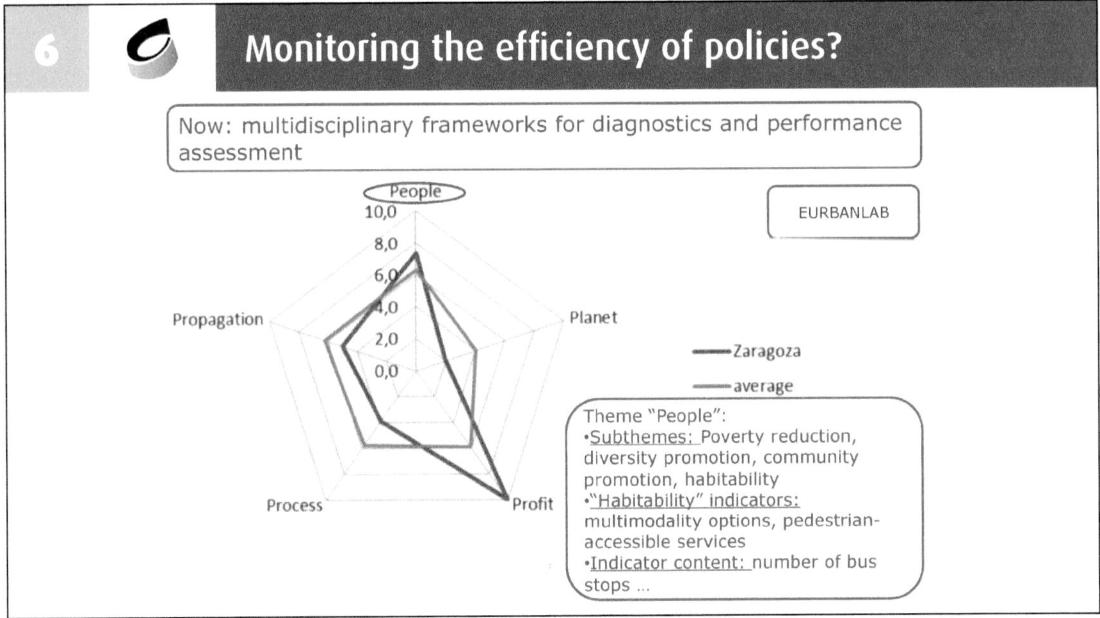

To evaluate the policies' efficiency, what is now needed is a multidisciplinary approach, coupling various factors from distinct fields of knowledge - from social to economic to environmental impacts. Consequently, multiple indicators are imperative to proper diagnostics before and after policy implementation. This example from EURBANLAB evaluation framework is an interesting example: this project in Zaragoza has a good performance for people and business, but is bad for environment.

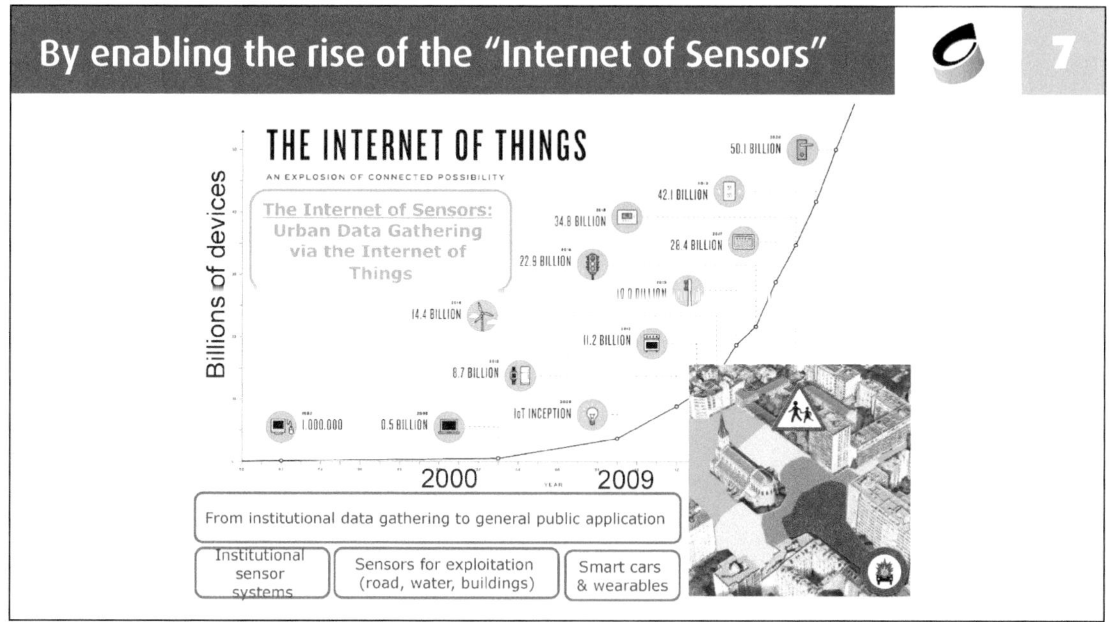

Given the need to collect a lot of data across all domains, new technological solutions are needed. The Internet of Things (IoT), with its remarkable and explosive growth in the last years, is a good candidate to meet the challenge. Based on this we proposed a subconcept, the Internet of Sensors (IoS), focusing on solutions for data gathering for urban applications. It includes institutional applications, data for urban operators (water networks, building energy performances) and applications for the general public (smart cars, smart wearables).

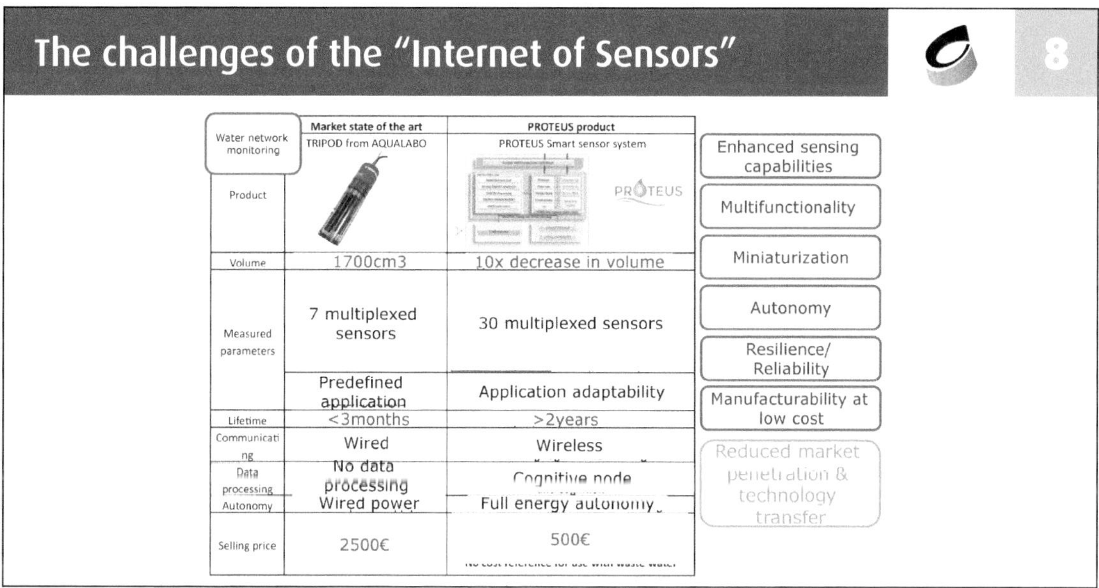

Water network monitoring	Market state of the art TRIPOD from AQUALABO	PROTEUS product PROTEUS Smart sensor system
Product		PROTEUS
Volume	1700cm3	10x decrease in volume
Measured parameters	7 multiplexed sensors	30 multiplexed sensors
	Predefined application	Application adaptability
Lifetime	<3months	>2years
Communicating	Wired	Wireless
Data processing	No data processing	Cognitive node
Autonomy	Wired power	Full energy autonomy
Selling price	2500€	500€

Despite the huge growth of the internet of sensors, it still features a lot of challenges, such as miniaturization, autonomy, multifunctionality, etc., despite the facts that this are crucial aspects to ensure market penetration. This table presents two different products: the first one, 'TRIPOD from AQUALABO' is a commercial product while the second 'Proteus' is, at this moment, a prototype developed along the course of the European project Proteus. This table exemplifies the margin of progress for commercial products to reach what the market actually needs (what is for instance proposed in project Proteus)..

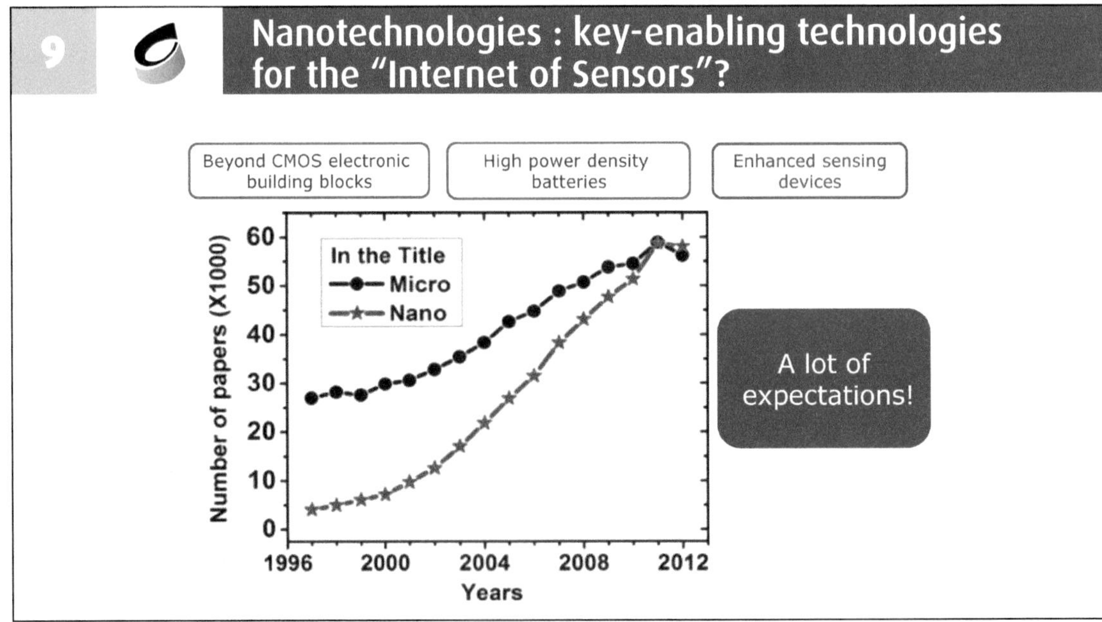

Nanotechnology has strongly developed in the last few years, as it is expected to provide solutions to various current issues in the Internet of Sensors, such as the next generation of electronic circuits or batteries. What will interest us in the next slides are the expectations in terms of sensing, notably for improved sensing capabilities or miniaturization and for multifunctionality.

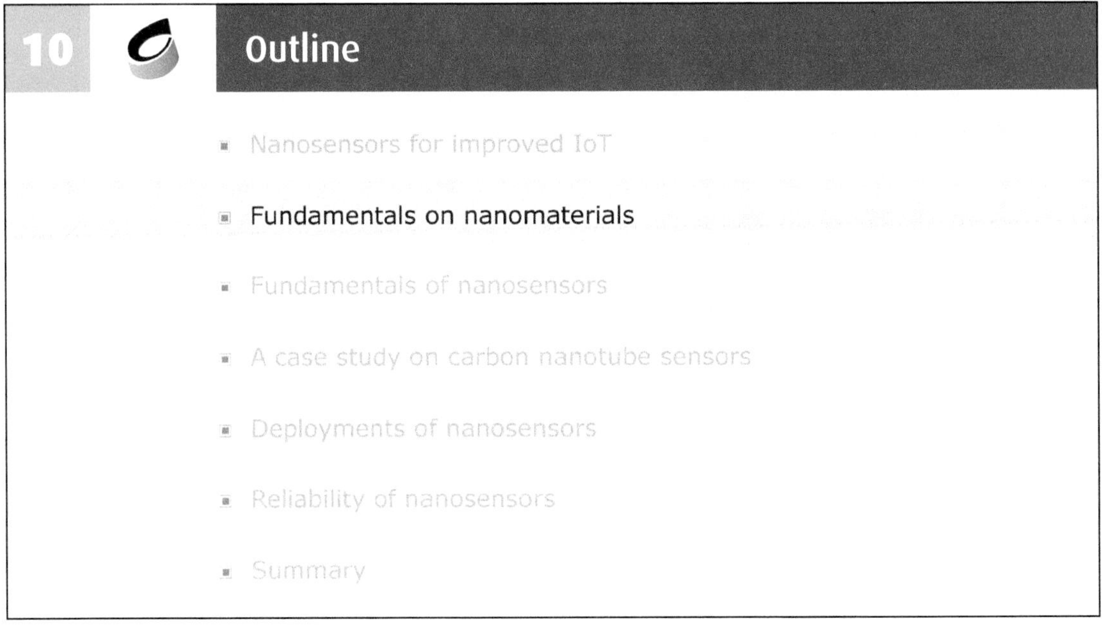

The following slides provides some basics about nanomaterials and nanotechnologies, before going into more specific topics.

Definition of « nano »

11

- *Nano* means *dwarf* in greek
- Nano=1e-9
  - Nano-*thing*→ any *thing* with one characteristic size below 100 nm

    Nanoparticles, nanosensors, nanocomposite

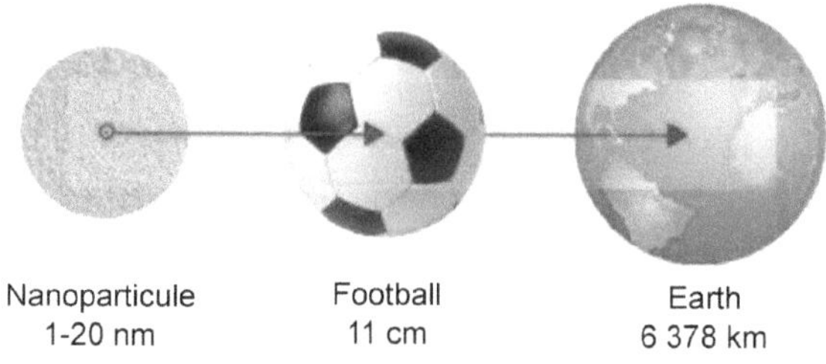

The word 'nano' means 'dwarf' in Greek, and is used today as a prefix meaning 1e-9. A 'nano-object' is an object with at least one characteristic size below 100 nm (for instance nanoparticles, nanoelectronics, nanostructures, etc). If all characteristic sizes are above 500nm then the micro prefix should be used.

Exploiting Nano: ancient practice

12

- Nanoparticles have been used since Roman antiquities to make color pigments

Left: Lycurgus' cup, British Museum, London
Right: Stained glass from La Saint Chapelle, Musée de Cluny, Paris

Romans were already using (unknowingly) nanoparticles for color pigments centuries ago, as shown by Lycurgus's cup. More recent use is observed in stained glass achievements from the Middle Age. The pigments were achieved by heavy grinding to dust and heating of the colored materials available in the environment, usually some forms of metal oxides (from quarries and mines).

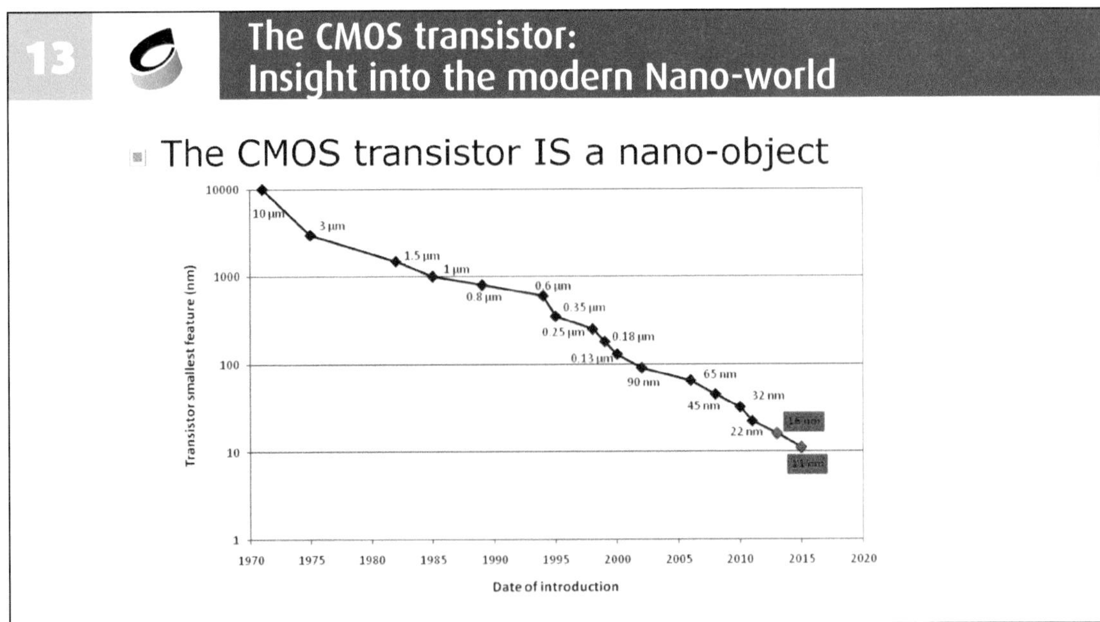

Let us start with a well-known, completely ubiquitous product of nanotechnologies: the CMOS transistor. Indeed, its channel length (that is, the technology node name) is currently below 100 nm. To keep following Moore's law, technology developers have to push heavily into nanotechnologies, reaching a channel size where the number of silicon atoms cannot be considered infinite anymore (leading to complex side effects). Let us remark that despite this, this field of science and technology is more currently referred to as microelectronics.

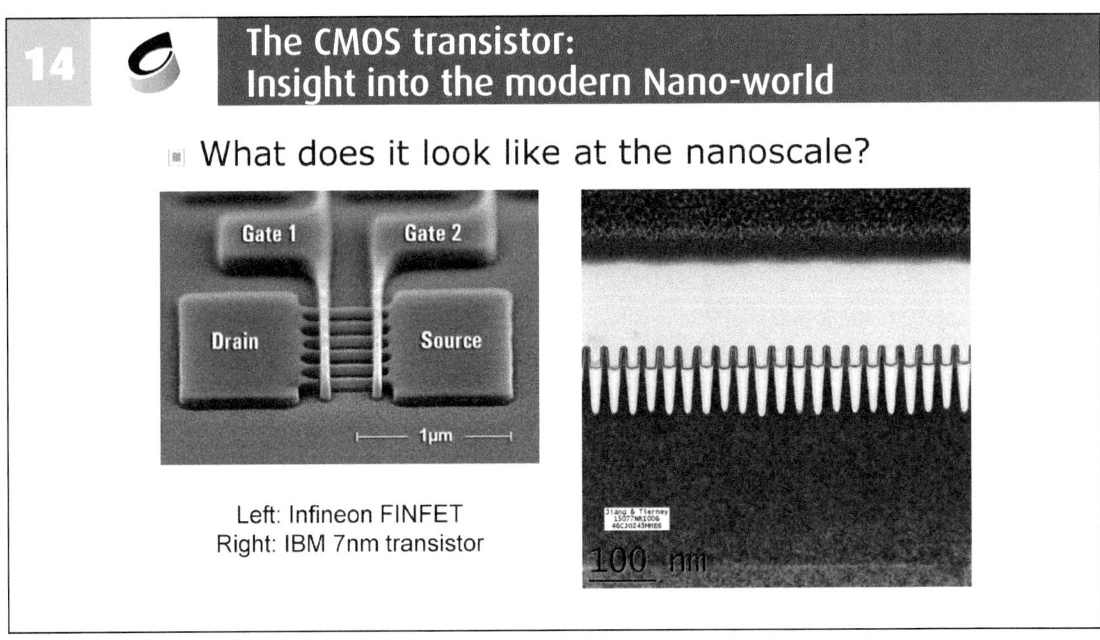

14
The CMOS transistor:
Insight into the modern Nano-world
What does it look like at the nanoscale?
Gate 1
Gate 2
Drain
Source
1µm
Left: Infineon FINFET
Right: IBM 7nm transistor
100 nm

These figures illustrate what kind of nanostructures the CMOS transistor industry is working with, here a FINFET by Infineon and a 7nm transistor by IBM. It is interesting to observe that at the nanoscale, devices often feature non-idealities compared to the theoretical structures: lack of uniformity, variability in pattern size, or small misalignments. This may leads to degraded device performances compared to predictions.

Classifying nanomaterials

15

- 2D nanomaterials: thin films technologies

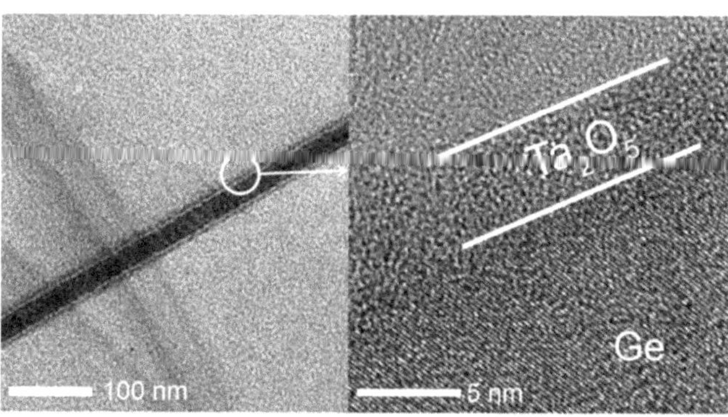

To understand the role that nanotechnologies play in sensing technologies (and more generally in the IoT), you have to understand a bit about how to categorize nanomaterials. A very popular way to classify them is according to their dimensionality (whether they feature a 2D, 1D or 0D structure). Thin films are widely used examples of 2D nanomaterials (sensors, transistors). In fact, all kind of materials (tantalum pentoxide, silicon dioxide, molybdenum, etc.) can be deposited into thin films, depending on the target application. A variety of criteria are available for their classification, such as thickness, crystalline structures, electronic properties... In the figure, one observes at different scales via TEM (transmission electron microscopy) a 5 nm Ta_2O_5 layer ontop of a thicker (20 nm) Ge layer. The right-hand side picture shows clearly the atomic planes of the Ge layer.

Classifying nanomaterials

16

- The ultimate 2D nanomaterial: the monolayer
- Example: graphene - single layer of graphite (carbon) – thickness 0.34nm

Science **306**, 666 (2004)

Electric Field Effect in Atomically Thin Carbon Films

http://www.observatorynano.eu/project/document/101/

The ultimate 2D thin film is the monolayer of atoms. Graphene is the most well-known example of a monolayer. It consists in just a single layer of graphite (that is, of what pencil tips are made). Mechanically, graphene has a higher Young's modulus than steel, with also very high thermal and electrical (ballistic) conductivities. The charge mobility in graphene is approximately 200 times the mobility in silicon (Si). The reference in the slide is the (Nobel-prize winning) paper published in Science 2004 by Geim and Novoselov that demonstrated the first isolation of graphene as a single layer of graphite, and its use in electronic and optical devices.

Classifying nanomaterials

- The future: new 2D nanomaterials
  - TMDC: transistion metal dichalcogenides

Nobel Lecture: Graphene: Materials in the Flatland

K. S. Novoselov

REVIEWS OF MODERN PHYSICS, VOLUME 83, JULY–SEPTEMBER 2011

BN **NbSe2** **MoS2**

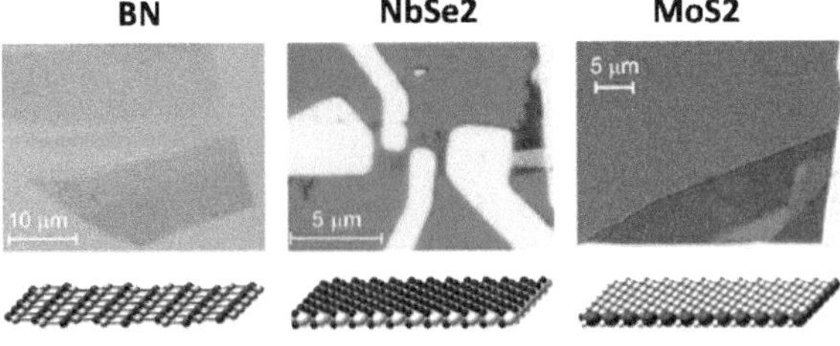

The next generation (after graphene) of monolayer thin, 2D materials are in the pipes: notably monolayers from transition metal dichalcogenide. These materials (a few popular examples are shown in the figures) feature interesting properties compared to graphene, which are attracting a lot of interests.

Classifying nanomaterials

- 1D nanomaterials:

-Nanotubes

-Nanowires

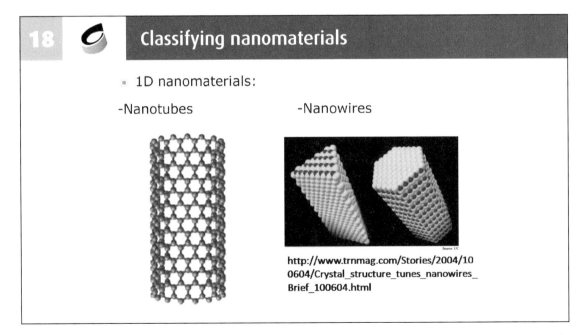

http://www.trnmag.com/Stories/2004/100604/Crystal_structure_tunes_nanowires_Brief_100604.html

Another type of nanomaterials is the 1D category consisting of either nanotubes (empty) or nanowires (filled). While the former presents a cylindrical nanostructure (empty as a tube), nanowires are nanostructures with atoms in the middle. The type of structure has a very strong impact in the properties.

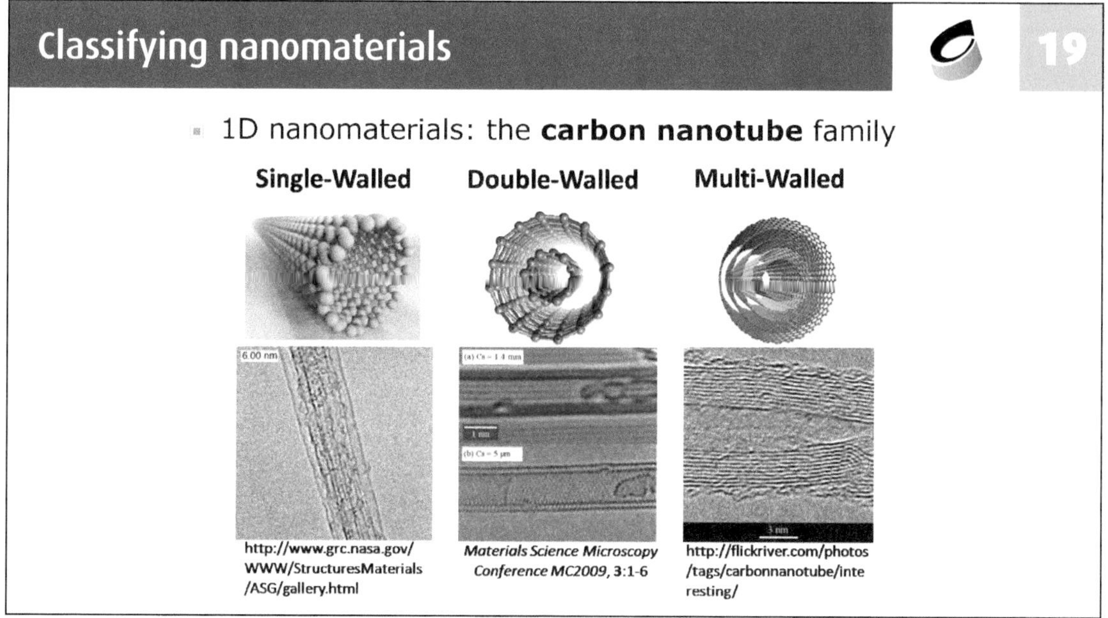

Regarding the carbon nanotubes (CNT) family (64000 publications in 2016 only), they are classified according to the number of 'walls': 'single-walled', 'double-walled' or 'multi-walled'. Their diameter ranges from 0.5nm to 100nm, while their length ranges from 500nm up to hundreds of micrometers. The number of walls impacts drastically the electronic properties (from semi-conducting to metallic). They are encountered in a wide range of applications such as electronics, energy, chemical processing, among others).

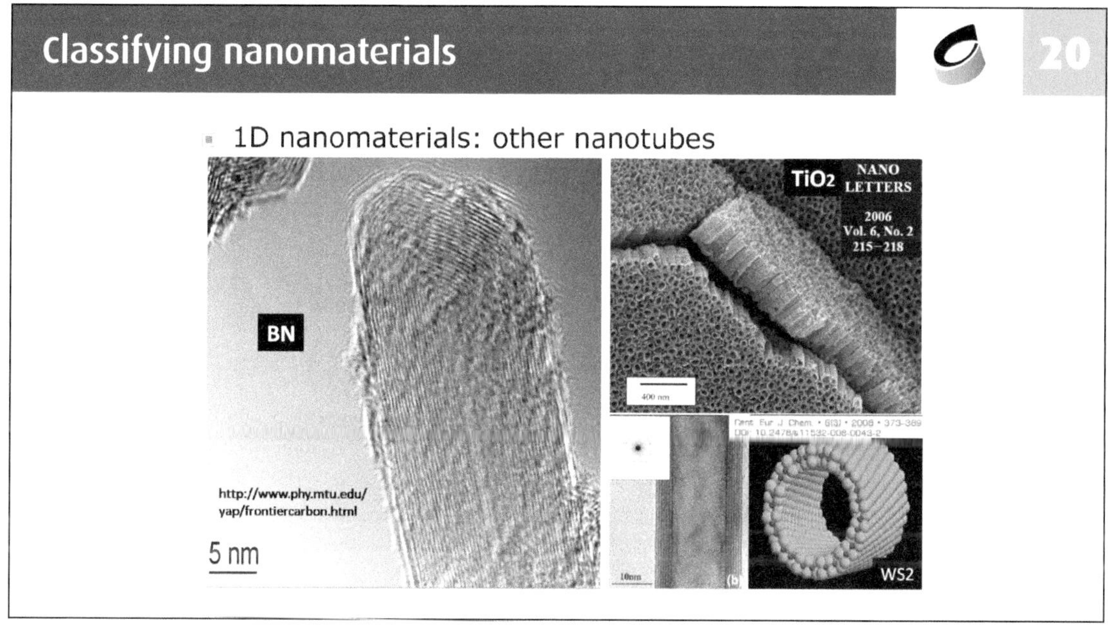

Beyond carbon nanotubes, the most popular form of nanotubes, there are other types of nanotubes such as from BN or from TiO_2.

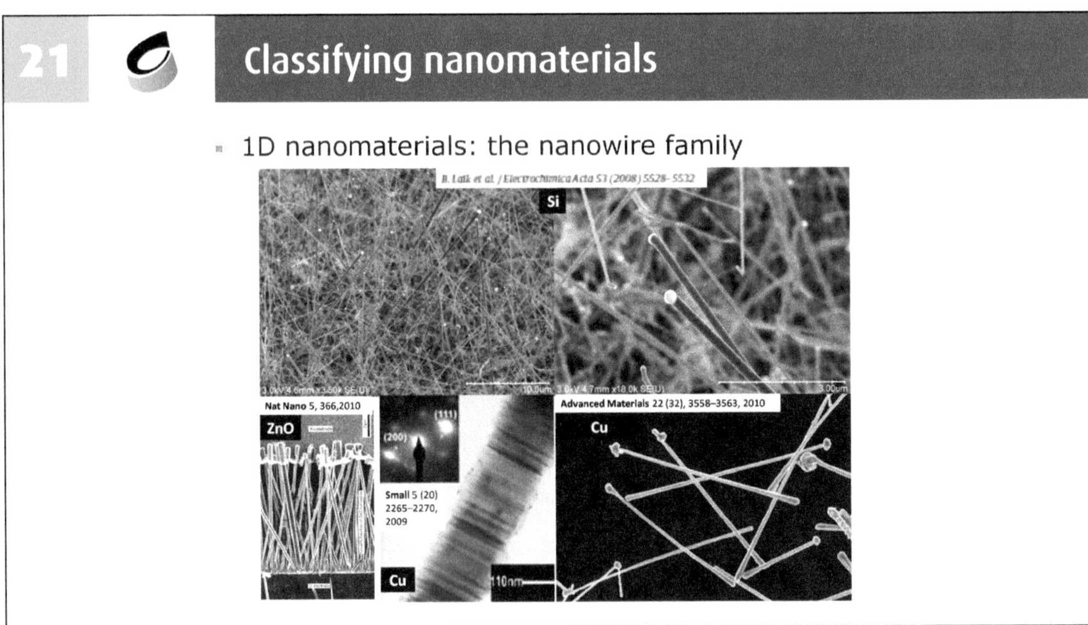

Various types of nanowires from different materials are shown in the figures, with quite a variety of outlook, and consequently a wide range of properties and potential applications.

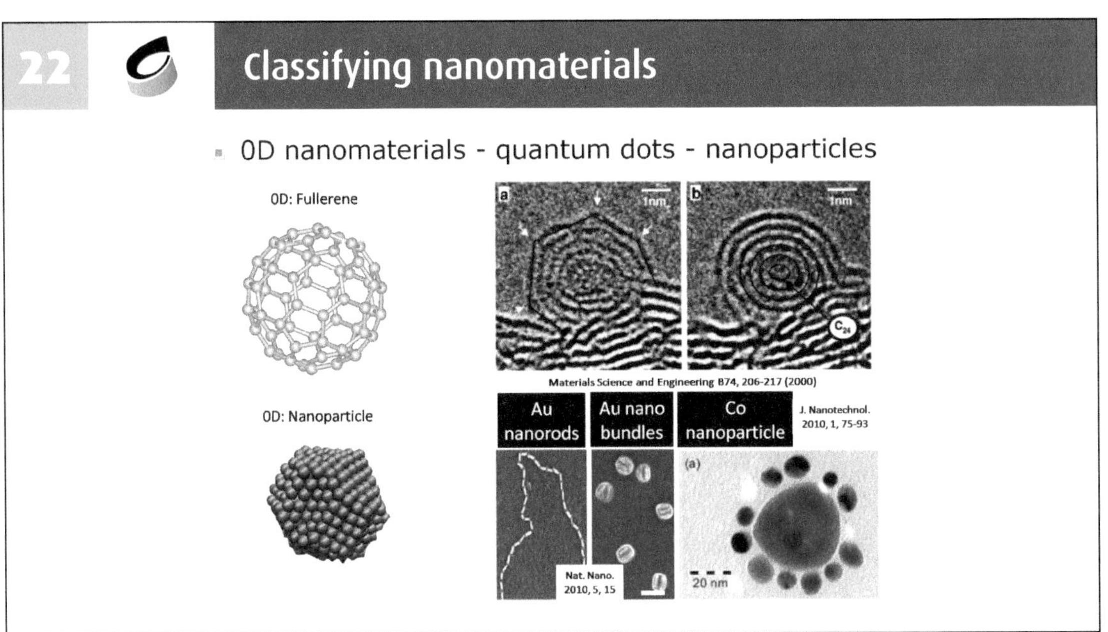

The last variety of nanomaterials is called quantum dots (0D nanomaterials). Their size is below 100nm in the three dimensions of space. They have highly different optical and electronic properties than their bulk counterparts. They are for instance popular in applications related to light absorption or emission, as the wavelength or even polarization of absorbed/emitted light can be adjusted by varying the shape and size of the quantum dots. Applications such as transistors and LEDs are being studied.

Physical/chemical differences between regular and nanoscale?

23

- Changes in band structure →impact on electronic, photonic and thermal behavior
  - Ex: Graphite is metallic, graphene is semi-metal, carbon nanotubes are semi-conducting or metallic
  - Ex: metal nanoparticle color depends on size
- Changes in mechanical properties (increased rigidity, increased limit strength)
- Increased reactivity and sensitivity (increased surface-over-volume ratio)

Overall, when decreasing size from regular to nano, physical and chemical properties change. There are changes in the band structure, which have impact in electronics, photonics and even thermal properties, and in the mechanical characteristics (for instance increases in rigidity). Furthermore, increasing the surface-over-volume ratio, the sensitivity and reactivity rise. One of the interests in going to nanometer size sensors is that the material becomes more sensitive to the environment for the same volume (and expectedly the same cost).

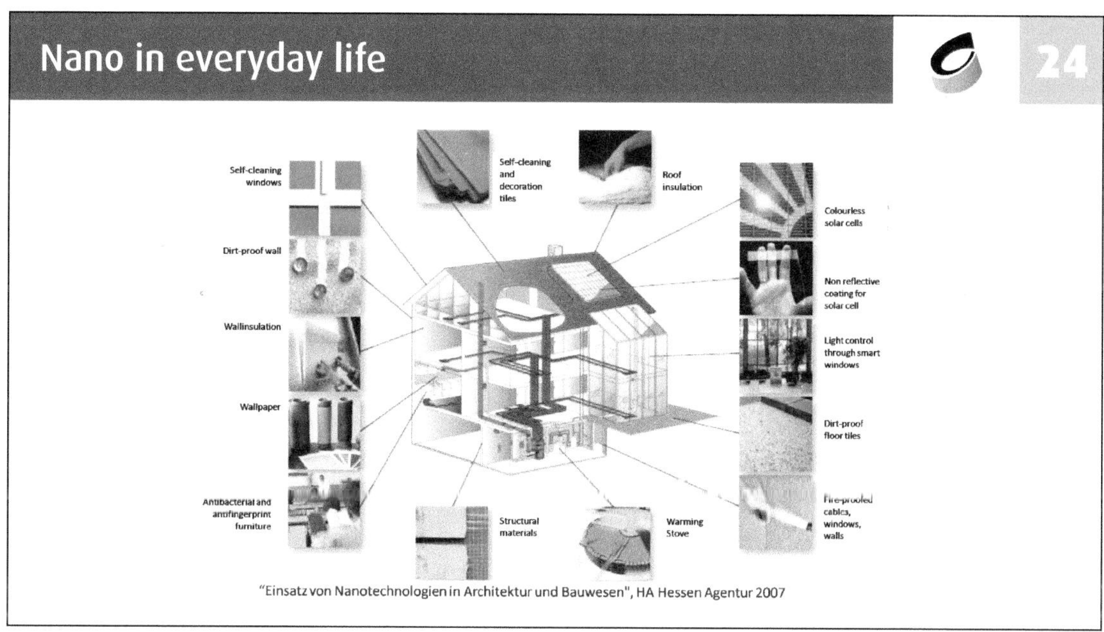

Looking around us, nanomaterials are present in a wide amount of everyday objects, in our house, car, smartphone, computers, sensors, etc..

25
Outline
Nanosensors for improved IoT
Fundamentals on nanomaterials
Fundamentals of nanosensors
A case study on carbon nanotube sensors
Deployments of nanosensors
Reliability of nanosensors
Summary

The following section focuses on nanosensors' fundamentals.

26
What is a nanosensor?
Nanosensor:
– sensor with characteristic size below 100nm
– sensor of any size integrating nanomaterials
Sensor?
– a electrical transducer (or sensing element) transforming a change in the environment into a electrically measurable quantity
– in the IoT field, sensor may also designate the full IoT node integrating various transducers
– Here, focus on the transducer only

This slide provides the different possible definitions for 'nanosensor' and 'sensor'. Please be careful that the term sensors is used in different ways depending on the community, either as only a transducer (that is, a sensing element), or as the full sensor node in the sense of IoT (a module that senses one or more parameters, does some computing and send the information to the user). Most frequently, nanosensor designates an electrical transducer incorporating nanomaterials. That is the meaning used in the rest of this talk.

The (potential) added value of nanosensors

27

- Sensitivity of a material to the environment depends on the surface area S
- Cost of a material depends on its volume V
- If the characteristic size L decreases, the surface-over-volume ratio S/V increases

→Nanomaterials are more sensitive that their bulk counterparts at equivalent cost.

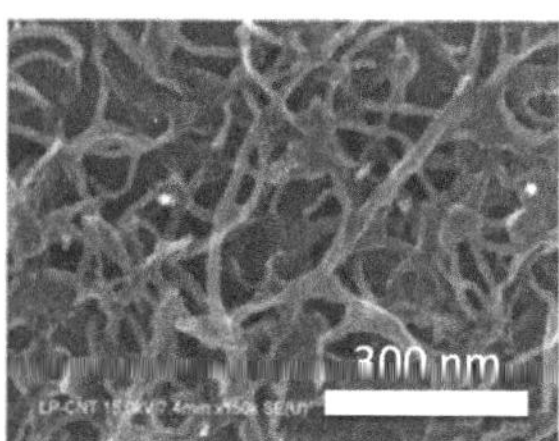

Why are we studying nanosensors? Taking into account that the sensitivity of a material is dependent on the surface area, and that its costs is related to its volume, the best sensor is achieved with highest surface-over-volume ratio (highest sensitivity with lowest cost). The surface-over-volume is roughly proportional to the inverse of the characteristic size; so the smaller the size, the better the sensor. Henceforth, let us use nanomaterials in sensors. Of course, in details, the sensing capabilities depends on on how the active nanomaterials are integrated into a transducer. The images show examples of nanomaterials with very intricate structures and high surface area.

The key features of a nanosensor

28

- The substrate (supporting material):
  - Flexible (polymer) or rigid (silicon, glass)
  - Low cost (polymer, glass) or high end (silicon)
- The transduction principle
  - Type of electron device: resistive, capacitive, transistor, logic gate, ...
  - Sensing principle: Chemical, mechanical, optical...
- The « nano » feature
  - Classification & mode of device integration for th active nanomaterial
  - Mode of contribution to transduction

The slide lists the features that are commonly used to describe a nanosensor. One of them is the substrate, that is the supporting material of the sensor, that can be flexible, or not, and made of different types of materials (glass, polymer, silicon, paper, etc.). The transduction principle is also crucial: it depends both on the sensing principle (chemical, electrical, optical, etc.) and on the type of electron device. These features impact the final use of the devices. For instance, depending whether the fabrication is done by printing on plastics, or if it is done in clean room (process on Silicon), the cost and reliability are different. Clean room fabrication is costlier, but it is more reliable. Nanosensors are classified by a lot of criteria in the literature, notably the way the nanomaterial is processed to fabricate the sensor, or the way it contributes to transduction

29 

Characterizing nanosensor performances

- Sensitivity & cross-sensitivity
  - Which electrical parameters vary as a function of which environmental parameters? (ex: strain sensor resistance depending on strain – desired sensitivity – and temperature –undesired sensitivity)
  - Is the dependance linear? Over which range of the desired parameter? Does it feature hysteresis?
  - Response curve: output parameter as a function of desired parameter
- Reproducibility: do two sensors out of the same fabrication process have the same electrical features and sensitivities
- Reliability:
  - Range of operation (min and max temperature, humidity, UV...)
  - how long does a sensor operate in regular conditions of use?

This slide describes a few of the major criteria used to characterize sensors (nano or not) and essential to determine which application they are suitable to.

30 

Preparing sensor integration

- Output signal:
  - Resistance, capacitance, current, voltage
  - Baseline value (output signal without loading)
  - Range of relative variations
- Mode of operation
  - Supply voltage
  - preheating
  - regeneration
- Level of integration
  - Raw analog output: No conditioning
  - Analog voltage in 0-5V
  - Digital signal
- Power consumption: μW/mW/W

This slide describes additional characteristics of a sensor (nano or not) which a user should look into before considering integrating it into a IoT device. For instance, a sensor that needs preheating for several seconds (usual in gas sensors based on metal oxides) will cause very high power consumption and is probably not suitable for energy-autonomous applications.

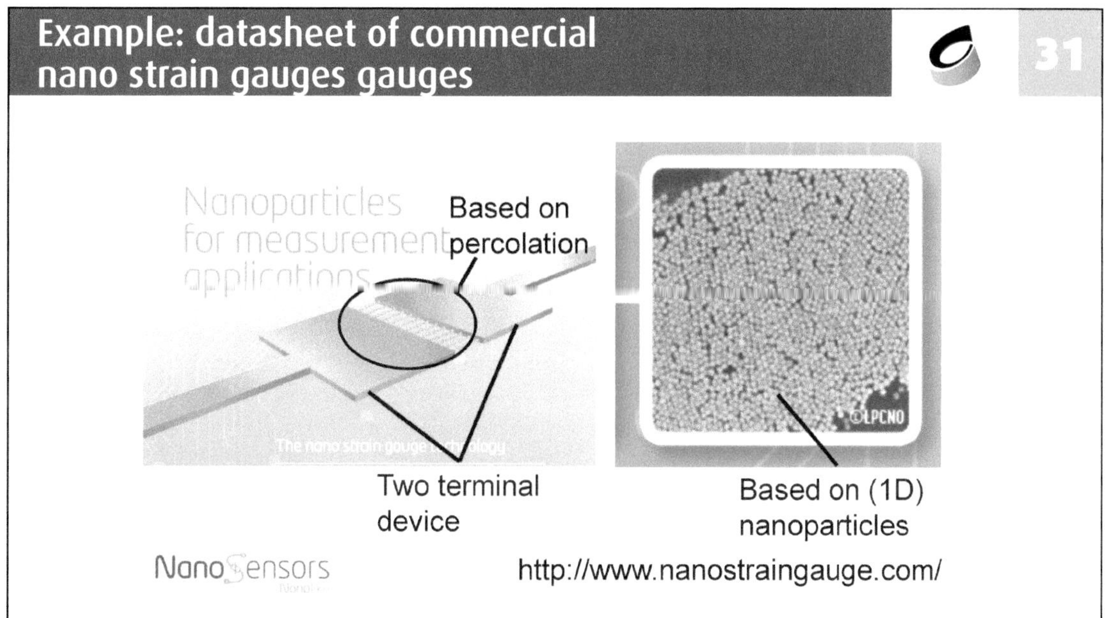

To get more familiar with these concepts, the next slides dissect the datasheet of commercial nano-enabled strain gauges sensors. This sensor solution operates by percolation effect in a network of gold nanoparticles lying between two gold electrodes (and forming percolating path). This sensor is able to measure small deformations, has a low size and low electrical consumption.

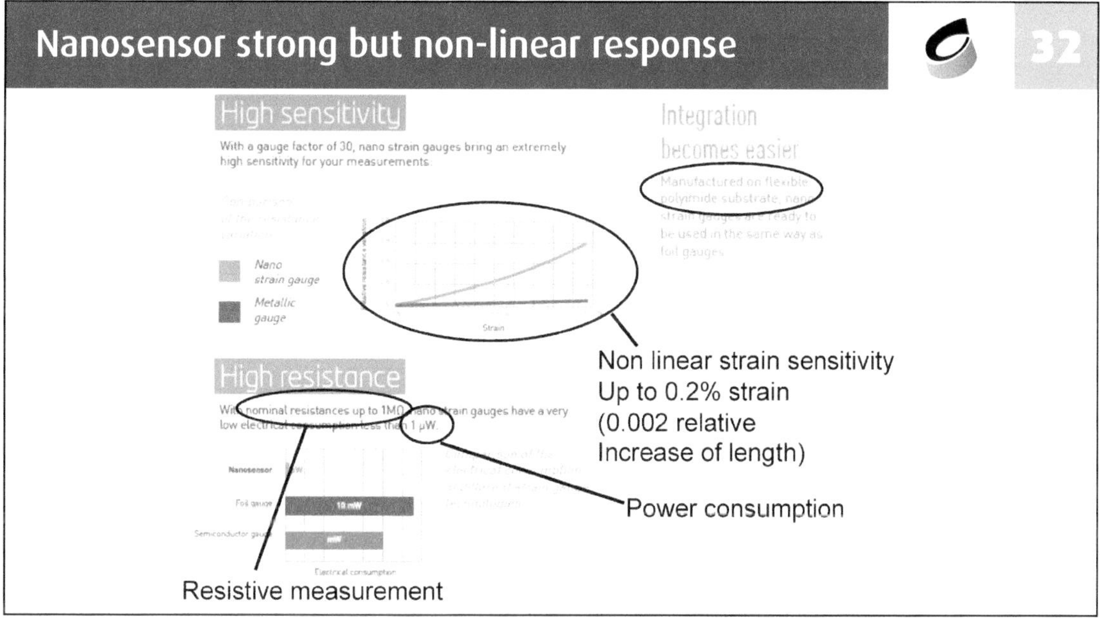

Analyzing the datasheet, it is possible to conclude that the nanosensor presents a strong but non-linear response and a comparison with metallic gauge is made. The drawback is that nonlinear sensors are tricky to use (and quite uncommon) in applications, One learns also that the substrate is flexible polyimide, The nominal resistance and power consumption are also provided. The MOhm-level of resistance is a problem, because current levels are quite low and the signal-over-noise-ratio (henceforth the sensitivity) may be low. Conversely the power consumption will be low. All these features matter when integrating devices into IoT devices, and should be checked in depth by IoT system designers.

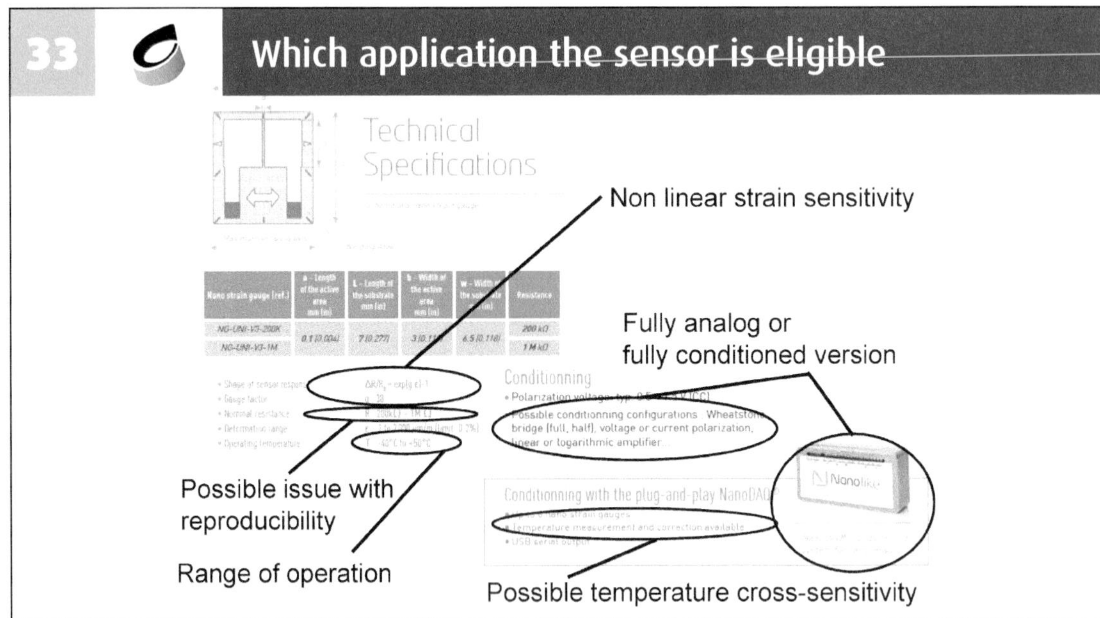

Regarding technical specifications, the datasheet gives information about the temperature range of operation, the sensor conditioning method (different options). A possible difficulty is mentioned regarding to reproducibility, since the nominal resistance has a relatively high range of variation. All of these characteristics are important to define which application the sensor is eligible for and how to integrate it into a IoT node. This shows the different aspects a user of nanosensor should be careful about while developing a new application or a new product.

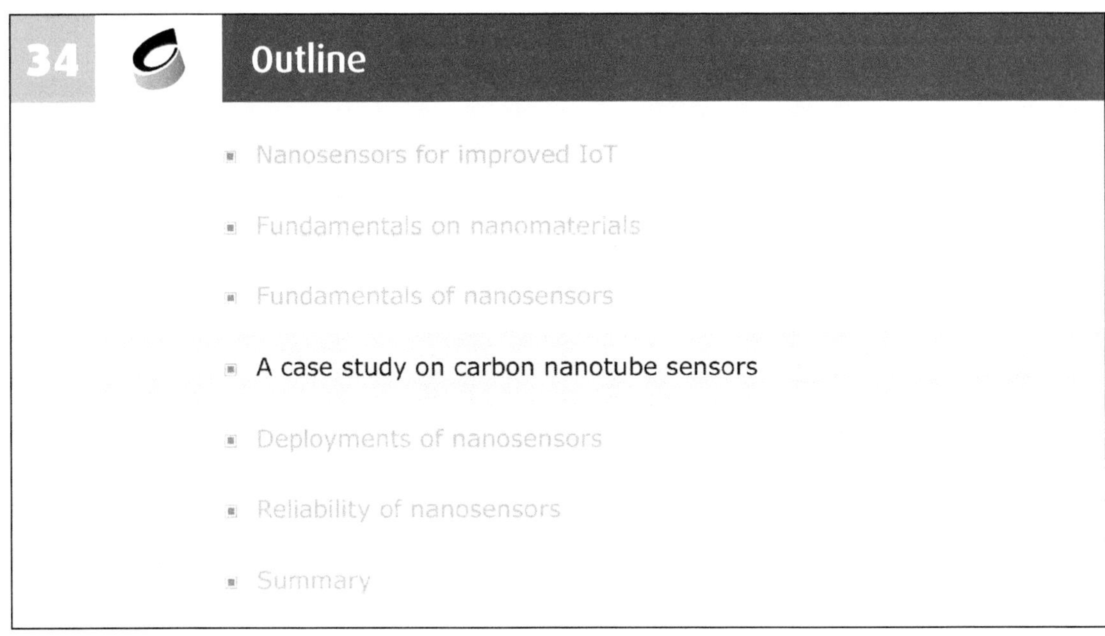

In the following section, an example on carbon nanotube sensors will be shown, based on work done by my team.. The goal is to give some sense of how nanosensor development works practically speaking.

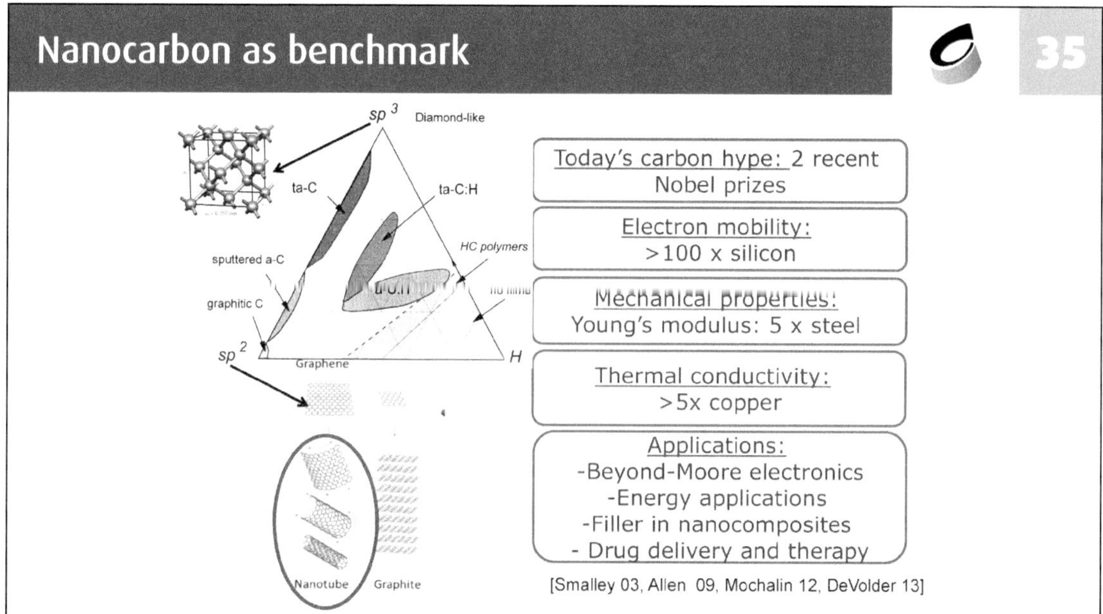

Nanocarbon as benchmark

Today's carbon hype: 2 recent Nobel prizes
Electron mobility: >100 x silicon

-Filler in nanocomposites
- Drug delivery and therapy
[Smalley 03, Allen 09, Mochalin 12, DeVolder 13]

Nanocarbon is being studied due to its attractive properties such as high electrical and thermal conductivities and good mechanical characteristics, characteristics which all depend on its crystalline configuration. As a consequence, different structures (specially a single layer of graphite, graphene, and carbon nanotubes) are being used in different applications in distinct fields of research.

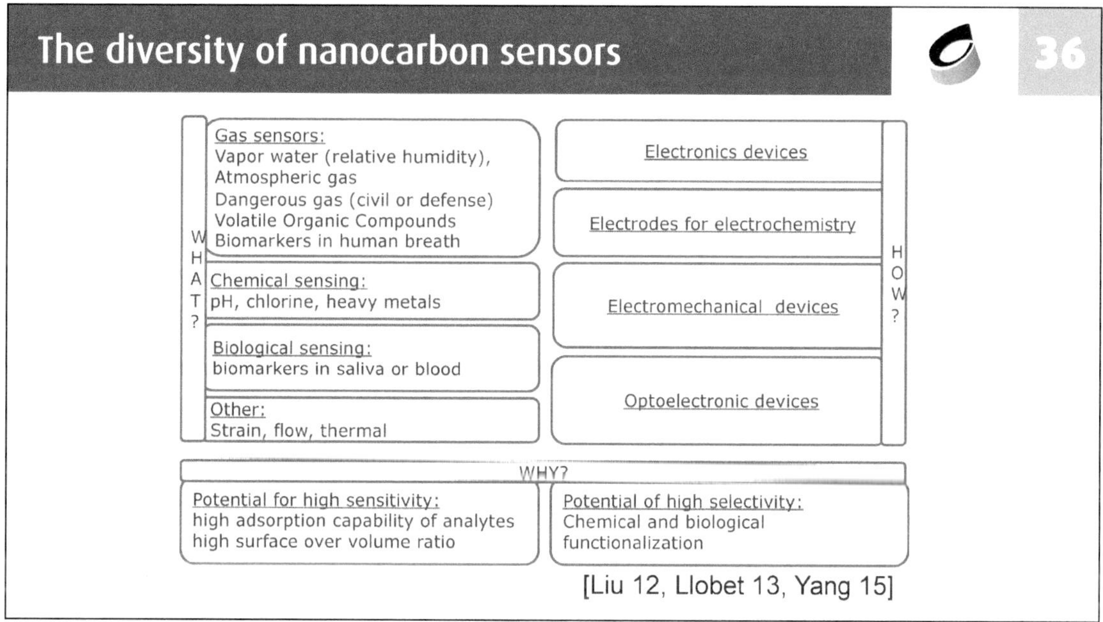

The diversity of nanocarbon sensors
36
WHAT?
Gas sensors:
Vapor water (relative humidity),
Atmospheric gas
Dangerous gas (civil or defense)
Volatile Organic Compounds
Biomarkers in human breath
Chemical sensing:
pH, chlorine, heavy metals
Biological sensing:
biomarkers in saliva or blood
Other:
Strain, flow, thermal
HOW?
Electronics devices
Electrodes for electrochemistry
Electromechanical devices
Optoelectronic devices
WHY?
Potential for high sensitivity:
high adsorption capability of analytes
high surface over volume ratio
Potential of high selectivity:
Chemical and biological functionalization
[Liu 12, Llobet 13, Yang 15]

As previously mentioned, the number of applications where nanocarbon structures are introduced is constantly growing. It is possible to emphasize gas, chemical, biological, mechanical and thermal sensors, where the sensitivity and the selectivity of nanocarbon materials play a relevant role.

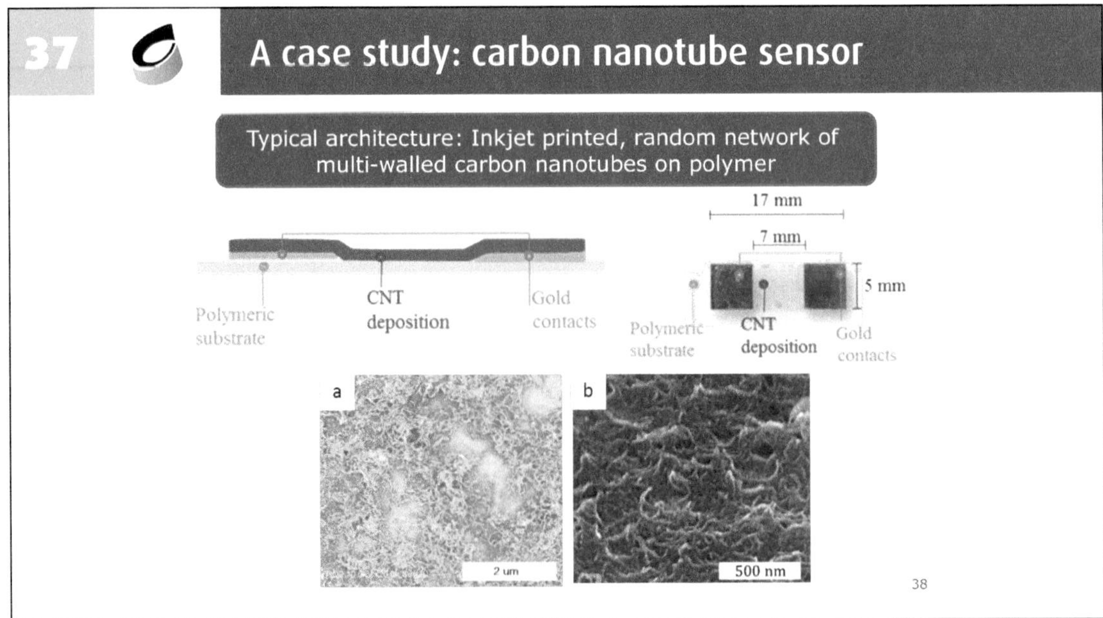

In this slide, the structure of the carbon nantoubes sensors made in my team is shown. The substrate is polymeric. Gold was deposited to create the electrodes. A random network of multi-walled carbon nanotubes (shown in the SEM figures at the bottom) was deposited by inkjet printing on the polymer. The overall device size is also demonstrated.

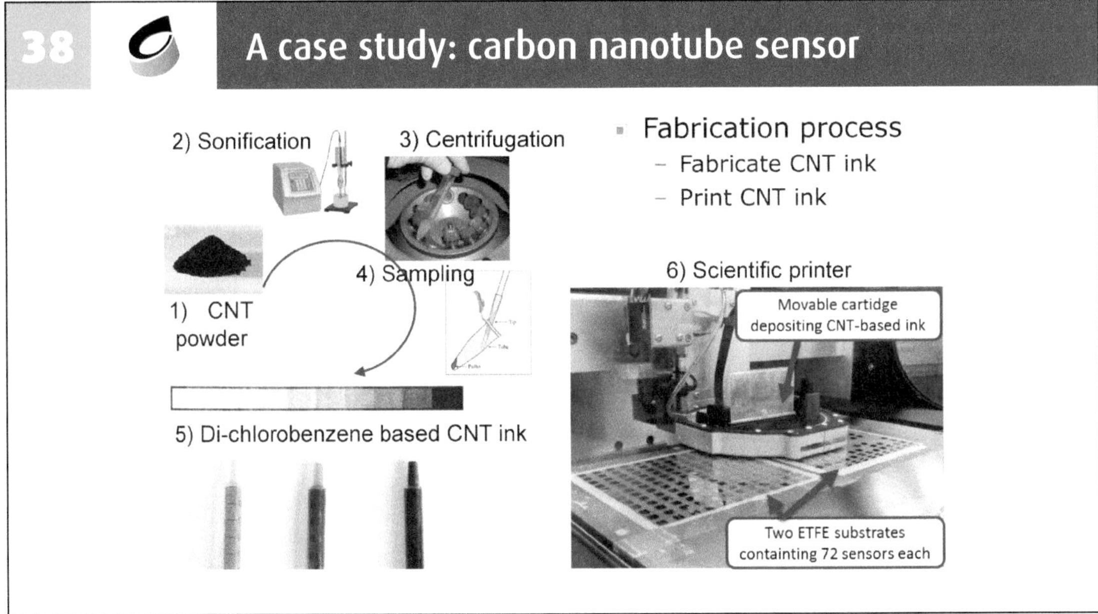

The carbon nanotubes deposition involves two main steps: the production of carbon nanotubes ink and then the printing of the ink. The former starts from carbon powder and finishes with a di-chlorobenzene based ink (the sequence and the details can be adapted depending on the desired ink characteristics) then, here with a scientific ink jet printer, the ink can be printed, creating the device structure and pattern.

Electrical characterization of the nanosensors?

39

- Probe station, Parametric analyzer, Low current source-meter ...

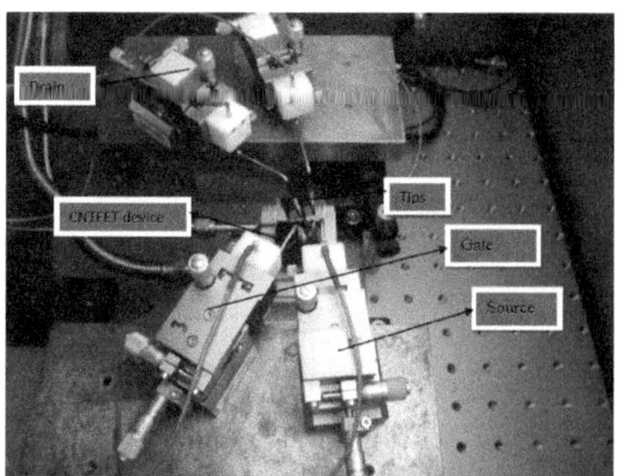

The device characterization is crucial to evaluate if the device is working as expected. Considering the electrical characterization, it can be done using a probe station, using a parametric analyzer and a low current source-meter. The probe station is mostly used at the early stage of device development; at later stages the device should rather be connected to a printed circuit board (PCB) by wire bonding.

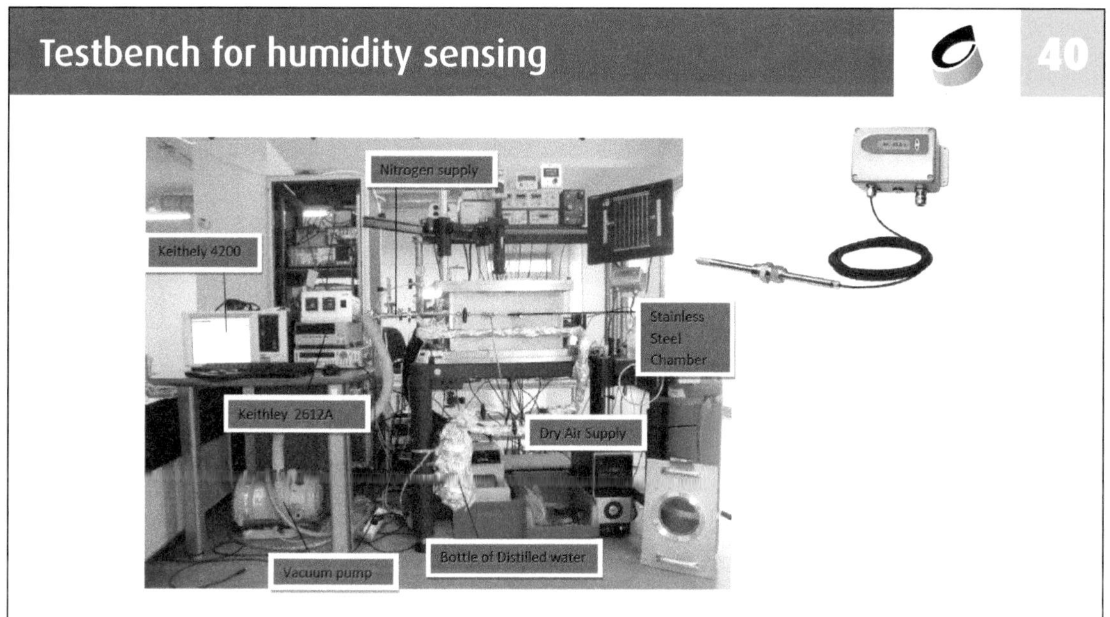

In this figure a test bench used for the analysis of humidity sensitivity of devices is shown and the equipment is detailed. More generally, it is quite complex to develop test benches to analyze device sensitivity. The important aspect are the controllability and reproducibility of the environmental conditions.

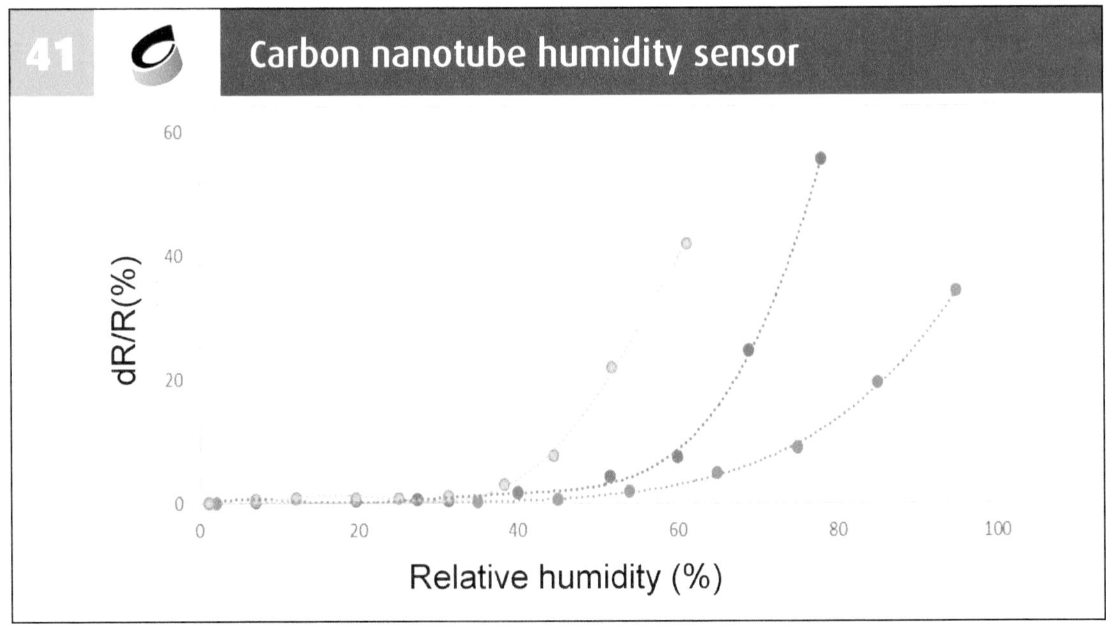

In this slide, the dependence on relative humidity is shown for three different versions of the carbon nanotube sensors. This is the usual data that one gets for this type of humidity sensor (stronger sensitivity at relative humidity above 40 to 60%).

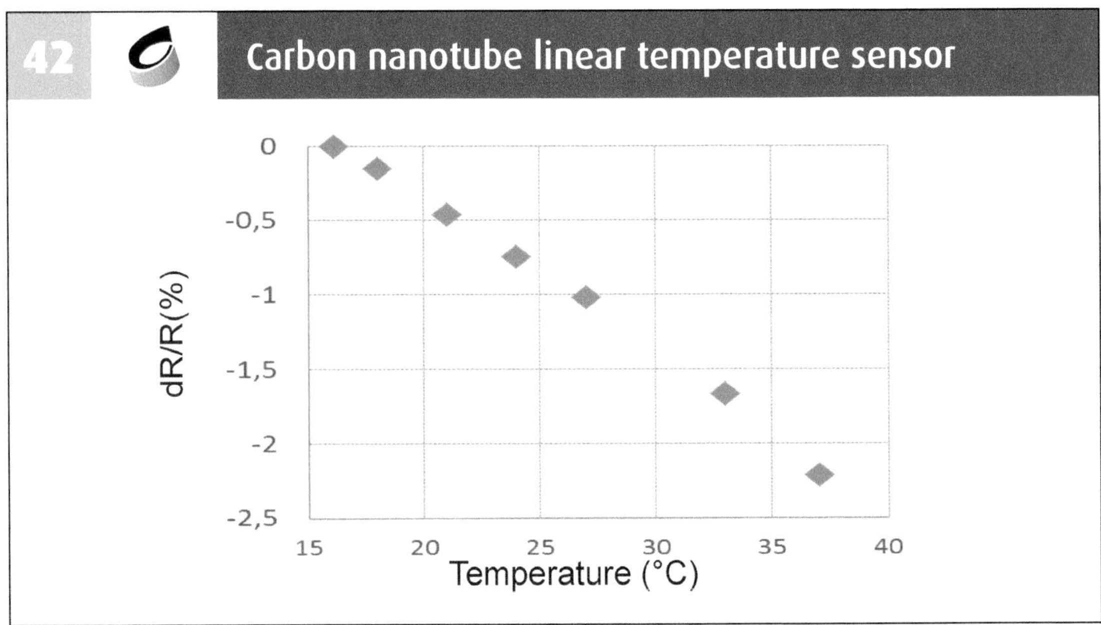

This graph demonstrates the carbon nanotube temperature sensor linear behavior. A linear variation is preferable since it is easier to manage by the user or by any algorithm that uses the measured data.

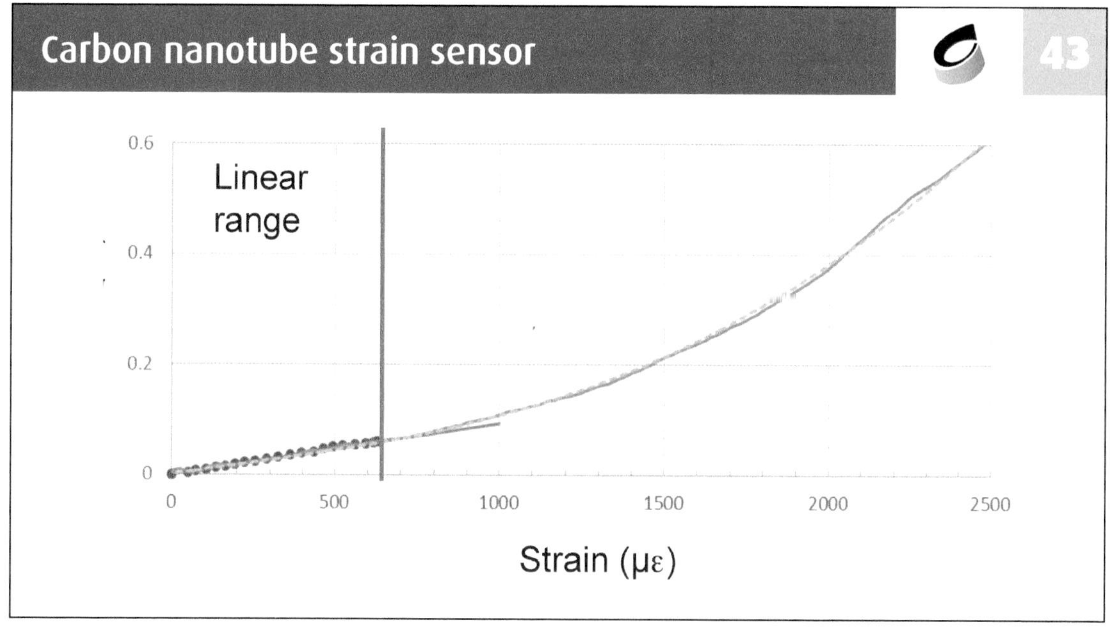

In this slide, the response of a carbon nanotube strain sensor with the strain is shown. The sensor response presents a linear and a non-linear region. In the former, the sensor response is reproducible between distinct devices in the same batch, and for several cycles of mechanical loading and unloading.

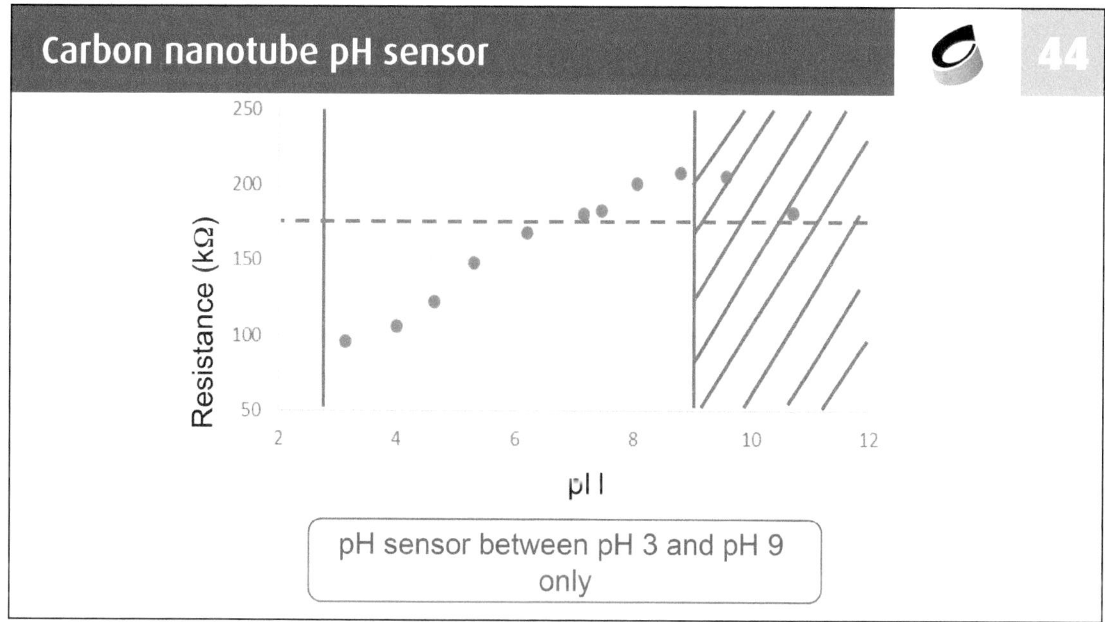

The dependence of resistance on pH is demonstrated for a specific carbon nanotube pH sensor. As it shown, this sensor can be used between pH 3 and 9 only, because beyond this range, the response is not monotonous anymore. That is, if the sensor yields a response at 175 kΩ, the user cannot know whether the pH is 7 or 11, unless the range of use is limited to below pH 9.

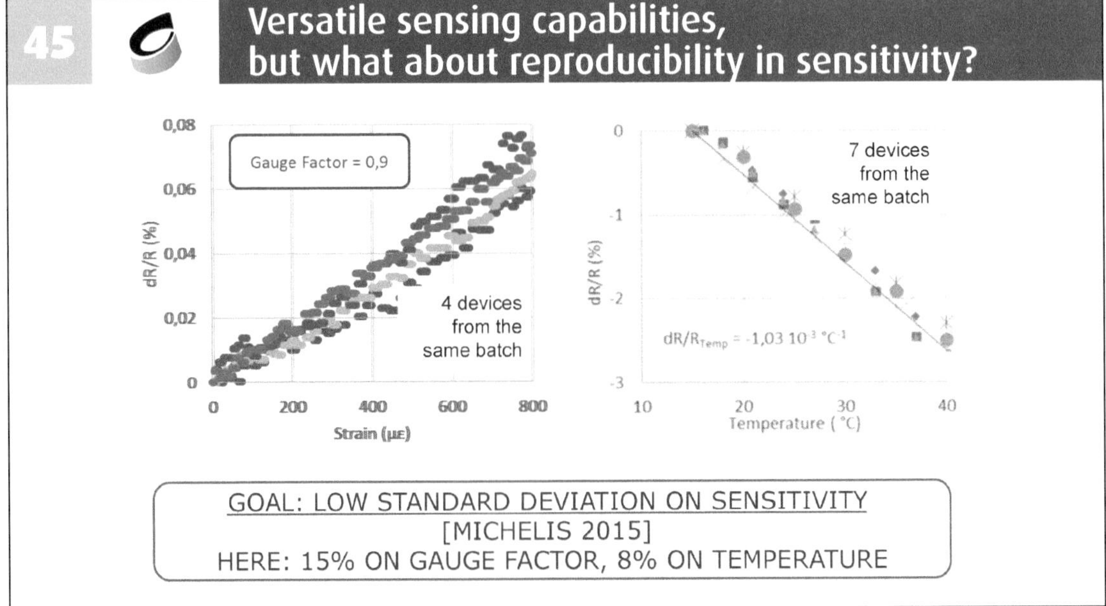

In the previous slides, we discussed the different sensitivity capabilities. However, for applications, what is primordial is the reproducibility in sensitivity. It means that the sensors have a similar response (low standard deviation betwenn the slope of the different response curves) to the same environment conditions. The presented graphs show the dependence on strain and temperature from different sensors from the same batch, with a standard deviation level comparable with low cost commercial products.

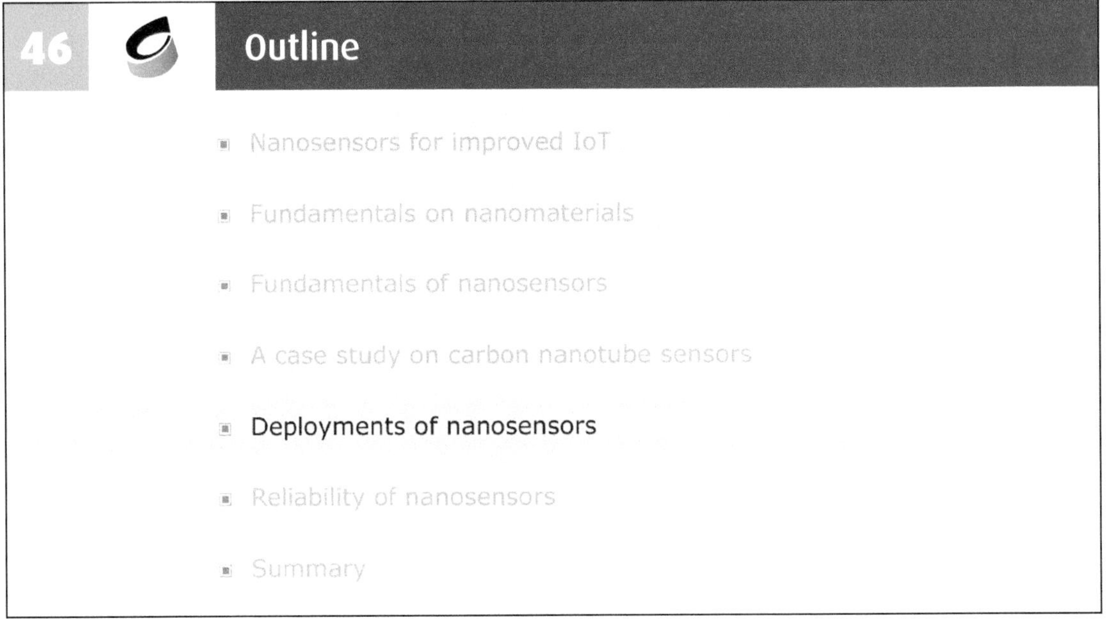

Now that we know what nanosensors can measure, we describe how nanosensors can be deployed in the field.

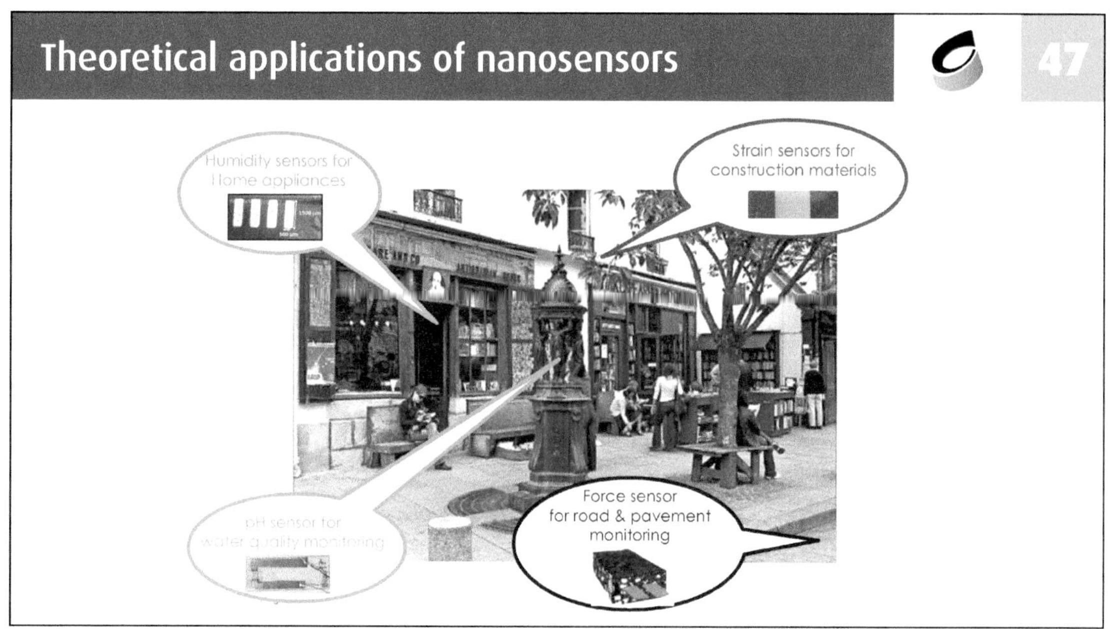

Theoretically, given the diversity of results available in the literature or even in my team's work, nanosensors could be applied in everything around us, in our everyday life. In this slide, you can see on a well-known image from Paris example of where concretely such sensors could be found, for instance to monitor traffic, to monitor the state of buildings, the quality of water or the indoor comfort. And these are just a few examples from our own results, using nanocarbon only. The possibilities are infinite.

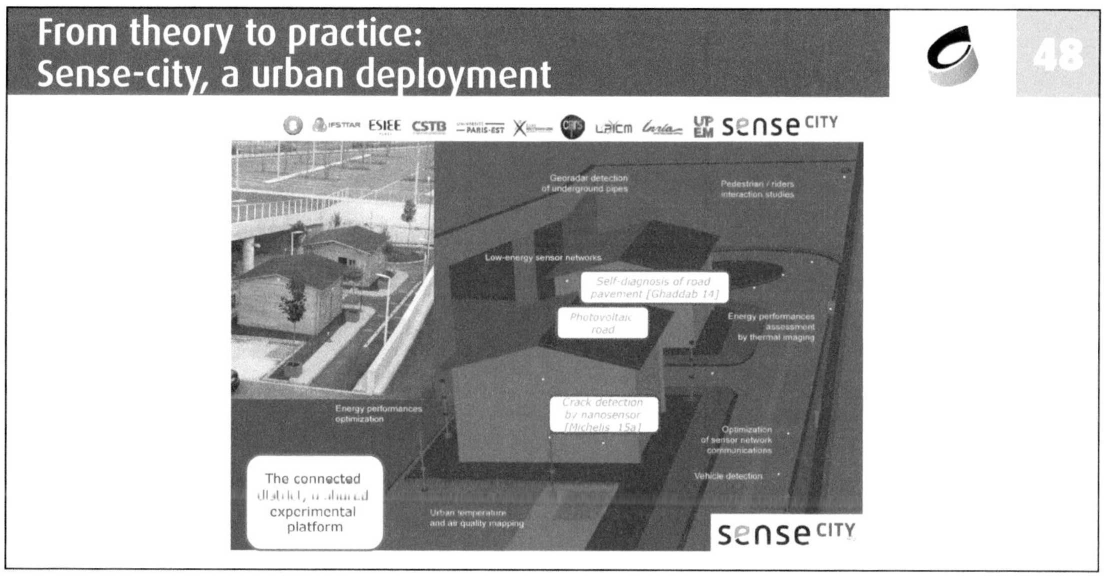

The previous slide says "theoretical" because proving sensitivity in the lab is not enough. What matters is whether the results hold in the field. The Sense-city project was designed to help out for this phase of sensor development. Sense-City relies on a small district (see inset to the left) that can be used to deploy all kinds of experimental technologies without the constraint of real life (no permanent traffic, inhabitants...). Examples of the experiments in progress on the platform are listed in the main picture, such as crack detection with nanosensors, photovoltaic road, self-diagnosis of road pavement, temperature and air quality mapping, low-energy sensor networks, among others...

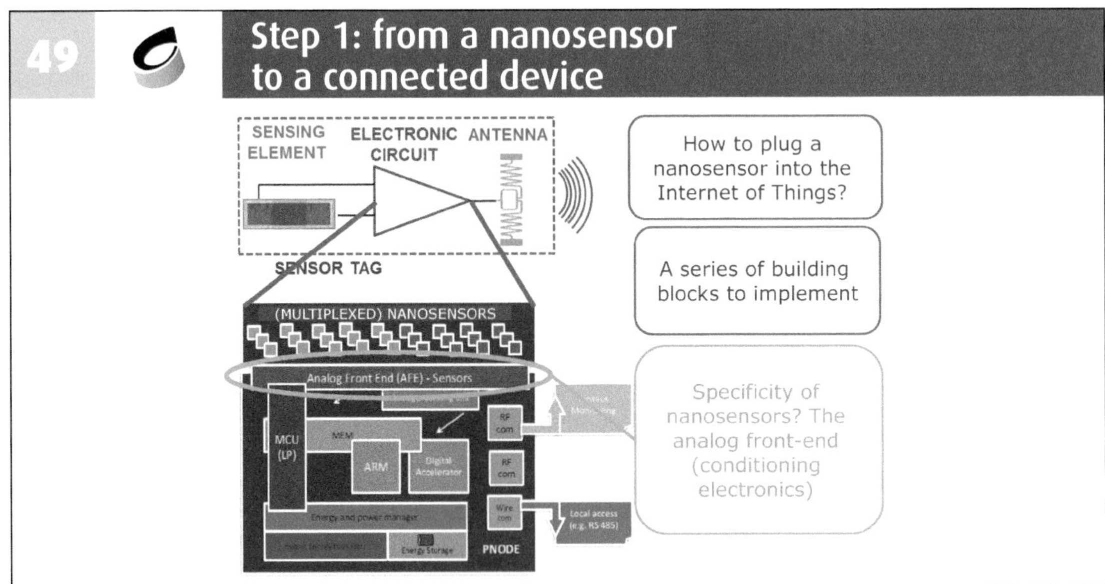

And to be able to deploy nanosensors in the field (and first in Sense-City platform), they need to be integrated into some connected device that will be able to read the information from the sensor and communicate them to the user. The connected device can be fairly typical of any regular IoT device (from the most simple – top left – to the most advanced - bottom left, IoT node developed in H2020 Proteus project on water quality monitoring with nano and MEMS sensor). Only the analog front-end needs to be specifically adapted to comply with the specificity of nanosensors. .

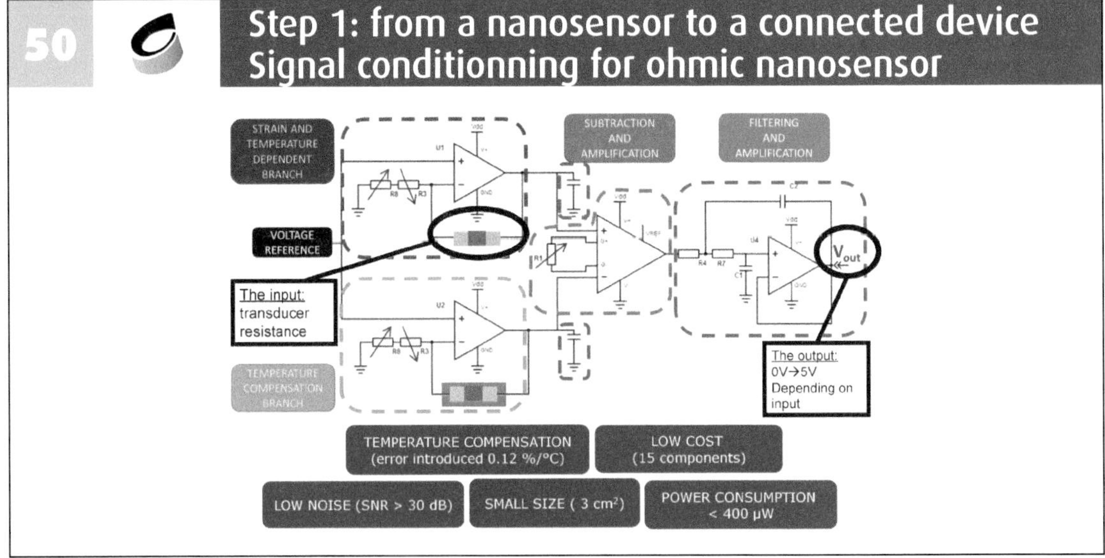

Notably, the signal conditioning should account for the specificity of the target nanosensor on a case by case basis. In the example presented here, the carbon nanotube ohmic strain sensor has a very strong cross-sensitivity to temperature (perturbing factor), so the signal conditioning integrates the temperature compensation at hardware level. A few performances of the architecture are listed here. The design should use low power, low cost components. It should be as small as possible. Of course, it will be considerably larger than any CMOS based design can possibly be, but it is enough for a proof of concept. Let us remark that in a lot of IoT applications today, the market is not yet large enough to justify the added costs of moving to integrated designs.

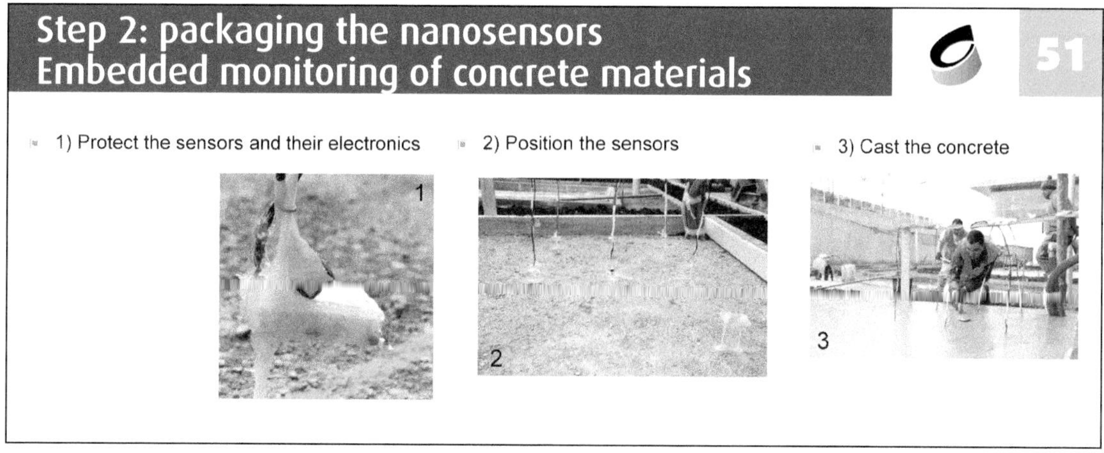

Once the electronic chain is ready, nanosensors can be deployed in the field. Here the example is for carbon nanotubes strain sensors embedded in a mortar slab in Sense-City (the goal of this research is to monitor the lifetime of infrastructures with embedded sensors). The protection of the sensor and its electronics are very important, especially in concrete which is very aggressive (highly basic pH, humid environment).

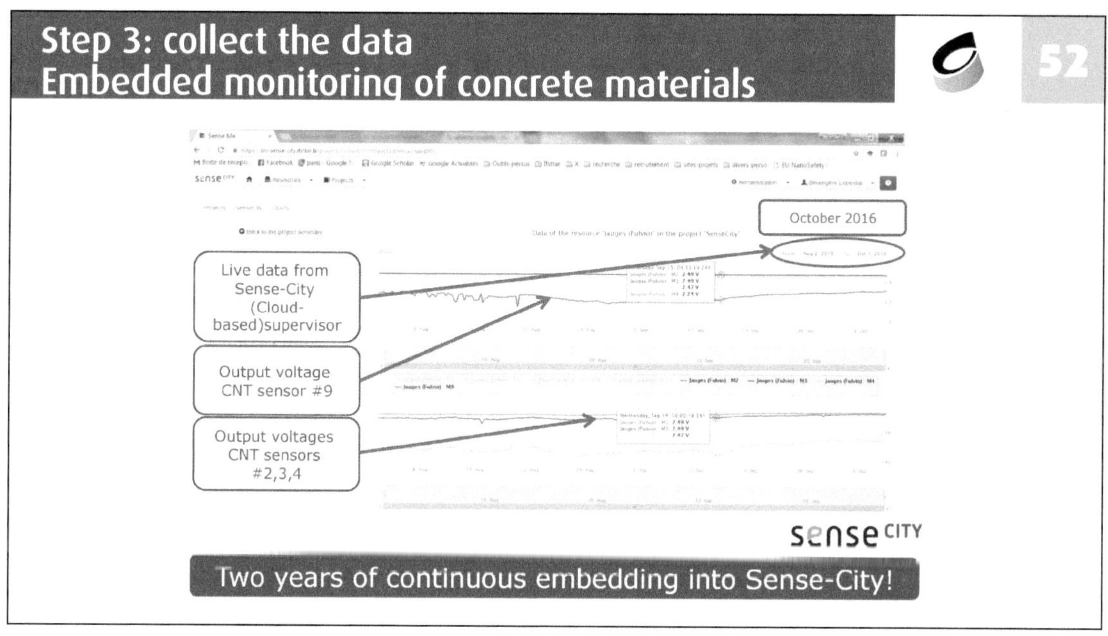

The strategy to collect of data is extremely important to understand what it is happening in real time (for instance, to understand if the sensor is alive and operating as expected, and then to analyze the data to understand the relevance of the information). In Sense-City, a cloud-based supervision system is proposed. Here we show the real time data from the carbon nanotube strain sensors (discussed in the 2 previous slides) accessible via Sense-City website. In late 2014 (when the sensors were put in mortar), it was the first example of embeddeding in concrete for carbon nanotube strain sensors. Two years later, our sensors are still alive in the mortar slab, as we can see on the website.

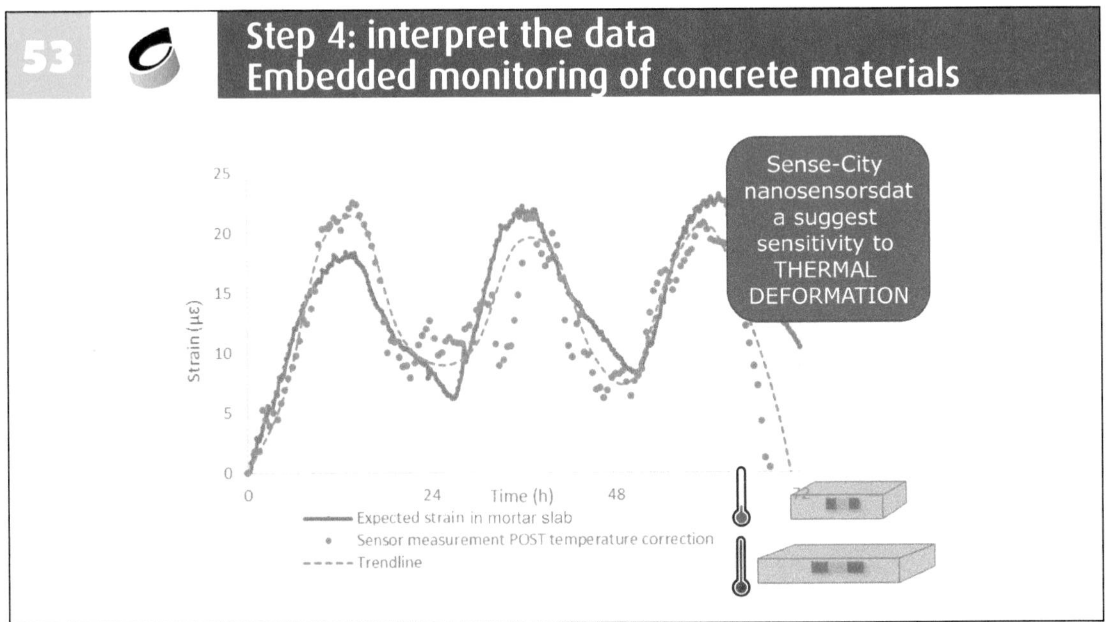

This slide zooms on three days of operation for the carbon nanotube sensor. Data suggests that the sensor response follows the thermally-induced deformation of the mortar slab. It is good, because it is the type of information that interests end-users (specialists of construction materials) and which is not readily available currently with market solutions.

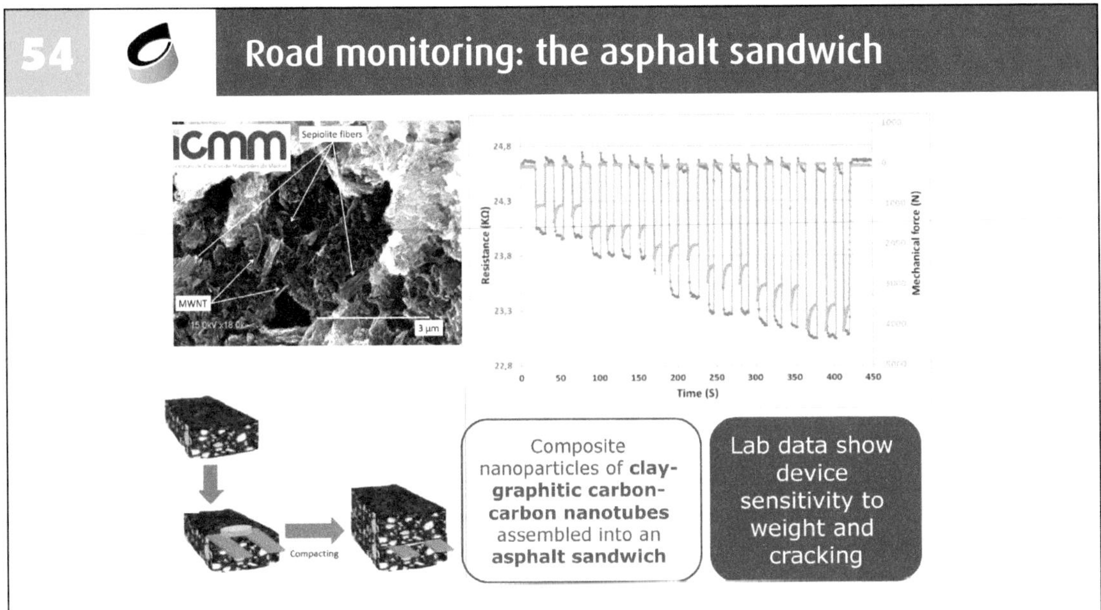

Another example of nanosensor development for road monitoring. A nanosensor composed by composite nanoparticles of clay-graphitic carbon-carbon nanotubes assembled into as asphalt sandwich (scheme presented in the figures) was created. As it shown in the graph, the nanosensor is sensitive to the force applied to it .

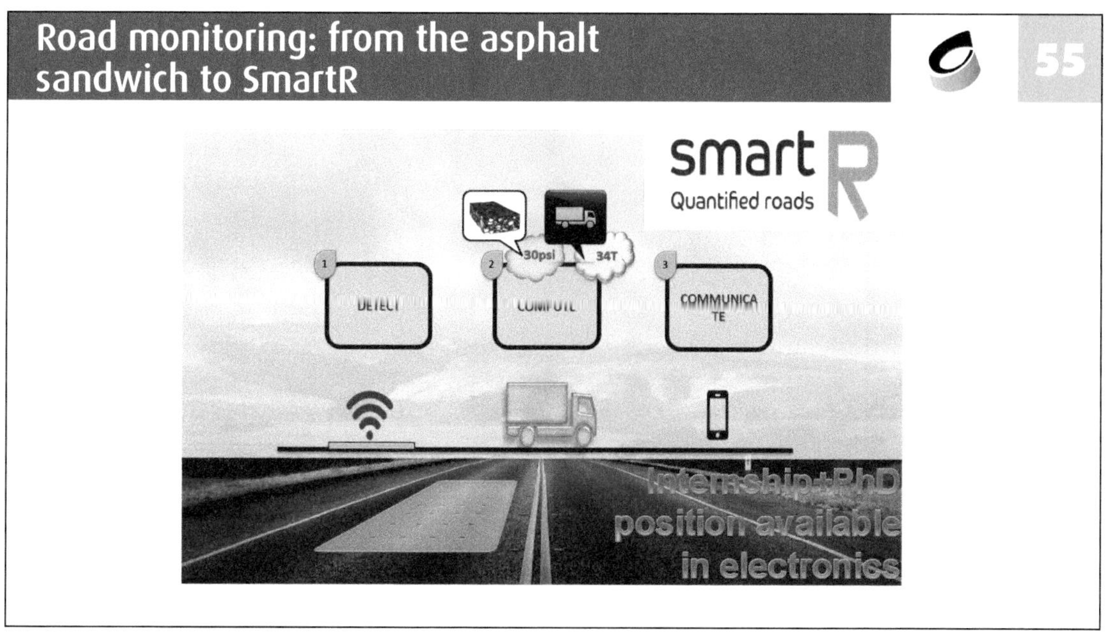

This project has been extended and is now being transferred to the start up SmartR. A version of the sensor described in slide 55 is being adapted to monitor the weight of vehicles and the state of the road.

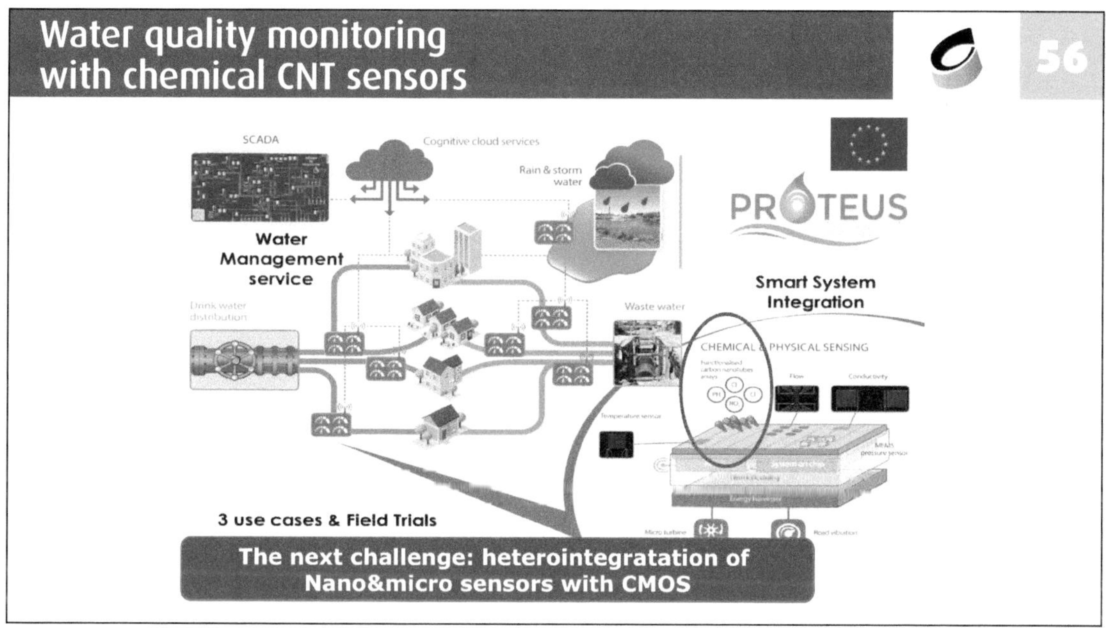

Another example related with nanosensors is the research project PROTEUS, to monitor the water quality with carbon nanotube sensors. This project aims at field deployment in different use cases in drink, waste and rain water. The sensor integrates on the same sensor nodes MEMS sensors, carbon nanotubes sensors and CMOS circuitry, as well as multi-source energy harvesting capabilities. The project involves distinct partnerships from academia to industry.

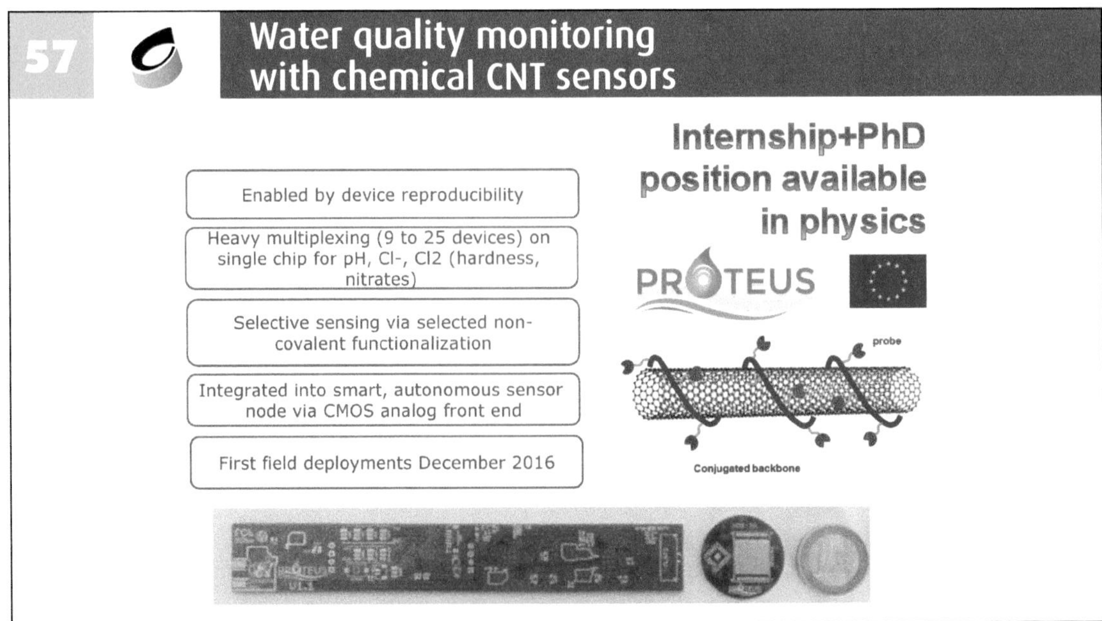

PROTEUS had its first field deployment in Sense-City in December 2016, showing the high level of integration of all the new technologies considered in the project. Regarding to the carbon nanotube part, there is heavy multiplexing (9 CNT sensors in the current version, 25 in the next version) of the chemical CNT sensors. CNT sensors are used for chemical sensing. To make the CNT sensitive to different chemicals, they are functionalized by specific polymer chains. In 2017, this system will be tested in Almada water network in Portugal.

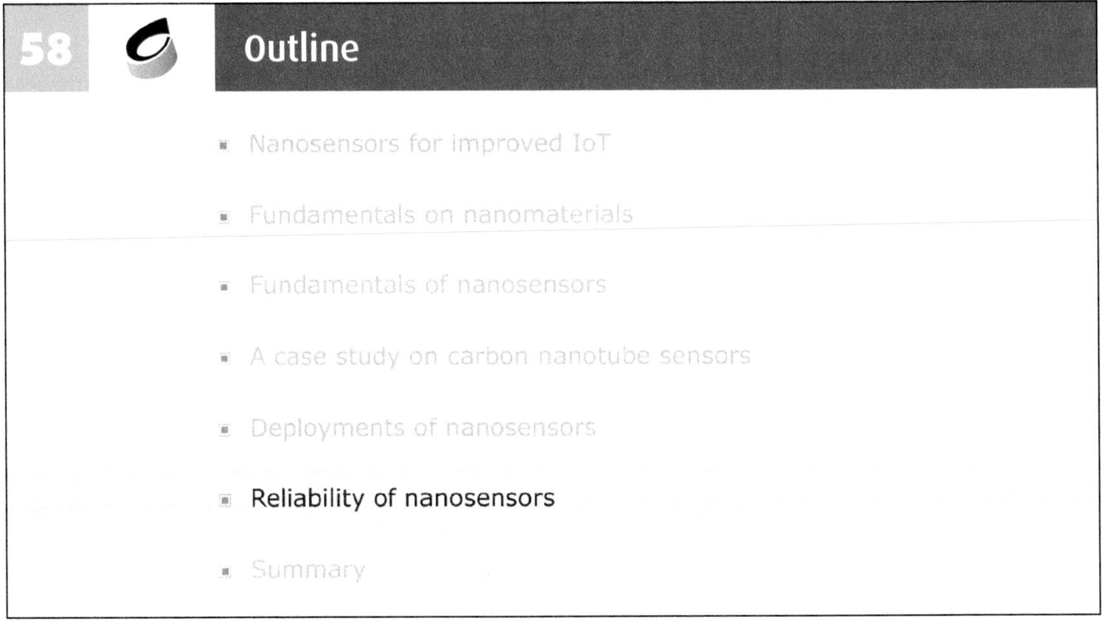

Now that we know that nanosensors can be deployed in the field, the question is how long they last. The following slides focus on the reliability of nanosensors.

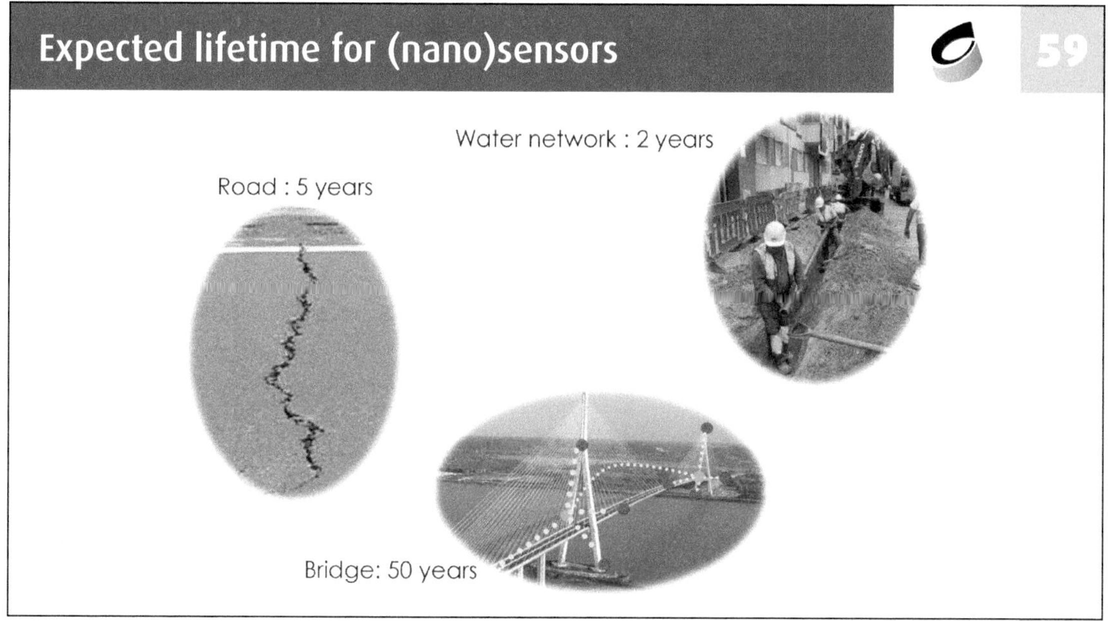

Depending on the application, the expectations regarding to sensor life time may change drastically. This is one of the problems of finding markets for nanosensors, as industrial partners do not trust that they can have a sufficient resiliece outside of lab conditions

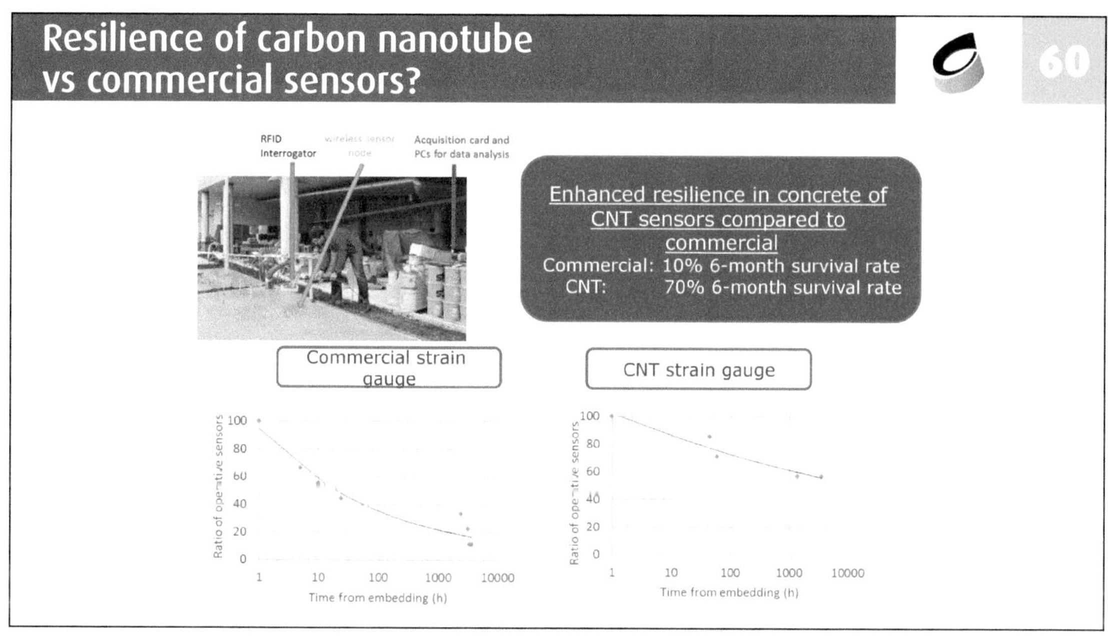

In the experiment that has been described before (the carbon nanotubes sensors deployed in mortar in Sense-City), we actually showed that our carbon nanotube strain sensors had a better survival rate in mortar than commercial sensors.

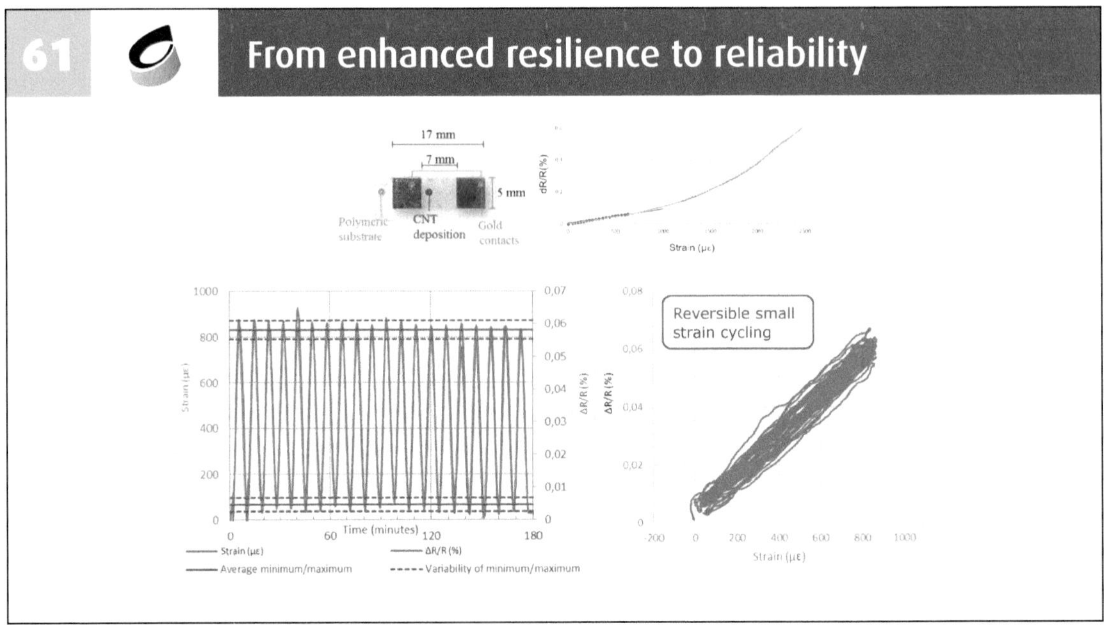

It is not enough to know that the sensor survives. You need to show that its sensivity remains stable. Continuing with the example of the carbon nanotube strain sensors discussed before, we studied the cyclic response of the device, to show that it remains stable in time (if strain levels remain low enough).

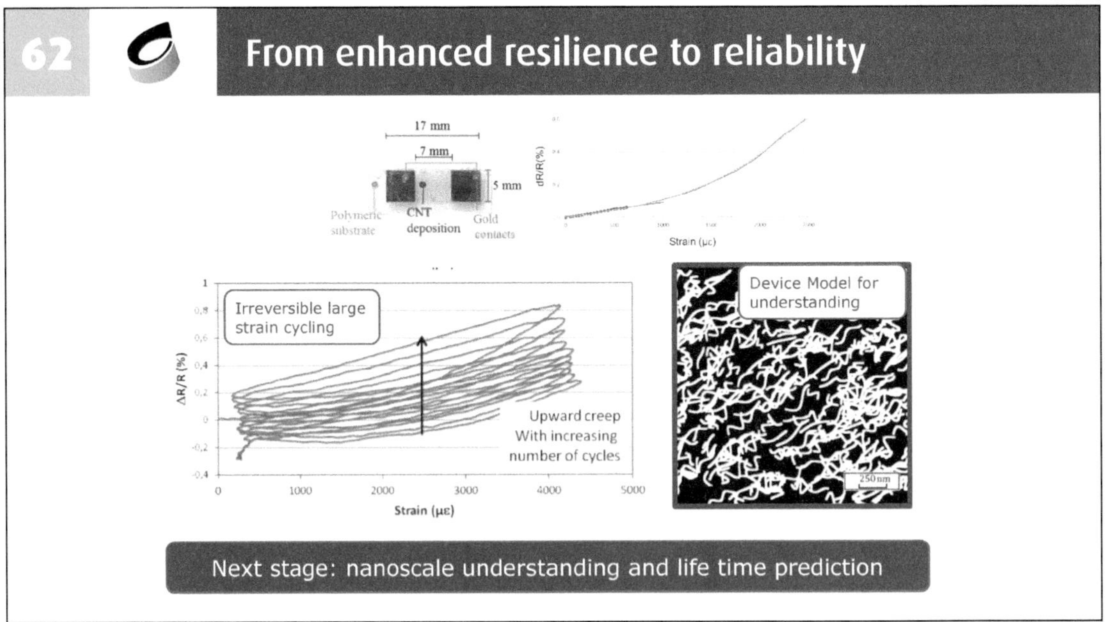

However, if strain levels increase, the sensor response starts to evolve with time (which is not good). This means that an appropriate application for this sensor is an application where the maximal strains applied to the sensor remain all the time below the threshold for reversible sensor operation. We are also building nanoscale numerical models of our devices to understand these features of the devices.

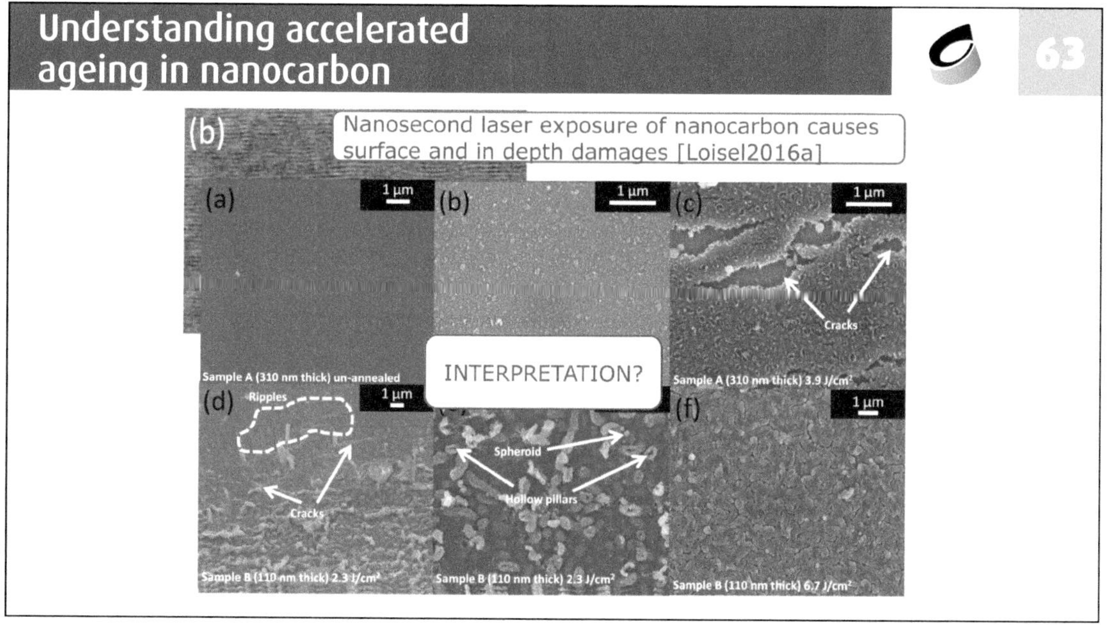

As a last example regarding to the topic of reliability and ageing in nanomaterials, this slide shows how the surface features of nanocarbon can evolve when it is exposed to a laser beam (accelerated ageing experiment). The challenge for us is to understand the mechanisms driving these changes.

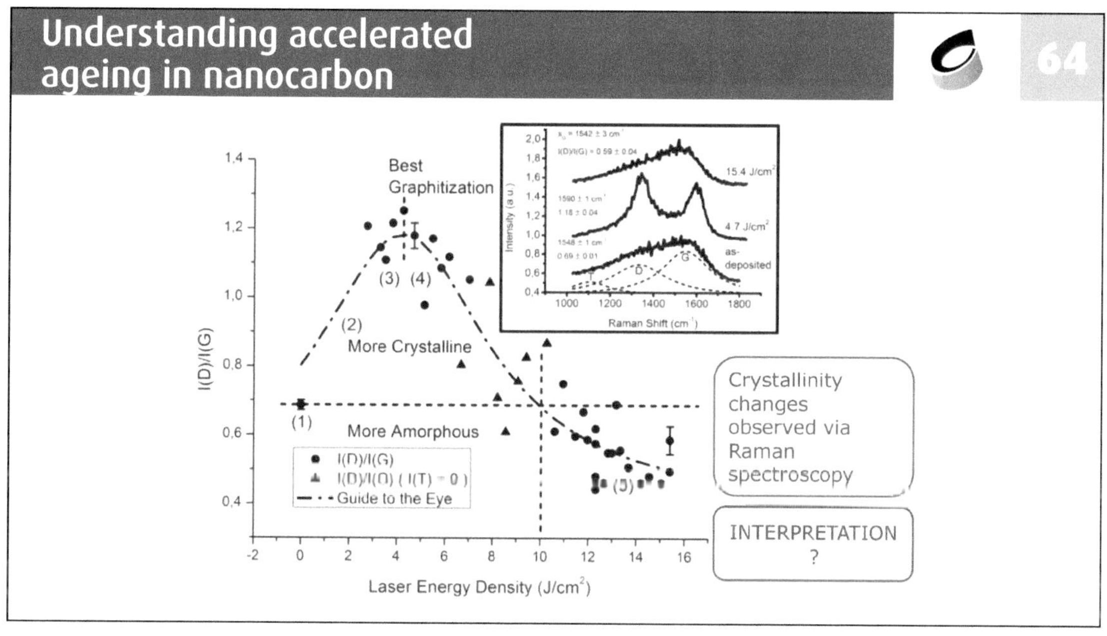

We have characterization tools that enable us to understand how these surface changes correlate with what happens at the nanoscale (in the crystalline structure of the material); here Raman spectroscopy shows the evolution of the material from more graphitic (highly organized) to more amorphous (highly random) depending on the intensity of the laser beam.

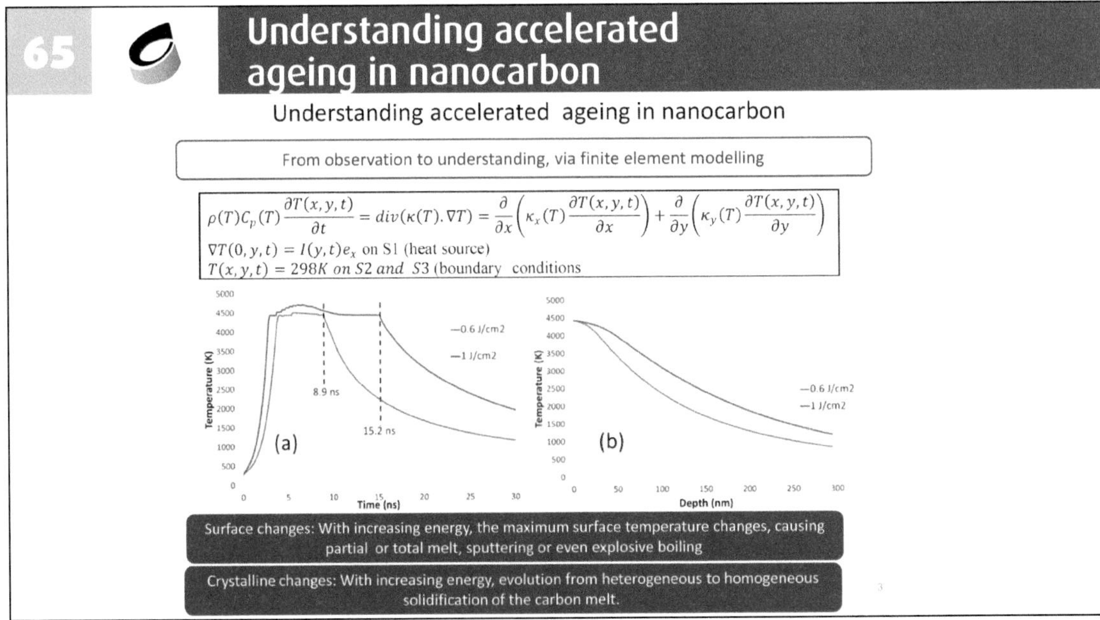

65
Understanding accelerated ageing in nanocarbon
Understanding accelerated ageing in nanocarbon
From observation to understanding, via finite element modelling
$\rho(T)C_p(T)\frac{\partial T(x,y,t)}{\partial t} = div(\kappa(T).\nabla T) = \frac{\partial}{\partial x}\left(\kappa_x(T)\frac{\partial T(x,y,t)}{\partial x}\right) + \frac{\partial}{\partial y}\left(\kappa_y(T)\frac{\partial T(x,y,t)}{\partial y}\right)$
$\nabla T(0,y,t) = I(y,t)e_x$ on S1 (heat source)
$T(x,y,t) = 298K$ on S2 and S3 (boundary conditions)
Temperature (K)
Time (ns)
8.9 ns
15.2 ns
—0.6 J/cm2
—1 J/cm2
(a)
(b)
Depth (nm)
Surface changes: With increasing energy, the maximum surface temperature changes, causing partial or total melt, sputtering or even explosive boiling
Crystalline changes: With increasing energy, evolution from heterogeneous to homogeneous solidification of the carbon melt.

But to understand what happens, we need to go further. Here we built a detail numerical model to calculate the temperature of the material when exposed to the laser beam. We show that the temperature may reach up to 4000K at fairly low beam fluence (4000K is the melting temperature of graphite; by comparison the surface temperature of the sun is only 5 800K and the melting point of silicon is around 1500K). The numerical study of thermal patterns in the sample explains all the features observed experimentally. The numerical model provides invaluable information, as the actual temperature pattern and evolution are very complex to measure experimentally in this experiment.

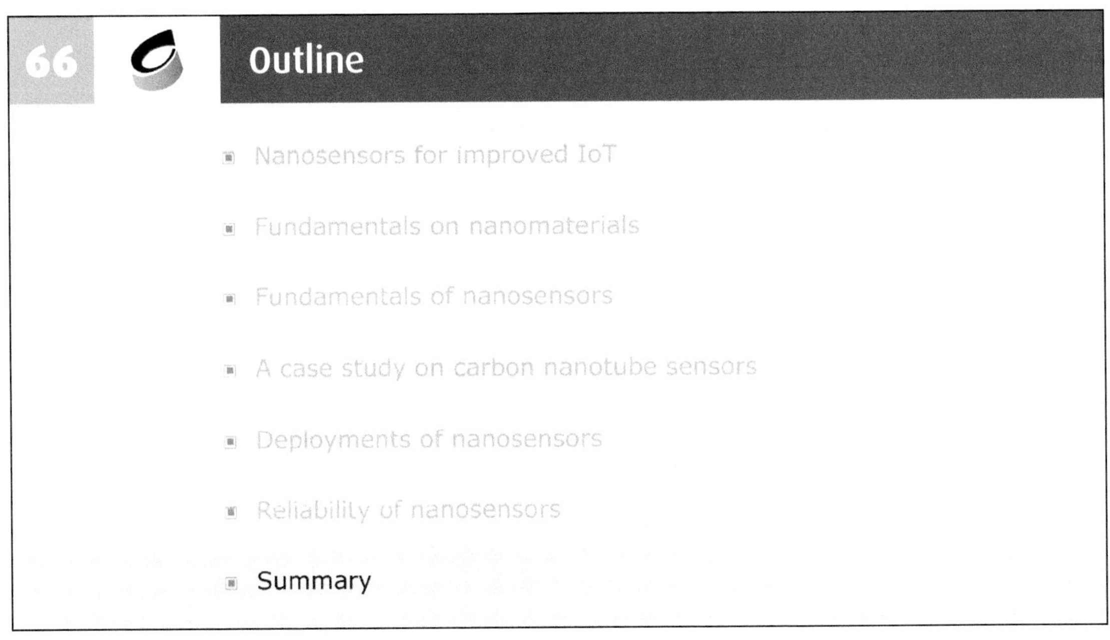

66
Outline
Nanosensors for improved IoT
Fundamentals on nanomaterials
Fundamentals of nanosensors
A case study on carbon nanotube sensors
Deployments of nanosensors
Reliability of nanosensors
Summary

Finally, a brief summary of the discussed topics is presented.

Summary

67

- Nanotechnologies are key-enabler for the Future of the Internet of Things
- A wide variety of nanomaterials now available commercially and in research, notably used in nanosensor applications
- Nanosensors operate like regular sensors, but cross sensitivity and reproducibility are to be checked systematically
- Deployments of nanosensors are possible, but show that reliability is a deep issue.

Nanotechnologies play a relevant role to prepare the future of the IoT. The wide variety of nanomaterials, with their different properties, motivates their use in various fields of the IoT, notably for nanosensors. Nanosensors behave in general like regular sensors, but may still have additional issues (sensitivity to perturbing factors, problems of reproducibility) which should be checked out by any end user. Deployment of nanosensors is also possible and yields interesting results. Now the challenge is to work on the reliability of nanosensors.

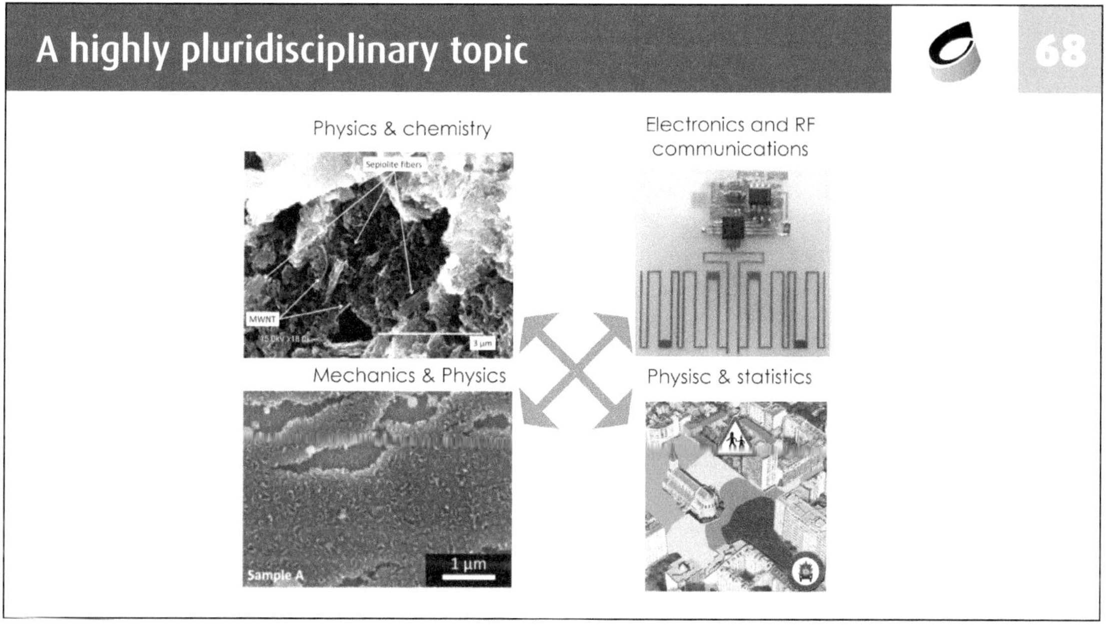

All in all, the development of nanosensors for IoT is a very pluridisciplinary topic, since it involves different areas of knowledge such as: electronics, chemistry, physics, materials science, statistics, and energy, among others. Consequently, a strong collaboration between these areas is crucial to meet with success.

About the Editor

João Goes

João Goes graduated from Instituto Superior Técnico (IST), Lisbon, in Electrical and Computer Engineering (ECE), in 1992. He obtained the M.Sc. and the Ph.D. degrees in ECE, in 1996 and 2000 respectively, from the Technical University of Lisbon, and the 'Agregado' degree ('Habilitation' degree) in Electronics, in 2012 from the Nova University of Lisbon (NOVA). He has been with the Department of Electrical Engineering (DEE) of the Faculty of Sciences and Technology (FCT) of NOVA, since April 1998 where he is currently an Associate Professor and, since July 2012, he is the Department Chair. Since 1998, he has been a Senior Researcher at the Centre for Technology and Systems (CTS) at UNINOVA and responsible for the Microelectronics Research Unit. In Nov. 2012 he also became the Director of CTS. In Sep. 2003 he co-founded and served, for 4 years, as the CTO (and Board member) of ACACIA Semiconductor SA, a Portuguese engineering company specialized in high-performance data converter and analog front-end products (acquired by Silicon and Software Systems, S3, in Oct. 2007). Since Nov. 2007 he lectures and performs his research with part-time consultancy for S3. From March 1997 until March 1998 he was Project Manager at Chipidea SA (now SYNOPSYS). From December 1993 to February 1997 he worked as a Senior Researcher at Integrated Circuits and Systems Group (GCSI) at IST doing research on data converters and analog filters. João Goes has supervised (graduated) 9 Ph.D. theses, 16 M.Sc. theses, and 9 graduation projects. He has published over 160 papers in international journals and leading IEEE conferences and he is co-author of 6 books. João Goes is a Senior Member of IEEE since 2009 (Member since 1995) and Member of the Circuits and Systems (CAS) and Solid-State Circuits (SSC) Societies. He was also the Chairman of the IEEE CAS Analog Signal Processing Technical Committee, ASPTC, (the largest within CASS) for the term 2014-2015. He is currently an Associate Editor of the IEEE Transactions on Circuits and Systems - II, TCAS-II, for the term 2016-2017.

Universidade NOVA de Lisboa,
NOVA School of Science and Technology
Portugal

About the Authors

Pieter Harpe

Pieter Harpe received the M.Sc. and Ph.D. degrees from the Eindhoven University of Technology, The Netherlands. In 2008, he joined Holst Centre / IMEC where he worked on low-power ADCs. In April 2011, he joined Eindhoven University of Technology as assistant professor on low-power mixed-signal circuits. His main interests include power-efficient and reconfigurable data converters and low-power analog design. He is a TPC member for ISSCC, ESSCIRC and AACD, and serves as Distinguished Lecturer for the IEEE Solid-State Circuits Society. He is recipient of the IEEE ISSCC 2015 Distinguished Technical Paper Award.

Bérengère Lebental

Bérengère Lebental graduated from Ecole Polytechnique (Paris Saclay University, France) Engineering Program in 2006, received two MSc in Physics and Nanotechnology from Ecole Polytechnique in 2007 and her PhD from Université Paris-Est, France, in 2010 (ENPC-PariTech PhD award in 2011). Since 2010, she is research scientist at IFSTTAR (French Institute of Science and Technology of Tranportat, Development and Network) and at LPICM (Laboratory of Physics of Interfaces and Thin Films, a joined research team between Ecole Polytechnique and CNRS). A physicist specialized in the nanoelectronics of carbon-based nanomaterials, her research focuses on the development of reproducible and reliable nanosensors for applications to urban sustainability, with a focus on micromechanical and chemical sensing. She coordinates the 9M€ French Equipment Program Sense-City and the 4M€ European research project Proteus on water quality monitoring.

Franco Maloberti

Franco Maloberti received the Laurea Degree in Physics (Summa cum Laude) from the University of Parma, Italy, and the Dr. Honoris Causa degree from Inaoe, Puebla, Mexico in 1996. He was a Visiting Professor at ETH-PEL, Zurich in 1993 and at EPFL-LEG, Lausanne in 2004. He was the TI/J. Kilby Analog Engineering Chair Professor at the Texas A&M University and the Microelectronic Chair Professor at University of Texas at Dallas. Currently he is Professor at the University of Pavia, Italy and Honorary Professor at the University of Macau, China. His professional expertise is in the design of analog integrated circuits and data converters. He has written more than 500 published papers, six books and holds 34 patents. He is the Chairman of the Academic Committee of the Microelectronics Key Lab. Macau, China. He is the President of the IEEE CAS Society, he was VP Region 8 of IEEE CAS (1995-1997), Associate Editor of IEEE-TCAS-II, President of the IEEE Sensor Council (2002-2003), IEEE CAS BoG member (2003-2005), VP Publications IEEE CAS (2007-2008), DL IEEE SSC Society and DL IEEE CAS Society. He received the 1999 IEEE CAS Society Meritorious Service Award, the 2000 CAS Society Golden Jubilee Medal, and the IEEE 2000 Millennium Medal, the 1996 IEE Fleming Premium, the ESSCIRC 2007 Best Paper Award and the IEEJ Workshop 2007 and 2010 Best Paper Award. He received the IEEE CAS Society 2013 Mac Van Valkenburg Award. He is an IEEE Life Fellow.

Augusto Marques

Augusto Marques is the CTO of Aura Semiconductor, where he is focused on the development of state of the art radio technology and timing products. Previously, he was a Fellow Advisor for RF and Analog Circuit Design with MediaTek. As an advisor, he was responsible for the development of high performance, high volume cellular and connectivity transceivers for mobile applications. Before that, he was Senior Fellow and VP of RF at NXP Semiconductors. Before the acquisition of the Wireless Business Unit of Silicon Laboratories, Dr. Marques was elected as a Silicon Laboratories Fellow and served as the Wireless Engineering Director supervising the development of GSM, GPRS, EDGE, and 3G products. During his tenure at Silicon Laboratories, Dr. Marques was an architect and/or principal designer for the world's first CMOS GSM/GPRS synthesizer, the world's first CMOS GSM/GPRS transceiver (Aero), the world's first CMOS GSM/GPRS transceiver with integrated DCXO (Aero +), and the world's first single-chip CMOS GSM/GPRS transceiver (Aero II). Cumulatively, he has been an architect and/or principal designer for products that have shipped in excess of 1 billion units and is an inventor on more than 20 US patents in the areas of analog and RF integrated circuits. Dr. Marques holds a Ph.D. in Microelectronics from the Katholieke Universiteit Leuven, Belgium with a Master's in Physics and a Bachelor's degree in Computer Engineering from the University of Coimbra, Portugal.

Luis B. Oliveira

Luis B. Oliveira was born in Lisbon, Portugal, in 1979. He graduated with a degree in electrical and computer engineering, and a Ph.D. degree, in 2002 and 2007 respectively, from Instituto Superior Técnico (IST), Technical University of Lisbon. Since 2001 he has been a member of the Analog and Mixed Signal Circuits Group at INESC-ID. Although his research work has been done mainly at INESC-ID, he has had intense collaboration with TUDelft, in The Netherlands, and University of Alberta, in Canada. In 2007, he joined the teaching staff of the Department of Electrical Engineering of Faculdade de Ciências e Tecnologia, Universidade Nova de Lisboa, and is currently a researcher at CTS-UNINOVA. Prof. Oliveira is an IEEE Senior Member and chair of the Broadcast Technology Society, Circuits and Systems Society, and Consumer Electronics Society, IEEE Joint Chapter. He is a member of the IEEE CASS Analog Signal Processing Technical Committee, ASPTC, (the largest within CASS), since 2008. He has more than 100 publications in International Journals and Conferences, and is co-author of two books: "Analysis and Design of Quadrature Oscillators" (Springer, 2008) and "Wideband CMOS Receivers" (Springer, 2015).

João P. Oliveira

João P. Oliveira was born in Paris, France, in 1969. He graduated in 1992 at Instituto Superior Técnico (IST) of Technical University of Lisbon. He has also received his M.Sc. degree in electrical engineering and computer science in 1996 from IST. He obtained the Ph.D. degree in 2010 from Universidade Nova de Lisboa (NOVA). He has been with Department of Electrical Engineering of the Faculdade de Ciências e Tecnologia (FCT) of NOVA as an Assistant Professor of Microelectronics, since 2003. Since 2004 he has worked as a Senior Researcher of the Centre for Technology and Systems (CTS) at UNINOVA. Since 2003 he has been a Co-founder of MOBBIT Systems SA, a specialized system design Portuguese engineering company. From 1996 to 2003 he worked in the telecommunications industry in the area of 2G and 3G radio terminal equipment and switched packet data. From 1992 and 1996, he was a research engineer in the Integrated Circuits and Systems Group (CGSI) at IST where he worked in the area of switched current-mode ADCs and filters. He is Member of IEEE and Member of the Circuits and Systems (CAS) and Solid-State Circuits (SSC) Societies.

Noel O'Riordan

Noel O'Riordan has over twenty years' experience in the field of mixed signal integrated circuits in the industrial, consumer, automotive and RF sectors, with S3 Group, Dublin Ireland. His work has covered circuitry for ADCs, Filters, Power Management, LNAs, DSP and full IC development, in technology processes from 28nm CMOS to HV SOI-BCD. He worked in the R&D Laboratories of ITALTEL in the area of Electromagnetics for telecommunication systems, in Milan Italy from 1991 to 1996. He is an Adjunct Senior Lecturer in Electronic Engineering at University College Dublin, Ireland. He obtained a BE and MEngSc degree in Electronic Engineering from University College Dublin, Ireland in 1991.

Nuno Paulino

Nuno Paulino was born in Beja, Portugal, in 1969. He graduated from Instituto Superior Técnico (IST), Lisbon, in 1992. He obtained the M.Sc. degree, in 1996 from the Technical University of Lisbon and obtained the Ph.D degree in 2008 from the Universidade Nova de Lisboa. He has been with the Department of Electrical Engineering (DEE) of the Faculdade de Ciências e Tecnologia, Universidade Nova de Lisboa (FCT NOVA), since 1999. Since 1999 he has been also working as a Senior Researcher of the Centre of Technology and Systems (CTS) at UNINOVA. In 2003 he co-founded ACACIA Semiconductor, a Portuguese engineering company specialized in high-performance data converter and analog front-end products, acquired by S3 in 2007. From 1996 to 1999 he worked as Analog Design Engineer at Rockwell Semiconductor, USA. He is a Member of the ASPTC of IEEE CAS Society.

Sandeep Perdoor

Sandeep Perdoor, graduated in Electronics and Communication from National Institute of Technology, Surathkal, India in 2002. He was awarded the university gold medal for the year 2002. In 2009, he received M.Sc cum Laude in Microelectronics from Delft University of Technology, Netherlands. From 2002 to 2007, he was with Texas Instruments, India, where he led the development of several pipelined ADC based products. From 2009 to 2011, he was with ST Ericsson, India, where he worked on the development of 3G transceivers. Since 2011, he is with Aura Semiconductor, where he has led the development of several timing and RF products. He is currently the RF business unit lead responsible for the development of BLE based solutions. He holds 5 patents.

Marcelino Santos

Marcelino B. dos Santos received the M.Sc. and Ph.D. degrees in Electrical and Computer Engineering from Instituto Superior Técnico, Technical University of Lisbon, Portugal, in 1994 and 2001 respectively. He teaches electronics and microelectronics at Instituto Superior Técnico since 1990. He advised eight PhD and more than thirty MSc thesis.

He is the responsible for the Quality, Test and Co-Design of HW/SW Systems research group of Instituto de Engenharia de Sistemas e Computadores R&D in Lisbon (INESC-ID). His research interests include power management, testability, ultra-low power mixed signal design and educational issues on microelectronics. He is a senior member of IEEE and he is the chair of the Portuguese join chapters of Industrial Electronics, Industrial Applications and Power Electronics. He published more than thirty journal papers and 130 conference papers. Marcelino Santos was co-founder of SILICONGATE LDA in 2008. He is, since then, the CTO of the company, today a word leading company in power management IP.

Rui Santos-Tavares

Rui Santos-Tavares was born in Oeiras, Portugal, in 1975. He graduated from Universidade Nova de Lisboa, Lisbon, in 1998. He obtained the M. Sc. and the PhD degrees, respectively, in 2001 and 2010, from the Universidade Nova de Lisboa.

Rui has been with the Department of Electrical Engineering (DEE) of the Faculdade de Ciências e Tecnologia (FCT) of Universidade Nova de Lisboa (UNL), since September 1998, where he is currently Assistant Professor. Since 1998 he has been also working as a Researcher of the Centre for Technology and Systems (CTS) at UNINOVA. From 1998 until 2001 he worked as an Assistant Researcher of the Interoperability Supported by Standards group (GRIS) at UNINOVA, where he has actively participated in several National and joint European cooperative projects in science and technology (e.g. ESPRIT IV 22056 - FunSTEP). Since 2001 he is with the Microelectronics, Design and Fractional Signal Processing Group (MESP) from CTS/UNINOVA. Professor Santos-Tavares has published several papers in international leading conferences, a book and has been serving as a reviewer for many IEEE Conferences. He is member of IEEE Circuits and Systems (CAS) Society since 2010. He participated in the organization board of IEEE ISCAS'2015, in Lisbon in May 2015.

Leonel Sousa

Leonel Sousa received a Ph.D. degree in Electrical and Computer Engineering from the Instituto Superior Técnico (IST), Universidade de Lisboa (UL), Lisbon, Portugal, in 1996, where he is currently Full Professor. He is also a Senior Researcher with the R&D Instituto de Engenharia de Sistemas e Computadores (INESC-ID). He has contributed to more than 200 papers in journals and international conferences, for which he got several awards - such as, DASIP'13 Best Paper Award, SAMOS'11 'Stamatis Vassiliadis' Best Paper Award, DASIP'10 Best Poster Award, and the Honorable Mention Award UTL/Santander Totta for the quality of the publications in 2009. He has edited two special issues of international journals, and he is currently Associate Editor of the IEEE Transactions on Multimedia, IEEE Transactions on Circuits and Systems for Video Technology, IEEE Access, IET Electronics Letters and Springer JRTIP, and Editor-in-Chief of the Eurasip Journal on Embedded Systems, from Springer. He is Fellow of the IET, a Distinguished Scientist of ACM and a Senior Member of IEEE, and Member of IFIP WG10.3 on concurrent systems.